THE ELLIPSE

THE ELLIPSE
A HISTORICAL AND
MATHEMATICAL JOURNEY

ARTHUR MAZER

A JOHN WILEY & SONS, INC., PUBLICATION

For general information on our other products and services or for technical support, please contact our Customer Care Department within the United States at (800) 762-2974, outside the United States at (317) 572-3993 or fax (317) 572-4002.

Wiley also publishes its books in a variety of electronic formats. Some content that appears in print may not be available in electronic formats. For more information about Wiley products, visit our web site at www.wiley.com.

Library of Congress Cataloging-in-Publication Data:

Mazer, Arthur, 1958-
The ellipse : a historical and mathematical journey / Arthur Mazer.
    p. cm.
  Includes bibliographical references and index.
  ISBN 978-0-470-58718-8 (pbk.)
1. Ellipse. 2. Mathematics–History. I. Title.
QA485.M39 2010
516'.152–dc22   2009035919

Printed in the United States of America

10 9 8 7 6 5 4 3 2 1

CONTENTS

PREFACE

Two of my passions are history and math. Historians often consider mathematics separate or at best tangential to their own discipline, while, by contrast, historians snuggle up with philosophy and the arts in an intimate embrace. Try this experiment: go to the library and randomly select a history book on ancient Greece. The book will describe the geopolitical landscape in which Greek culture emerged, the incessant feuding between the city states, the wars with Persia, the Peloponnesian war, and the Macedonian conquest. Also included in the book will be a section on the influential Greek philosophers and philosophical schools. And equally likely is an analysis of the artwork that provides a reflection of the times. Most likely, there is no reference to mathematical and scientific achievements, and the rare book that does mention mathematics and science is very stingy in its offerings. The reader is left to conclude that philosophical ideals are the drivers of historical change, the evolution of which can be seen in the arts. Mathematical and scientific achievements are mere outcomes of the philosophical drivers and not worth mentioning in a book on history.

There is of course the opposite argument in which one exchanges the positions of the mathematician with that of the philosophers. That is, mathematics and science are the drivers of historical evolution and in Darwinian fashion philosophies and political entities that promote scientific excellence flourish, while those that do not fade away. This latter argument provides the perspective for this book.

The seventeenth century was the bridge between the sixteenth century's counterreformation and the eighteenth century's enlightenment. It was the mathematicians who built that bridge as their efforts to settle the geocentric versus heliocentric debate over the universal order resulted in Newton's and Leibniz' invention of calculus along with Newton's laws of motion. The mathematicians concluded the debate with their demonstration that the planets revolve around the sun along elliptic pathways. In a broader context, the outcome of the argument was a scientific breakthrough that altered European philosophies so that their nations could utilize their newly found scientific prowess. *The Ellipse* relates the story from the beginnings of the geocentric versus heliocentric debate to its conclusion.

The impact of the debate is sufficient to warrant a retelling of the story. But this is not only a story of tremendous political, philosophical, and not to mention scientific and mathematical consequences, it is also one heck of a story that rivals any Hollywood production. Were we not taken in by Humphrey Bogart and Katherine Hepburn's dedication to a seemingly impossible mission in *The African Queen*? Johannes Kepler launched himself on a mission impossible that he pursued with fierce dedication as it consumed 8 years of his life. Were we not enthralled by Abigail Breslin as her fresh honesty disarmed the pretentious organizers of the Sunshine Pageant in *Little Miss Sunshine*? In the face of the Inquisition as they condemned Bruno to death at the pyre, Bruno exposed the hypocrisy of his sentencers stating, "You give this sentence with more fear than I receive it." Such are the elements of this story that it is not only significant but also compelling.

Those somewhat familiar with this story might launch a protest. Given its centrality to man's development, this story has been picked over by many outstanding individuals. The result is that there are already many fine accessible books on the topic, such as Arthur Koestler's *The Sleepwalkers*. What does *The Ellipse* offer? There are two offerings. First, the premise above that mathematics and science are the drivers of historical evolution directs the historical narrative. There is a true exchange of the roles of philosophers and mathematicians from what is evident in the standard historical literature. As with standard history books, this book describes the geopolitical environment. But philosophers are given a scant role, while mathematicians assume the center stage. Second, this is predominantly a math book with a specific objective. The objective is to take the reader through all of the mathematics necessary to derive the ellipse as the shape of a planet's path about the sun. The historical narrative accompanies the mathematics providing background music.

Throughout the book, the ellipse remains the goal, but it receives little attention until the very last mathematical section. Most of the book sets the stage, and the mathematical props of geometry, algebra, trigonometry, and calculus are put in place. Presenting these topics allows for the participation of a wide audience. Basic topics are available for those who may not as of yet had an introduction to one or more of the foundational subjects. And for those who have allowed their mathematical knowledge to dissipate due to lack of practice over several years, a review of the topics allows for a reacquaintance. Finally, for those who are well versed and find the exercise of deriving the ellipse trivial, enjoy the accompanying narrative.

Apart from devoting quite a few pages to history, the presentation is unconventional in several respects. The style is informal with a focus on intuition as opposed to concrete proof. Additionally, the book includes topics that are not covered in a standard curriculum, that is, fractals, four-dimensional spheres, and constructing a pentagon. (I particularly want to provide supplementary material to teachers having students with a keen interest in mathematics.) Finally, I include linear algebra as a part of the chapter that addresses high school algebra. Normally, this material follows calculus. Nevertheless, calculus is not a prerequisite for linear algebra, and by keeping the presentation at an appropriate level, the ideas are accessible to a high school student. Once this tool is available, the scope of problems that one can address expands into new dimensions, literally.

There are prefaces in which the author claims their writing experience was filled with only joy and that the words came so naturally that the book nearly wrote itself. I am jealous for my experience has certainly been different. There were joyous moments, but difficulties visited me as well. The challenge of maintaining technical soundness within an informal writing style blanketed the project from its inception to the final word. Setting a balance between storytelling and mathematics has been equally confounding, as has been determining the information that I should park in these two zones. Fortunately, I have had the advice of many a good-natured friend to assist me with these challenges. I would like to acknowledge my high school geometry teacher, Joseph Triebsch, who first introduced me to Euclid and advised me to address the above-mentioned challenges head on. Others who have assisted include Alejandro Aceves, Ted Gooley, David Halpern, and Tudor Ratiu. Their willingness to take time from their quite busy schedules and provide honest feedback is greatly appreciated. Should the reader judge that I have not adequately met the aforementioned challenges, it is not due to my not having been forewarned and equally not due to a lack of alternative approaches as suggested by my friends. The project did allow me to get in touch with old friends, all of whom I have not been in contact with for many years. This experience was filled with only joy and more than compensated for the difficulties that surfaced during the writing.

I must also acknowledge my family, Lijuan, Julius, and Amelia, for putting up with me. For over a year around the dinner table, they were absolutely cheery while listening to my discourses on *The Ellipse*. I still cannot discern whether they actually enjoyed my hijacking of the normal family conversation time and conversion of it to lecture sessions or were just indulging their clueless old man. Either way, I am lucky and in their debt.

CHAPTER 1

INTRODUCTION

My first teaching job did not start out too smoothly. I would feverishly spend my evenings preparing material that I thought would excite the students. Then the next day I would watch the expression on my students' faces as they sat through my lecture. Their expressions were similar to that on my Uncle Moe's face when he once recalled an experience on the Bataan Death March. How could the lectures that I painstakingly prepared with the hope of instilling excitement have been as tortuous as the Bataan Death March? To find an answer to this question, I went to the source. I asked the students what was going wrong. After 16 years, with the exception of one suggestion, I have vague recollection of the students' feedback. After 16 years, with the exception of one individual, I cannot remember the faces behind any of the suggestions. Concerning the one individual, not only do I have clarity concerning her face and suggestion, but I also have perfect recollection of my response.

The individual suggested that I deliver the lectures in storylike fashion and have a story behind the mathematics that was being taught. My response that I kept to myself was "you have got to be joking." My feeling was that mathematics was the story; the story cannot be changed to something else to accommodate someone's lack of appreciation for the subject. This was one suggestion that I did not oblige. And while for the most part the other students responded positively to the changes that I did make, this student sat through the entire semester with her tortured expression intact.

It is difficult to recall the specifics of something that was said over 16 years ago, the contents of a normal conversation remain in the past while we move on. Despite my reaction, there must have been some meaning that resonated and continued doing so, otherwise I would have long ago forgotten the conversation. Now I see the student's suggestion as brilliant and right on target. By not taking her suggestion, I blew the chance to get more students excited by mathematics through compelling and human stories that are at the heart of mathematics. At the time, I just did not have the vision to see what she was getting at. After 16 years, I have once more given it some thought and this book is the resulting vision. This is a mathematical story and a true one at that.

The story follows man's pursuit of the ellipse. The ellipse is the shape of a planet's path as it orbits the sun. The ellipse is special because it is a demonstration of man's successful efforts to describe his natural environment using mathematics and this mathematical revelation paved the pathway from the Counter Reformation

The Ellipse: A Historical and Mathematical Journey   by Arthur Mazer
Copyright © 2010 by John Wiley & Sons, Inc.

to the Enlightenment. Man pursued the ellipse in a dogged manner as if a mission to find it had been seeded into his genetic code. Through wars, enlightened times, book burnings, religious persecution, imprisonment, vanquished empires, centuries of ignorance, more wars, plagues, fear of being ridiculed, the Renaissance, the Reformation, the Counter Reformation, excommunication, witchcraft trials, the Inquisition, more wars, and more plagues, man leaf by leaf nurtured a mathematical beanstalk toward the ellipse. This book examines the development and fabric of the beanstalk. It describes the creation of geometry, algebra, trigonometry, and finally calculus, all targeted toward the ellipse.

What are the ingredients that make up a good story? *Heroes*: They are in this story as the book presents a glimpse of the lives of several mathematicians from Aristarchus to Leibniz who made significant contributions to the beanstalk. *Villains*: The story of men threatened by progress and doing their best to thwart—being central to the story of the ellipse. *Struggles*: The problem of planetary motion is sufficiently vexing to assure some mathematical difficulty, and as the previous paragraph indicates, additional struggles result from a tormented history. *Dedication*: The dedicated effort of the contributors is at once admirable and inspiring. *Uncertainty*: While the book reconstructs mathematical history with the certainty that man arrives at the ellipse, many contributors had absolutely no premonition of where their contributions would lead. This uncertainty is germane to our story. *Character flaws*: Our heroes were not perfect and their mistakes are part of the story. *Tragedy*: Getting speared in the back while contemplating geometry, a victim of one's own insecurity, a burning at the stake as a victim of the Inquisition—these are a small sampling of personal tragedies that unfold as we follow the ellipse. *Triumph*: After a tortuous path, this story triumphantly ends at the ellipse. What else is in a good story? I dare not get explicit, but it is in there.

With such a great story, one would think that someone had told it before. Indeed, the story has been told; the most comprehensive historical presentation is Arthur Koestler's distinguished book, *The Sleepwalkers*. In addition, there are history books and excellent biographies of the main contributors, mathematical history books, and books covering the various mathematical topics that are contained in this book. So what is different about this book? Simply put, the history books only address the history, the math books only address the math, and the mathematical history books only address the mathematical history. This book is a math book covering the topics of geometry, algebra, trigonometry, and calculus which contains a historical narrative that sets the context for the mathematical developments. Following my belief that separating the disciplines of the history of mathematics and science from general history is an unnatural amputation, the narrative weaves the mathematical history into the broader history of the times while focusing along the main thread of uncovering the ellipse.

There is a final category of book that readers of this book may be interested in, popular books that explain mathematical and scientific theory—books explaining general relativity, quantum mechanics, chaos theory, and string theory abound for those without the requisite mathematical background. Of necessity, the core is missing in these books, the mathematics. Just as love binds two humans in true intimacy, mathematics binds the theorist with evidence. It is difficult to have a true appreciation of the theory without the mathematics, which is unfortunate because it keeps the general public at a distance from theory. This book takes the reader through all the mathematical developments needed to uncover the ellipse, and the reader will become truly intimate with the theory. The book delves into the subjects of geometry, algebra, trigonometry, and calculus, and once the mathematical machinery is finally assembled, we stock the ellipse.

Mathematicians are explorers. They follow their imagination into new territory and map out their findings. Then their discoveries become gateways for other mathematicians who can push the path into further unexplored territories. Unlike the great sea-going explorers of the fifteenth and sixteenth centuries who were exploring the surface of a finite earth, the domain of the mathematician is infinite. The subject will never be exhausted, mathematical knowledge will continue to expand, and the beanstalk will keep growing. However, like the great explorers of the fifteenth and sixteenth centuries, mathematical journeys may target a specific objective (akin to Magellan's circumnavigation of the world) or the consequences of mathematical journeys may be fully unrelated to their intentions (akin to Columbus' accidental discovery of a new continent). We can even go one step further; it is possible that some mathematical journeys have no intent whatsoever other than to amuse the journeying mathematician.

This book presents mathematics as a journey. There is the intended pathway toward the ellipse and there are sojourns along bifurcating branches of the beanstalk that are unrelated to the ellipse. The journey passes through the normal high school curriculum and calculus. By placing all the subject matter together, it is possible to demonstrate relations between what are normally taught as separate disciplines. For example, the area of an ellipse, a geometric concept, is finally arrived at only after developing concepts in linear algebra and trigonometry; the approach highlights the interplay of all the disciplines toward an applied problem. In addition, setting the objective of uncovering the ellipse motivates the mathematics. For example, studies of motion motivate the presentation of calculus and the fundamental theorem of calculus is presented as a statement of the relation between displacement and velocity. The sojourns with no apparent relation to the ellipse are undertaken solely because they are irresistible.

The book allows you as a reader to plot your own course in accordance with your own purpose. Readers with excellent proficiency in calculus will certainly plot their way through the book differently from those who may be a little out of touch with their high school mathematics and calculus. And those entirely unfamiliar with one or more of the subjects will plot another course altogether. The first section of Chapter 2 hosts the main narrative and tells the story of man's pursuit of the ellipse beginning with Aristarchus, the first known heliocentrist, and ending with Newton's successful unveiling nearly two millennia after Aristarchus. Each subsequent chapter begins with a narrative that is pertinent to the mathematical material in the chapter. By and large the mathematical material is included for one or more of the following reasons. The material is necessary to understand the topic of the chapter and will be used in subsequent chapters, or the material presents concrete examples of relevant concepts, or I have just indulged my own fancy and included material that I find fun. Sections containing material that falls solely within the last category are clearly marked as excursions and may be skipped without compromising your understanding of the remaining material. As for the remaining material, plot your course in accordance with your own purpose. You may grasp the high-level concepts and move on, or for those who want to go through the nitty-gritty, it is in there. Enjoy your journey.

CHAPTER 2

THE TRAIL: STARTING OUT

2.1 A STICKY MATTER

CLASSMATE: *Be careful. Take such stands in the classroom only. If you speak like that in public, you could be called a heretic.*

KEPLER: *My beliefs are my beliefs. I will make no secret of them.*

Kepler and Galileo lived during a time of transition. The church had lost much of its authority during the Reformation and answered with the Counter Reformation in an attempt to recover its former position. There were several factors contributing to the Reformation: nationalism, taxation, and a wayward clergy. The central method of the Counter Reformation was that that the church had honed over its 1000-year reign of power, fear.

For centuries the church could afford its excesses. Its position as the sole interpreter of scriptures allowed it to control human activity with the threat of eternal damnation. The message was simple and not subtle—follow the church's dogma toward eternal salvation or suffer unimaginable consequences, not only for the short period of your life on earth, but for eternity. And the church proffered vivid descriptions of what the consequences would be so that the unimaginable became images that were seared into the minds of medieval Europe. Demons thrusting pitch forks into screaming victims, deformed beasts pursuing their victims without mercy, and rings of fire forever scorching its victims—these images of hell had been painted in medieval churches across Europe. Through fear, the church stifled intellectual development throughout the Dark Ages.

The church maintained its monopoly as the sole interpreter of scriptures through two methods. First, the predominant avenue to an education was through church seminaries or church-sponsored universities; there were few independent secular educational institutions. Second, Latin, which was only taught in the seminaries and universities, was the language of the Bible. There were no translations into local languages, so the majority of Europeans could only rely upon the church's interpretation. In 1439, Johannes Gutenberg invented a simple device that would challenge the church's monopoly on intellectual activity, the printing press. Soon the Bible would be printed and distributed in local languages and the masses would be free to read and interpret scriptures for themselves. The Reformation was born, and after recovering its

The Ellipse: A Historical and Mathematical Journey   by Arthur Mazer
Copyright © 2010 by John Wiley & Sons, Inc.

footing, the church responded; it launched the Counter Reformation and unleashed the Inquisition.

The result was chaos as Protestants responded to the Counter Reformation with war. The Germanic states that comprised the Holy Roman Empire launched a revolt against the church-supported Habsburg dynasty. This spawned the Thirty Years' War between Christians. Each side required discipline from their followers. In Italy, the church accepted no challenges to its authority and enforced its dogma with the Inquisition. Protestants followed suit, enforcing discipline the only way that they knew how, with fear. Those who did not agree with the dogma of the leading Protestant clergymen were excommunicated. It was in this environment that the Lutheran Kepler and the Catholic Galileo initiated modern science and mathematics, and it was in this environment that both were punished for their remarkable accomplishments.

Is the sun fixed, with the earth and its sister planets revolving about the sun, or is the earth fixed with all that is in heaven revolving about the earth? This seems to be an innocent question, certainly not a question that would lead to censorship, excommunication, imprisonment, torture, and execution on the pyre, with all of these indignities sponsored by an institution claiming to show humanity the way to salvation. And yet, the quest to answer this seemingly innocent question catalyzed all of these responses within the church. In those times, the church was far more politically consumed than the present-day church and political motives engendered these ugly responses. On the scientific side, the quest to determine the path of the planets catalyzed the development of calculus and brought science and mathematics into the modern era. This chapter follows the history of the quest in a narrative that addresses both political and scientific dimensions. The mathematics presented later in the book follows the narrative.

While there are many potential points to begin this story, we choose to begin with Aristarchus (310–230 B.C.), a Greek astronomer and mathematician from Samos. Aristarchus is the first individual known to have proposed heliocentricity based upon geometric analysis. The analysis contained two components: a method for calculating the relative size of the sun and a proposition that distance explains the fixed path of the stars from the perspective of a moving earth. This latter proposition explicitly addresses what is known as the parallax problem. Detractors of heliocentrism state that the stars would not daily appear in the same position as the earth revolves around the sun if the earth were to do so. In short order, their argument goes, the stars do appear in the same position, so the earth must be stationary. Aristarchus retorts that even though the earth moves, the stars appear fixed because the distance between the earth and stars is orders of magnitude greater than the comparatively small distances that the earth moves. With this argument, Aristarchus confronts man with the scale of the universe and how little we are within it, not a very popular notion.

Aristarchus makes another scaling argument, this one a bit more quantitative with his estimate of the relative size of the sun. This estimate demonstrates Aristarchus' grasp of geometry while at the same time illustrating the limitations of the instruments used to take astronomical observations. The geometric argument is flawless, providing a correct equation, but the measurement of an angle required by the equation is far off base. Placing his poor measurement into the formula, Aristarchus calculated that the diameter of the sun was about 20 times that of the earth, whereas the sun's actual diameter is on the order of 300 times that of the earth's. Nevertheless, Aristarchus is the first to propose that the sun is the significantly larger body, likewise not a popular notion. Aristarchus' work in which he proposes the heliocentric model did not survive. We can only conjecture that he found it more reasonable that the smaller body should orbit the larger body, not vice versa.

The prefix *geo*, which finds great use in the English language, has its origins in the Greek word *ge*, meaning "earth." A slight permutation of geo yields *ego*, which is the Latin word for *I*. The heliocentric universe that Aristarchus proposed was much less ego friendly than the geocentric universe that had been accepted since Aristotle. The earth lies precisely in the middle of Aristotle's universe and it is dominant everything else, much lighter than the earth, revolves around the earth in perfect circles. Given man's collective ego, not even the finest snake oil salesman in history could have sold Aristarchus' view in Aristarchus' time. The church's response to the heliocentric view nearly 1800 years later echoes a response by a contemporary of Aristarchus, Cleanthes. Cleanthes was so affronted by Aristarchus that he wrote a treatise entitled *Against Aristarchus* in which he states that "it was the duty of Greeks to indict Aristarchus of Samos on the charge of impropriety." The charge of impropriety is eerily similar to the charge of heresy that the church would accuse adherents of heliocentrism of at a later time.

Despite Cleanthes' appeals, there is no evidence of court action against Aristarchus. In fact, Aristarchus' proposal was firmly rooted in Greek tradition, one that respects not only knowledge but also the quest for knowledge. Chaos often begets intellectual activity and the percolating cauldron that was Greece fits this pattern. Prior to Alexander, there was not an empire or even a monolithic civilization known as Greece. On the contrary, Greece was a constellation of city states, each with its own distinct culture. Some were ruled by tyrants and others by assembly. Some stressed military values, while others stressed arts and learning. What held them together as distinctively Greek was a common polytheistic religion, a common language, and geographic proximity. Another element binding the Greeks was the common threat of Persia, which would place them in temporary alliance. More often than not, when external threats diminished, the Greek states would war with one another.

It was not until Philip of Macedonia united the Greeks that a national entity emerged. When Philip's son, Alexander, assumed power and established his empire, it was the Athenian culture of arts and knowledge that he exported and transplanted. This culture had an unusual tolerance for individual expression, one that resonated well with the indigenous inhabitants of the lands that Alexander conquered. The Athenian theater tradition demonstrates the high esteem that Athenians held for the right of self-expression, provided of course that you were a citizen as opposed to a woman, foreigner, or slave. (The latter category was not an insignificant portion of the population; at one time slaves comprised 30% of the Athenian population.)

The Athenians delighted in theater. At the festival of Dionysus, there was a tradition of sponsoring four playwrights to showcase their work. A playwright whose work was selected for sponsorship received much prestige and the competition to be selected was fierce. Even in the midst of war, the Athenians would celebrate the festival of Dionysus. During the Peloponnesian War, which was poorly managed by

Athenian politicians and drained the city's economy, the playwright Aristophanes lampooned the Athenian leadership in his comedy *Lysistrata* (414 B.C.). In the play, the women of both Athens and Sparta, Athens' nemesis, unite to bring an end to the senseless war. Their plan is simple; they would deprive their respective males of carnal pleasure by going on a sex strike. The men are unable to withstand this denial and are driven to a peace treaty. There would be no other political entity until the advent of modern democracy that would allow its citizens to openly mock the policies of its leadership, particularly when the entity is at war.

The Greek valued knowledge and learning centers were established throughout the lands conquered by Alexander. These learning centers inherited the Athenian respect for self-expression. Foremost among all learning centers was the university at Alexandria; the city established by and named for the famous conqueror is in Egypt. The debt of mathematics to the university at Alexandria cannot be understated. Shortly after Aristarchus, Euclid (circa 300 B.C.) wrote his incomparable work *The Elements*.

The Elements would be the standard text for mathematics training throughout the Middle East and Europe over the next 2000 years. At the age of 40, Abraham Lincoln undertook the study of *The Elements* to exercise his mind and much of our modern-day high school mathematics curriculum draws from Euclid's *The Elements*. It is *The Elements* that cements the axiomatic deductive process that lies at the core of mathematics. *The Elements* begins with a set of definitions that are used throughout the book. Axioms follow the definitions. Afterward, the text branches out in a tree of propositions that engender yet more propositions, but all are derived from *The Elements'* axiomatic roots.

Aside from Euclid, there were others of tremendous intellectual stature associated with the university at Alexandria who made lasting contributions to mathematics. Central to the quest of an understanding of the juxtaposition of the stars, sun, planets, and earth are Archimedes (287–212 B.C.), Apollonius (262–190 B.C.), and Ptolemy (83–168 A.D.). All of these outstanding mathematicians and scientists were thoroughly educated in Euclid's *Elements* and it permeates their work. All made significant contributions beyond *The Elements*. All were dead wrong in their assessment of Aristarchus' thesis.

The story predates the university at Alexandria. Aristotle had posited the commonly held view of the universe's structure. According to Aristotle, the stars, planets, sun, and moon float above the earth as these entities are lighter than earth and they circulate about the earth through an invisible medium that he coined the ether. The earth itself, being the heaviest of all objects, is immovable, fixed within the ether. As with many scientific theses that Aristotle posited, this was more of a product of fanciful imagination than an actual scientific investigation. And as with many of Aristotle's physical theses, it has been thoroughly discredited. Despite the fact that he wrote a significant amount on topics that he knew nothing about, Aristotle's scientifically vacant musings became dogma over a 2000-year span. The idolization of Aristotle would not happen under the Greeks; indeed, while Greek philosophers may have frequently cited Aristotle, the Greek scientists give little mention of him.

Aside from his theory on the structure of the universe, another of Aristotle's theories has significance in the development of calculus. That is his incorrect view concerning the motion of falling objects. Aristotle's view is that a heavy object falls faster than a light object. An argument against Aristotle's theory accessible to contemporary Greeks was that two objects could be combined into one by fusing them together; the combined object would not gain speed in its descent from the separate objects. Indeed, just tie a rope around two objects. If Aristotle is correct, the new object joined by the rope will fall faster than the separate objects, but it does not.

Because of Aristotle's prominence in history, it is worthwhile to examine his role as a scientist. Aristotle is frequently credited with the development of the scientific method, subjecting a hypothesis to rigorous testing. If this is the case, Aristotle never applied the scientific method to his views on the dimensions of the universe, the composition of the universe, or the motion of falling objects. Concerning the dimensions and composition of the universe, he had neither the knowledge nor the means to subject his views to testing. Concerning the motion of falling objects, an experiment whereby objects of different shapes and weights are dropped repeatedly from a fixed height allowing the scientist to observe which, if any, objects fall at greater speed could have easily been performed. Such an experiment would have shown Aristotle's hypothesis to be incorrect. But Aristotle never performed the experiment. As with his views on the dimension and composition of the universe, Aristotle proposed them without evidence.

By contrast, Aristarchus, a near contemporary of Aristotle's, only 70 years younger, performed a careful geometric analysis and then subjected his analysis to experimental measurement. The evidence he gave for his correct conclusion that the sun is larger than the earth was accepted and endorsed by the most capable of the Greek scientists, Archimedes. We have little record of the more personal aspects of his life, and yet personal stories about Archimedes have become a part of mathematical folklore, a tribute to his well-deserved legendary status. The mathematical achievements of this man are staggering. The originality of his work and the scope of subjects that he investigated placed him years ahead of his contemporaries and science would not catch up for another 1800 years. Archimedes left for posterity 12 works that we know of. He was a consummate problem solver developing brilliant methods. Archimedes did not formalize his solutions into theory. But the theory is recognizable and the depth of it is amazing. We briefly describe three works.

In *On Floating Bodies*, Archimedes determines stable configurations of floating bodies. A stable configuration is one that does not drift away from its equilibrium position under a disturbance. As an example, a pendulum with its weight directly above the pendulum's pivot is in equilibrium but not stable since the weight will swing downward upon being disturbed. The pendulum with its weight directly below the pivot is in equilibrium and stable. Archimedes examined equilibrium configurations for bodies floating in water and then determined which were stable, that is, which would not flip over in the presence of a disturbance. Archimedes correctly asserts that the stable configuration of a floating object is the one with the lowest energy level. He then proceeds to apply this principle to nontrivial shapes and determines their stable configuration. The physical assertion of the stable configuration does not explicitly use the term *energy* for the concept of energy had not yet been discovered. Rather, Archimedes poses his work in terms of centers of gravity. The very thought of attempting a stability analysis at this stage of intellectual development demonstrates his creativity and daring. One would not find comparable works for another

1900 years, well after the advent of calculus. And arguably, one could cite this work to justify Archimedes as the father of integral calculus because he develops methods of integral calculus to calculate centers of gravity for nontrivial shapes.

Again, Archimedes demonstrates his pioneering efforts in physics and calculus in his work *On Spirals*. In this work, Archimedes calculates the velocity of an object moving along a prescribed spiral pathway, relates the velocity to lengths of arcs, and determines the area that the particle sweeps out between the spiral and a coordinate axis. With this work, one could cite Archimedes as the father of differential calculus and even go one step further. While he does not formalize it, Archimedes uses the fundamental theorem of calculus to relate velocity to length and area.

Archimedes judges his work by a different standard than those who review it with a historical perspective. The work that Archimedes was most proud of is *On the Sphere and Cylinder*, where he determines the formula for the volume of a sphere. Perhaps this is because it is such a difficult problem and Archimedes is a consummate problem solver, rightfully proud of his problem-solving capacity. The method is exceptionally original and once again demonstrates Archimedes' comfort with calculus. Indeed, Archimedes also relates the volume of the sphere to the surface area, giving another example of the application of the fundamental theorem of calculus.

It is not too difficult to imagine a historical scenario in which scholars following Archimedes' works formalize his problem-solving methods into theory and develop calculus long before Newton. But this is not how history happened. On the scientific side, Archimedes was far ahead of his time. The number system that Archimedes used did not have the counterpart of a zero, making calculations tedious and difficult to follow. Even more, algebra had not yet been formalized, and the many complex algebraic manipulations that Archimedes executes are difficult to communicate in the geometric language of Archimedes' times. In addition, formalization of Archimedes' results requires the general concept of a function as well as the concept of a coordinate axis system, neither of which was developed in Europe until Rene Descartes in the seventeenth century. Archimedes' works do not receive the attention that they merit. Perhaps, sadly, they served no other purpose than to dazzle modern historians by the stunning capacity of this man to be not only centuries but nearly two millennia ahead of his time. Then again, perhaps, as we shall see, they played a more inspiring role.

One further work of Archimedes requires our attention for it is germane to the topic of this book. In *The Sandreckoner*, Archimedes supports the geocentric vision of the universe and opposes Aristarchus. Indeed, it is from Archimedes' response to Aristarchus that we are aware of Aristarchus' works; as previously noted, the original works of Aristarchus in which he expounds his heliocentric theory have been lost. It is a pity that Archimedes placed his opposition to Aristarchus in writing. Certainly, Archimedes carried significant authority in the intellectual world. His refutation may well have influenced others to turn away from Aristarchus.

The historical route to calculus and the structure of the universe bypasses Archimedes and flows through a contemporary, Apollonius. Apollonius was a lecturer and researcher at Alexandria. Apollonius' initial impact upon the search for the structure of the universe was to point in the wrong direction with a very persuasive finger. He applied his ingenuity toward correcting a flaw in the geocentric universe. The stars follow a daily circular motion and maintain their relationship with each other. But the Greeks identified objects with curious behavior and assigned the Greek word planetoid, meaning wanderer, to these bodies. Aristotle's model does not give a convincing description of planetary motion.

The planets float among the stars, shifting their position. A first attempt to describe this drifting behavior was to hypothesize that they follow their circular orbits around the earth at different speeds than the stars. This appears to be almost correct, but there are anomalies. In particular, the planets appear to not be moving at a constant speed and the planets shift their direction on occasion. How could one account for this observation? Apollonius salvages the Aristotelian view with a proposition that each planet moves in a small circle that rotates about a larger circle. Picture a giant circular arch across the sky and a planet being carried across by an invisible wheel that rolls over the arch. As the wheel moves along the dominant archway, its motion creates a secondary circular path known as an epicycle. The planet follows these composite motions. With the correct diameters for the larger arch and smaller wheel, as well as correct speeds around each circle, the anomalies could be explained. This is what Apollonius had in mind, but he did not carry out the calculations in detail.

Apollonius is best known for his studies of conic sections. These are curves that result from the intersection of a plane with a cone. The intersection generates a circle, ellipse, hyperbola, or parabola dependent upon the manner in which the intersection occurs. Apollonius was not the first Greek to investigate conic sections, but he presents the most thorough accounting of their properties. Apollonius also calculated the value of pi (so did Archimedes), providing a necessary constant for finding the lengths and areas of both circles and ellipses.

In addition, in his investigations of conics, Apollonius determines the tangent line to the surface of the curves, the objective of differential calculus. This places Apollonius as perhaps, not the father of differential calculus, but certainly an inspirer. Apollonius' interest is purely abstract as he does not show applications for any of his works. He could not imagine that the ellipse would supplant his own epicycles as the correct description of planetary motion, but that would be a millennium and a smattering of centuries away.

The weight of authority from both Archimedes and Apollonius was enough to crush further investigations into a heliocentric system. Nobody from Alexandria followed the direction indicated by Aristarchus. A Babylonian, Seleucus, took up the cause of Aristarchus two centuries later (circa 190 B.C.) and presented additional arguments in favor of a heliocentric system. However, Seleucus' works received little attention. It is noteworthy that Seleucus had adopted Greek culture although he was not racially Greek. The acceptance of Greek culture by inhabitants of the lands ruled by the Greeks was common and many of the intellectuals in the Greek universities were not in fact Greek.

Ptolemy (83–161 A.D.) was another great ancient scientist who received a Greek education but was not Greek. Ptolemy endowed the West with the longitudinal and latitudinal coordinate system that is used to locate points on the earth's surface; this achievement cements Ptolemy's reputation in the West as the father of geography.

His works were widely read and quoted for centuries, particularly by mapmakers and navigators. Ptolemy was also a fine mathematician who developed extremely accurate trigonometric tables and in doing so demonstrated the level of sophistication in the area of trigonometry during his time.

Central to our story, Ptolemy turned his attention to astronomy and calculated the specifics of motion along the larger and smaller wheels as proposed by Apollonius. Unfortunately, he found that no matter how he chose dimensions and speeds, he could not precisely replicate the motion of the planets as measured and logged at the astronomical observatory in Alexandria. The stubbornness and ingenuity of mankind in pursuing what is dead wrong are mind boggling. Ptolemy responded by introducing two devices, the eccentric and the equant. Ptolemy ever so slightly displaced the earth from the center of the universe by assigning the center of a planet's orbit to a point that is distant from the earth. Another device was to introduce an equant for each planet. The planet's angular velocity about its equant remains constant but as the equant is not centered on the earth, the planet's speed about the earth is not uniform. Ptolemy's universe is like an organ with its rotating gears moving in precision to produce a harmonic outcome. As the observations are not quite so harmonic, take a monkey wrench to the machinery and make corrections. One point of note is that Ptolemy's mastery of trigonometry was essential to carry out the calculations. And over the centuries the calculations became unwieldy. By the time of Copernicus, the Ptolemaic universe consisted of 40 epicycles.

While at first glance this does not look good for Ptolemy, let us look at his approach in hindsight from a different perspective. In the early nineteenth century, the mathematician Fourier proposed a solution to a difficult set of equations known as the heat equations. His solution was an infinite composition of trigonometric functions. This caused much controversy because it was unknown if an infinite combination of functions had any meaning. As always, controversy in mathematics begets progress and Fourier was vindicated; indeed, one can construct a meaningful solution using an infinite composition of trigonometric functions. Ptolemy, with his wheel on a wheel on a wheel construction of the pathways of planets, was the first individual to attempt this. So even though he was dead wrong about the universe, he was years ahead of his time in creating functional series expansions.

Returning to our story, it is in the Ptolemaic universe that Western thought stagnates for 1500 years. No further contributions are made toward the question of the structure of the universe and planetary motions. Western science is frozen in Aristotelian musings, while Aristarchus, Seleucus, and the far-reaching works of Archimedes are ignored. What forces are responsible for these sad circumstances? The force most immediately responsible for the decline in the pace of scientific progress is the Romans. Their march to power did not bode well for the sciences; a Roman soldier killed Archimedes. The historian Polybius (200–118 B.C.) recorded the story of Archimedes' death and it has become folklore that is retold in nearly every text that mentions Archimedes. So here it is.

It is believed that Archimedes was educated at Alexandria. While in Egypt, Archimedes demonstrated his mechanical creativity for there he invented the Archimedean screw, a water-pumping device that is still used today. Archimedes' gift with mechanics would be integral to his legendary status. Archimedes did not remain in Egypt but left to return to his native city of Syracuse, one of a series of cities that the Greeks had established throughout the Mediterranean prior to the days of Alexander.

During the life of Archimedes, a century after Alexander's death, Rome was establishing itself as a power within the Mediterranean. Another contender, Carthage, battled with Rome for supremacy throughout the Punic Wars. Syracuse was situated in between these two powers, trying to ally itself with the unknown victor. At first Syracuse lined up behind Rome. Then under Hannibal the Carthaginians gained the upper hand; Hannibal led an army that included warrior elephants across the Alps and defeated the Romans. At this point, Syracuse as well as other Sicilian cities realigned themselves with Carthage. This set the stage for Rome to besiege the cities of Sicily when Rome reestablished itself as the stronger power.

At the time Syracuse was a city state under the leadership of King Hiero. Hiero instructed Archimedes to draw up a defense plan against the Roman assault. In this task, Archimedes exhibited the same brilliance that he did in his scholarly works. He had the advantage of a well-protected city as the city walls had been well placed. Facing the sea, the wall abutted the Mediterranean. The remainder of the wall enclosed the city along a roughly semicircular path. The semicircular wall was predominantly built upon cliffs with few places of safe access.

Archimedes applied his mechanical ingenuity, which was evident when he invented the Archimedean screw. Archimedes designed catapults of various sizes, large, mid, and small sized, optimally tuned for different ranges. Midsized and smaller sized catapults allowed for quicker reloading as they held lighter projectiles. Aside from the range-specific catapults, Archimedes built several cranelike structures that could be used to hoist large and heavy objects. The hoisting mechanism consisted of a beam that rotated on a pivot. A heavy object was hooked and connected to one end of the beam by a rope. A counterweight on the other side of the beam hoisted the object using leverage. Archimedes designed a hooking mechanism that has been likened to a claw for it could grapple objects that would then be hoisted by the cranes. Finally, Archimedes designed short-range weapons that could fire multiple darts with a single shot, a sort of ancient machine gun.

Surveying the land and recognizing that there were few places where the wall was accessible, Marcellus, the commanding Roman general, at first decided to attack Syracuse by sea. His plan was to breach the walls with a series of ladders carried and secured by specially designed boats. The Romans had used this technique on cities similar to Syracuse and were confident of easily breaching the city walls by reason of overwhelming manpower. Once within the city walls, Roman soldiers with spears would operate according to standard procedure: pillage, burn, kill, and rape at will.

On the day of the attack, the Romans approached Syracuse by sea and Archimedes' long-range catapults bombarded Marcellus' navy. The vessels that survived the initial barrage were then targets for Archimedes' mid- and short-range catapults. If a boat reached the city wall, Archimedes' claw grappled the vessel and hoisted the vessel into a dangling vertical position, whereupon the vessel was released and smashed into the rocky seashore. All the while, the Roman soldiers were targets for shorter range weapons. The result was a resounding victory for the Greeks.

This was discomfiting to Marcellus and Marcellus decided to attack by land. Unfortunately for Marcellus, his analysis that initially caused him to forgo a land assault proved to be correct. Archimedes lined up his firepower in a few areas where the wall was approachable. The firepower was concentrated, overwhelming, and accurate as the Roman army had to pass through the same series of long-range, midrange, and short-range weaponry that earlier greeted their navy. A barrage of iron and stone projectiles as well as arrows greeted the Roman soldiers who came close to the wall. In addition, the claw grappled individual soldiers and dropped them into the rocks below. The overall picture for the Romans was very demoralizing.

Marcellus became prudent; he decided to forgo an assault and dug in for a siege embargoing the city and starving it into submission. This was not the preferred approach since it pinned down many Roman resources that could otherwise be used in further conquest. Furthermore, one never knew how long the supplies of the besieged city would last and the uncertainty of the duration of the siege, 6 months, 1 year, 2 years, or longer, would not be good for one's career. However, Marcellus saw no other choice and he set up camp.

Two and a half years into the siege, the embargo had taken its toll and the Syracusans resolve weakened. Marcellus launched an attack along with an order to spare Archimedes. The assault was successful, but the order to spare Archimedes was not obeyed. One legend has it that Archimedes was steeped in concentration, observing geometric figures that he had drawn in a sandbox, when the attack came. A Roman soldier entered his home and Archimedes snapped, "Don't disturb my circles." Not knowing the man was Archimedes, the soldier executed him by impaling Archimedes with a spear. Marcellus honored Archimedes by burying him in accordance with Archimedes' wishes. A sphere and a cylinder were engraved on the tombstone to commemorate Archimedes' most proud discovery of the formula for the volume of a sphere.

If Archimedes had been at the university, perhaps he could have developed a school of followers who would further develop his ideas. Instead, he did the next best thing. He sent copies of his works to Alexandria and most likely these were shared with other universities. His works were revolutionary and difficult to comprehend. The wealth of material that could have been formalized into theory and text was not given the attention it merited. Eventually, mathematics would catch up with Archimedes, but that would not be for another millennium and a smattering of centuries.

The Romans cobbled an empire together through fear. Local leaders understood that they would submit to Roman authority or suffer retribution. And the Romans were not subtle about the form that the retribution would take. Indeed, the standard infantry operating procedure described above, burn, pillage, kill, and rape, was meant as an intentional warning to those who might waiver in their commitment to Rome.

In the early days of the Roman Empire, as long as there was willing submission to Roman authority and taxes were paid, local leaders could maintain their autonomy. Under this arrangement, local religious and cultural traditions continued. If the culture had a tradition of learning, as in the case of Alexandria, it could locally support a university. Ptolemy lived under Roman rule and the university at Alexandria continued until 415 A.D. But learning and science were never germane to Roman culture. Indeed, on the cultural front, the Romans offered next to nothing over the 500-year span of their empire. The Roman Empire was an exercise in power. Whereas the Greeks bequeathed the university system along with a formidable accumulation of knowledge, the Romans bequeathed their profane and unenlightened credo of power for the sake of power. Within this cultural vacuum, Romans adopted the Christian religion.

Upon adopting Christianity in 312, the Romans assumed a less charitable attitude toward their territories. They reversed their previous tolerance for local traditions, and in 391 the Roman emperor Theodosius prohibited the practice of any religion other than Christianity. The prohibition sparked significant unrest as local populations were unwilling to forgo their traditional ways. The Roman response to this unrest was harsh. What occurred next at Alexandria was emblematic of what was happening across the empire.

The university's library was within the city's pagan temple complex, a place where the Alexandrian pagan population sought sanctuary from marauding Roman Christians. The library housed the most complete collection of literature within the Roman Empire. The library consisted of scrolls that were painstakingly transcribed one letter at a time by a professional cadre of scribes. At its largest, the library is rumored to have contained 500,000 scrolls. This was before its original burning by Mark Antony (48 B.C.), an accident that was brought on by Roman soldiers under Antony's command. During the battle between Marc Antony and Cleopatra's rival to the throne, her brother Ptolemy, the library was a victim which in modern-day jargon would be described as collateral damage. Mark Antony made it up to Cleopatra by pilfering 200,000 scrolls from another library, at Pergamum; unfortunately for Pergamum, Mark Antony did not find such an enchanting spirit as Cleopatra there. Theodosius was not a patron of knowledge or beauty. His decree of 391 instigated christians mobs to destroy pagan institutions. The library at Alexandria and the scrolls were not spared. While dedicated teachers tried to continue lecturing at the university, the staff eventually disbanded due to harassment through prosecution on the charges of heresy. In 415, the university's last remaining scholar-lecturer, a woman by the name of Hypatia, was accused of teaching heretical philosophies, assassinated, and cremated.

The mix of politics with faith has been an explosive combination that has turned disastrous throughout history. Christianity had moved into the cultural vacuum of the Roman Empire and with the fall of the Roman Empire the church would assume an even larger political role. Throughout its life at the center of European politics, the church would attempt to reconcile its religious inheritance, the message of salvation, with its political inheritance, power for the sake of power. All too often, the message of salvation became subservient to politics.

The church was a principal author of the post-Roman political order. The church struck a bargain with influential chieftains throughout Europe, and the church would legitimize their authority, conferring recognition of divine authority for a ruling class of nobility. In exchange the nobility would accept the Christian religion, furnish the church with lands and material support for cathedrals and monasteries through taxation, and allow the church to establish its own separate governing authority upon church-owned lands. In constructing this political edifice, the church had no compunction about sacrificing its moral authority. It explicitly accepted the indentured servitude of a landless class to the nobility by legitimizing the institute of serfdom, an institute rife with abuses that are contrary to Christian philosophy. Heredity perpetuated the arrangement as an individual's social status was conferred at birth. Through its political dealings, the church had become politically dominant, and more so than the Romans for from the fall of the Romans until the Reformation the reign of the church lasted 800 years.

Mirroring the social fixed point that the church in league with the ruling class had established, the church also established spiritual and philosophical fixed points to create a triangle of stability. Spiritually, the church offered the inspirational teachings of the Gospels, but intellectually the church inherited little direction. In the thirteenth century, the Europeans rediscovered Aristotle, which created a stir among a previously intellectually comatose population. Thomas Aquinas was the best known promoter of Aristotelian reason. The initial church response was predictable. Aristotle was summarily rejected by Pope John I, who in no uncertain terms condemned Aristotelian philosophy. But the church could not keep intellectual curiosity at bay and the church adapted. By accepting Aristotle's scientific principles and reinterpreting his political and ethical philosophies, the church coopted Aristotle and completed the final fixed point for their triangle of stability. What was it about Aristotelian science as opposed to the far-reaching ideas of Archimedes and Aristarchus that the church found attractive? Archimedes and Aristarchus had profound ideas that required further investigation. An institution that wishes to maintain the status quo does not want individuals following their own inquiry and certainly does not promote such a philosophy. Alternatively, Aristotle offered a comprehensible knowledge of everything. For church leadership, Aristotle was simple and intuitive and stifled independent inquiry. In one word, this was perfect. The effect on mathematics and science is known to all; there was little European progress in either mathematics or science under church domination.

Throughout the Middle Ages, the church maintained a grip on intellectual activity through its influence on university education. The university evolved to meet church requirements for an educated clergy that could address complex theological issues as well as administrative and financial issues. An additional impetus was the needs of a growing mercantile class who had educational requirements that the church could not fulfill. Both of these sectors worked together to meet their corresponding educational requirements.

Student-funded universities developed as the mercantile class sought knowledge beyond theological concerns. In the twelfth century, citizens would seek out a knowledgeable individual to teach a topic of interest in exchange for pay. At first, the process was ad hoc, and there was no formal administration or location. A natural place to hold classes was the old monastery schools that had been established during the Gregorian reforms, so the universities evolved out of the church's infrastructure. The process expanded both administratively and physically. Buildings were rented for the purpose of holding classes and later were owned as universities became their own entities.

As the infrastructure for the universities was created from church resources, the church had significant influence upon university administration that was later enhanced. Indeed, several incidents of conflict between local authorities and university students caused the student-sponsored universities to seek a form of church protection. The most famous incident occurred at the University of Paris in 1229. The incident began as a dispute between a small group of students and the proprietor of a tavern over a tab and eventually snowballed into a full-scale riot and strike. After exchanging words, the proprietor and his workers manhandled the students and threw them out onto the streets. The next day, the students returned with a larger force. A riot ensued and several establishments were damaged or destroyed. After the riot, the city police were called upon to punish the offending students and things went from bad to worse. The police used unwarranted tactics associated with a vigilante force. Rather than conducting an investigation, they chanced upon a group of students and killed a few. There was widespread belief among the larger student body that the attack was set upon completely innocent students and the students called an immediate strike. The university closed its doors for 2 years.

The strike prompted university officials to seek church protection from local authority; officials believed that students would return with such protection. The church responded favorably and upped the ante. Pope Gregory IX, an alumnus of the university, issued a papal bull that placed the university under papal authority, and from that time onward local officials had no authority over affairs concerning the university or its students. In addition, the church financed the university, providing the salaries of the staff. As noted above, incidents at other student-sponsored universities caused administrators to seek similar protection from the church; the Italian universities at Padua and Bologna are examples. These universities remained student funded with a significant sphere of influence from the church.

By the end of the fourteenth century, there were approximately 50 universities in Europe. The curriculum at the universities centered on theology, law, medicine, Aristotelian science, and mathematics. Concerning mathematics, as long as mathematical advances did not conflict with either biblical writings or Aristotelian science, there was a significant degree of allowable independent thought. During the twelfth and thirteenth centuries Europe played catch up. Indeed, from the ninth century through the thirteenth century Arab culture had spawned progress in mathematics, while Europeans made no contributions to the field. Translations of Arabic texts, some of which had been translated from the original Greek, afforded Europeans access to ancient Greek works, contemporary Arab works, and the Hindu Arabic numeric system. Euclid's *The Elements* and Ptolemy's *Almagest* were among the translated texts.

The method of translation is of interest. The translations were a multicultural effort between individuals of diverse backgrounds. Muslims, Jews, and Christians sat side by side rendering an Arab text into a local language and then into Latin. An Arab would read a text to a Jew who was perhaps conversational but not literate in Arabic. The Jew would then translate the text into the local language (often Spanish) and pass it on to a Catholic clergyman. The final translation into Latin was performed almost exclusively by a Catholic clergyman. In this manner, the Christian world was brought up to speed in mathematical and scientific endeavors. In Spain, the Jews were repaid for their assistance by centuries of discrimination, an Inquisition, and forced conversions that finally culminated in the confiscation of Jewish property and their wholesale eviction from Spain on the eve of the voyage of Christopher Columbus. Indeed, the seeds of the twentieth-century holocaust were planted in the Middle Ages.

Once up to speed, Europeans began to make contributions of their own. Most relevant to the study of motion is the work of Nicola Oresme (1323–1382). Oresme studied theology at the University of Paris. He led the life of an academic, an adviser to the King, and a clergyman, eventually becoming the Bishop of Liseaux. Oresme expanded upon the concept of a coordinate system to associate quantities of physical interest, such as position and temperature or position and time. In doing so, Oresme proposed the concept of a mathematical function and anticipated the work of Rene Descartes who in the seventeenth century would popularize Cartesian coordinates. Oresme also discovered the law of motion for an object under a constant force, a discovery that is generally attributed to Galileo at the end of the sixteenth century. The motion is described by a parabola and Galileo's proof is identical to that of Oresme.

There may have been an opportunity to follow up on the work of Oresme and initiate modern science. But Europe would have to wait for Kepler and Galileo two and a half centuries after Oresme. Europe was preoccupied by a disastrous fourteenth century. During this time, instead of pursuing learning, Europe pursued war. The Hundred Years' War was a conflict between two households for the throne of France, an eastern household and a western household that ruled over England as well as parts of Normandy and Brittany. The war was a series of battles with intermittent intervals of peace; the first battle was fought in 1337 and the last battle was fought in 1453. And aside from the self-inflicted wounds of war, the plague of the 1340s ravaged all of Europe. Estimates are that the population of Europe fell by between 50 and 70% as a result of the plague. It would take over a century for Europe to recover its population. As war and plague consumed all of Europe, Oresme's works, like those of Aristarchus and Archimedes, were orphaned and abandoned.

Political division was not confined to the monarchies of France and England. While the war between rival claimants to the throne unfolded in France, a similar internecine feud consumed the church. For a brief period between 1305 and 1377, the papacy resided in Avignon, France. Pope Gregory XI restored the office to its original location in Rome, but not before a substantial French influence had taken hold. Upon the death of Gregory XI in 1378, a riotous mob demanded that an Italian be elected to the office of pope and the cardinals acquiesced. They elected the Neapolitan Bartolomeo Prignano, who assumed the name Pope Urban VI. Regret at the choice set in and a substantial number of the cardinals who participated in the election of Pope Urban VI returned to Avignon and elected a rival pope, Pope Clement VII. The awkward situation in which there were two men who laid claim to the position of the sole and rightful interlocutor between God and man persisted for nearly 40 years. The nations of Europe were divided in their allegiance until the church finally reunited behind a single pope in 1415.

Plague, wars, and political intrigue were not enough to release Europe from the church's triangle of stability. But chaos begot more chaos and the church compounded its problems by leading its flock with complete dereliction. There were two directions from which the church was politically vulnerable, nationalism and reason. Nationalistic feelings had always been latent within Europe. With the advent of the Hundred Years' War, nationalistic sentiment crystallized and allegiance to a nation began to displace allegiance to the church. Nationalism would erode the church's political fixed point, while reason would call into question church doctrine and erode the church's scientific fixed point. Only the church's spiritual message would remain intact, but the message would not necessarily be spread by the Catholic church. Kings, queens, aristocrats, merchants, and yeomen were at the forefront of the nationalistic movement, while a budding university intelligentsia often of the church's own clergymen would lubricate the slope away from a church-dominated Europe. The greatest impetus for the movement was the church's own delinquency.

In 1460, a young Cardinal Borgia received a letter of rebuke from Pope Pius. The cause of concern was a party that the young cardinal had hosted; his guest list included the most enchanting ladies of Siena but curiously did not include their brothers, husbands, fathers, or any other male escorts. The cardinal and his fellow church brethren were free to enjoy their guests' company. The intercession of the pope did not cause Cardinal Borgia to uphold his vows of celibacy as it was publicly known that he fathered several children, including one of his own grandchildren.

Cardinal Borgia became Pope Alexander VI and brought his partying ways to Rome. A diarist recorded a party in which the finest harlots of Rome were brought in for entertainment. To add merriment through friendly competition, guests were encouraged to publicly fornicate and the pope dispensed awards for competitors with the most ejaculations. The church had returned Rome full circle to the orgiastic days of the Roman Empire with ill portent.

Another arena in which the church followed Roman precedent was in its policy of taxation. For centuries, the church imposed a tax across all of Europe. Taxation was unpopular, particularly among the mercantile class, and became a focal point around which Martin Luther attacked the church. To be fair, not all the collected taxes were spent on debauchery and war or dispensed to garner favor. The impressive cathedrals of the Renaissance along with the artworks of Botticelli, Michelangelo, Leonardo da Vinci, Rafael, and other accomplished artists were financed through church taxes. As impressive as these works are, they afforded little benefit to the citizens of Europe whose hard labor financed the efforts.

The excesses at the top were mirrored throughout the church's hierarchy. It was known that priests would proffer absolution in exchange for sex. Imagine a maiden presenting herself to her priest to confess a carnal sin and for the father to absolve the sin and assure the sinner a pathway to heaven through ecclesiastic intimacy. In addition to satiating sexual desire, clergymen also abused their authority to enrich themselves. Pursuing an all-too-common intersection between sex and money, more than one nunnery had the reputation of being a whorehouse. To augment their income, clergymen also sold absolution for any possible crime. Indeed, it was the public selling of absolution by the swindler Johann Tetzel, a Dominican friar, that inspired Martin Luther to denounce the church.

Johann Tetzel traveled between German towns announcing his arrival using the methods of a circus publicist. Tetzel would enter the town square alongside a team of performing acrobats and men lugging an enormous cross. In addition, Tetzel's entourage included an accountant from the church's banker and a sack full of indulgences with the pope's seal. An indulgence guaranteed its purchaser a place in heaven regardless of past as well as future crimes committed. Tetzel was a fine marketer and most likely he understood the need for price differentiation for his different policies; naturally, an indulgence covering future crimes was more expensive than one covering past crimes. One could also purchase an indulgence that would present a deceased loved one with the key to heaven. The profits from Tetzel's operation were split between local clergymen, the church, and the church's bankers, a Catholic German family named Fugger.

The church was so brazen that its activities were public knowledge. A showdown between reformers and the church was inevitable. Leading the reformers was Martin Luther, who took on the rather out-of-touch Pope Leo X. Luther railed against the corruption of the church and the taxes and rallied around nationalism. From a twenty-first-century perspective, one that is acutely aware of the negative consequences of German nationalism, Luther's admonitions to stir German pride are troubling. His voice resonated and many potentates of the hodgepodge of Germanic states that made up the Holy Roman Empire happily agreed to stop paying taxes to Rome and join the Protestant cause. Wholesale massacres of Catholic clergymen along with confiscation of property accompanied the Protestant revolution that occurred in the 1520s.

One consequence of the change in allegiance is that Protestant mercenaries who had joined the Spanish king's army, Charles V, in a church-inspired effort against the French under King Francis, turned against the church and sacked Rome. The sacking occurred in 1527 and these Protestants' ancestors, who had sacked Rome a millennium earlier, the Goths, and Teutonics, would have been proud of their progeny. Women were raped, buildings were burned, and wealth was stolen. The sacking of Rome was the symbol of the end of church dominance in Europe.

Luther was a populist. While his counterparts on the side of the church pronounced their edicts in Latin, Luther published in Latin as well as vernacular German. Of great assistance to Luther was the invention of the printing press and the maturing of a publishing community. Prior to the technical capacity for spreading one's message, it is doubtful that a challenge to the church was possible, even with the church in such a sordid state.

There are times when humanity collectively feels something in the air—that something big is going to happen. Those remembering the end of the cold war might know the feeling, a sense of anticipation that there will be a momentous shift in the international order. The national choir of the Soviet Union, the Godless state that was America's nemesis for nearly a century, came to entertain the nation's elite in Washington's Lincoln Center and sang, "God Bless America." Members of the audience at first stared in a stupor of disbelief and then cried; the cold war ended that night. Europe began the sixteenth century with a similar sense of anticipation. Renaissance painters brought the arts to its greatest level. The age of explorers was in its prime. Vasco da Gama pursued Africa to its bottom and opened a route to the Orient.

Christopher Columbus discovered a new continent to the west and soon Magellan's men would circumnavigate the globe. The church was lax and the Reformation would stir. In the year 1500 Nicolaus Copernicus, while completing his doctoral studies in law, followed his true passion with astronomy by observing a lunar eclipse. He then proceeded to eclipse Aristotle and illuminate the true scientific heritage of the Greeks. This was the most momentous and revolutionary achievement of the sixteenth century; henceforth, science would be where man would search for answers.

Nicolaus Coppernic (1473–1543), better known by his Latinized name Copernicus, was the first European to join Aristarchus. At the age of 10, Copernicus' maternal uncle became the legal guardian of the young Nicolas and his three orphaned siblings. Copernicus' father left behind sufficient funds to assure that the children were not wanting and the uncle, Lucas Waltzenrode, was a supportive man. Few men in Europe could have afforded the education that Copernicus received. In Poland, he studied at the University of Krakow. Upon completing his studies, good old Uncle Waltzenrode secured Nicolaus a lifetime appointment as church canon. The appointment included a comfortable stipend. In 1496, Copernicus went for further studies in Italy at Bologna, Padua, and Ferrara. Copernicus was a role model for the modern-day professional student, spending perhaps 14 years in higher education. In Renaissance fashion, Copernicus studied broadly, taking courses in law, the classics, and medicine, but it was his studies in mathematics and astronomy that were his passion.

Upon departing Italy in 1503, Copernicus became a secretary to his uncle, who was by then the bishop and governor of Warmia, an autonomous state within Polish Prussia. Copernicus' duties included judging legal cases and performing administrative functions for the church. Copernicus also spent time attending to the sick, often without compensation, a testimony to his good will and desire to do meaningful work. Copernicus' uncle, Bishop Waltzenrode, had paved a path for Copernicus to assume the position of bishop. But Copernicus refused to walk that road and maintained his much less esteemed rank. To a contemporary, the only explanation would be that the job was a comfortable job offering security for one with little ambition. And yet it was this man of seemingly little ambition who would move the earth and still the sun.

Along with caring for the ill, Copernicus' position afforded sufficient time to devote to his passion, astronomy. That astronomy consumed both Copernicus' head and heart was evident during his graduate years when Copernicus preferred eclipse observation to other forms of entertainment available to a 20 some year old in Rome. Having been educated in Greek and astronomy and possessing a sharp mathematical mind, Copernicus was well versed in Ptolemaic theory, understanding all the details of the calculation of the epicycles, eccentricities, and equants that are required to match observations. The Ptolemaic universe is unsightly. It is a jerry-rigged structure built on a wobbly foundation with struts added at points of notable weakness; the struts appear as out-of-place artifices that only accentuate the folly of the core. Both mathematicians and physicists have a grasp of theoretical aesthetics and more than once a theory has been assailed on the grounds that it is ugly.

Around 1610, Copernicus first reveals his heliocentric vision. In an unpublished manuscript known as *Commentariolus* (*Little Commentary*), Copernicus presents his view of the Ptolemaic system:

Yet the planetary theories of Ptolemy and most other astronomers, although consistent with the numerical data, seemed likewise to present no small difficulty. For these theories were not adequate unless certain equants were also conceived; it then appeared that a planet moved with uniform velocity neither on its deferent nor about the center of its epicycle. Hence a system of this sort seemed neither sufficiently absolute nor sufficiently pleasing to the mind.

After stating his objections, that the theory was not pleasing (it is ugly), Copernicus goes on to propose his alternative. Echoing Aristarchus, whom Copernicus would later cite, he supersizes the universe to explain parallax and proposes a heliocentric system. Going beyond what is known of Aristarchus, he explains the apparent motion of the heavens by motions of the earth. The short manuscript is a teaser, a preview to a larger piece of work that within the lines of *Commentariolus* Copernicus promises is forthcoming.

Copernicus circulated *Commentariolus* among a small circle of intimates. It would be about three decades before the larger body of work, *On Revolutions*, would be published in 1543. We can conjecture why it was that Copernicus waited until he was on his deathbed to publish a piece of work that he had finished long before. An obvious conjecture is the times, which mirrored the tile of Copernicus' book. It was indeed a revolutionary time as the polemics of Martin Luther, a volatile mix of politics and religion, wove itself into a festering regional dispute between the Teutonic knights and the King of Poland. Politicians, like dogs sniffing one another, seek out partners who may offer some advantage. Martin Luther inserted himself into the dispute proposing that the leader of the Teutonic knights, Albert of the House of Hohenzollern, convert to Lutheranism; Albert certainly saw the advantages to this proposal. The success of the Protestant revolt across central Europe was enough to scare the Catholic crown of Poland into conceding Prussia as a Protestant duchy to Albert in exchange for an oath of allegiance to Poland's King Sigismund I. Copernicus was in the thick of the dispute as he took command of the defenses of the Castle at Olsztyn on behalf of the Polish crown during the Teutonic War (1519–1521) and later worked through monetary reforms across Poland's religiously divided territories.

These pressing matters certainly interfered with Copernicus' true calling, but it is believed that Copernicus had completed at least a first draft of *On Revolutions* by 1530. For another decade he dallied on, making a revision here and there. Given that the contents could possibly lead to charges of heresy, one could conjecture that Copernicus delayed publication out of fear of church-sanctioned prosecution. But the church's attitude was lax. It had far greater concerns with Luther and the Reformation. Indeed, by 1533 Rome and Pope Clement VI had come to learn of Copernicus' heliocentrism through a series of lectures delivered by the pope's secretary. Not only did this elicit no negative response from Rome, but also Cardinal Schonberg, a well-respected authority who was a frequent visitor to Rome, encouraged Copernicus by requesting a more complete description of Copernicus' theory. Cardinal Schonberg's letter is punctuated by glowing accolades for Copernicus' achievement. Copernicus was assured that he would not have to bear the weight of religiously inspired persecution and responded by not responding to Cardinal Schonberg's request.

There is no reason to further conjecture why it was that Copernicus delayed publication of his life's work. Copernicus explains himself within the opening paragraph of his preface to *On Revolutions*: "the scorn which I had reason to fear on account of the novelty and absurdity of my opinion almost induced me to abandon completely the work which I had undertaken." At the age of 70, with a desire to leave a legacy, Copernicus agreed to publish his work knowing that his impending death would relieve him of the scorn he feared.

Copernicus left the publishing of *On Revolutions* to George Rheticus, an admirer who prodded Copernicus toward his decision to publish his works. In turn, Rheticus left the work with a Lutheran priest, Andreas Osiander, to oversee the publishing. Although there were no hotheads among the few who were up to date on Copernicus' universe, Osiander understood that religious passions could take root and cast Copernicus' heliocentrism into a fiery hell. As a precautionary measure that would enhance the probability of the book's survival, Osiander, without the permission or knowledge of Copernicus, included a foreword remarking that Copernicus' theory is a useful mathematical device to explain observation. Perhaps it was the reading of this foreword that instigated Copernicus' death, for to Copernicus his theory was no device but illuminated the actual structure of the universe. At any rate, the ruse served its purpose; in later years, *On Revolutions* would be saved from the church's pyre precisely on the grounds that its contents were a mere mathematical device. Osiander's feared backlash came soon enough.

The church initially stumbled in its response to the Reformation while the Lutheran wolf stole half its flock. The church had to recover, refocus, and deliver a coherent message. A wave of conservatism overcame the church; the church enshrined its conservative principles in the Council of Trent, which convened 25 sessions between 1545 and 1563. The council drew distinctions between Lutheran and Catholic doctrines issuing a series of dogmatic decrees. The issues addressed included the celibacy of the clergyman (Catholics yeah, Lutherans neigh); transubstantiation, the presence of Christ's body in the form of a wafer and Christ's blood in the form of wine during communion (Catholics yeah, Lutherans neigh); penance through contrition, confession, and priest-directed obligation (Catholics yeah, Lutherans yeah on contrition neigh to the rest); and individuals' right to interpret scriptures (Catholics neigh, Lutherans yeah). With the Lutherans championing the Reformation, the church championed the Counter Reformation.

The chaos of the Renaissance liberated arts, humanism, and Copernicus. The church's Counter Reformation was designed with the intention of returning Europe's intellect to the catacombs of orthodox scriptures; needless to say Copernican astronomy fell outside the domain of orthodox scriptures. Leading the assault on liberty was Pope Julius III. One cannot fault his desire to stamp out the excesses of church arrogance by curbing the clergy's impropriety as well as the sale of indulgences. But his zealotry morphed into bigotry as he and his successor, Pope Pius IV, burned homosexuals, expelled Jews from Catholic states, and unleashed the Office of the Inquisition to discipline heretics. One of the better known victims of the Inquisition is Giordano Bruno (1548–1600). In 1576, Bruno, an ordained priest, had captured the attention of the Inquisition with his promotion of Copernican ideas among other philosophies that were considered contrary to Catholicism. Fearing persecution Bruno traveled about Europe for 25 years. At the age of 43, in 1591, Bruno returned to his homeland believing that an unstated statute of limitations had passed. In 1592, he was arrested by the Inquisition. Among the members of the Inquisition's committee was Cardinal Bellarmino, who would later acquaint himself with Galileo. Bruno's trial on several counts of heresy was conducted at Rome over a 7-year period; during that time, Bruno endured a rather harsh imprisonment. The Inquisition found Bruno guilty and sentenced him to execution. Bruno's response to his inquisitors speaks for itself, " You pass this sentence with greater fear than I receive it." On execution day, a muzzlelike device was placed on Bruno's jaw and a spike pierced his tongue. He was marched naked to a public square and burned at the stake.

At this crucial point, science is represented by two remarkable men, Kepler and Galileo. Both were brilliant scientists and believers in their faith. It is well known that they both paved the way for the development of calculus and modern science but lesser known that they both espoused philosophies of the role of religious institutions separate from those of the state. Their remarkably similar philosophies were adapted by Christians of all denominations and Christianity has been strengthened as a result. While similar in philosophy, they were quite dissimilar in character. It is their similarities and differences that immortalize these men as fathers of modern science.

Kepler was a Lutheran. He received his education in Tubingen, originally established as a Catholic theological seminary but converted to a Lutheran seminary. As the quote at the beginning of this chapter shows, at Tubingen Kepler was known for letting his opinion be known. He actively participated in university-sponsored debates over a wide range of topics, where he displayed his unequaled intellect. In debate, Kepler was formidable. Like *Star Trek*'s Spock, no matter what argument was thrown at him, he could in an instant analyze it, pinpoint its weaknesses, and counter the argument with impeccable logic. It is little wonder that the irritating know-it-all garnered his fair share of enemies at Tubingen.

Kepler was a multifaceted character, certainly not schizophrenic, but with capacities beyond his prowess at logic. Adjectives to describe Kepler would be seemingly contradictory yet accurate. He was stubborn in his beliefs, flexible, brilliant, naive, wise, loyal, witty, self-deprecating, energetic, idealistic, pragmatic, logical, ill-tempered, and paranoid. Of course, he did not display all these temperaments at once; he could not contemporaneously appear as Lassie and Don Rickles, but at various times he could exhibit this wide range of behavior. There is one important facet to Kepler's character, that is, he was a constant motivator for his works throughout his life. In his writings, Kepler reveals a spiritual side that is akin to Buddha. Indeed, Kepler's initial dream was to become a clergyman in the Lutheran church. He was set on a profession in mathematics by university elders who perhaps believed that such an independent-thinking man would not be suitable in the pulpit. While the elders of the Lutheran church would commit their fair share of mistakes in their dealings with Kepler, this time their judgment was accurate. Kepler would gain well-deserved immortality as a mathematician; as a priest he would have most likely remained an unknown. Kepler, without much choice in the matter, assumed a position of teacher in a Lutheran community ensconced in a Catholic state close to Prague. He came to love mathematics, but as a partner to his spiritual compassion. For Kepler, science and religion were not separate endeavors; both efforts were directed at uncovering the way of God.

After several years as a teacher, Kepler became the assistant to the Habsburg's imperial mathematician and astronomer, Tycho Brahe (1546–1601). The meeting of these two men was a random event. The favor that King Frederick of Denmark had for Tycho Brahe was not shared by Fredrick's son, Christian and when the latter assumed the throne, Tycho was dispossessed. Tycho left his native Denmark and accepted an offer as the imperial mathematician and astronomer to the Habsburg's. At the same time that arrangements between the Habsburg emperor, Rudolf, and Tycho were being finalized, Ferdinand, Archdoke of Austria and Rudolf's brother issued a decree expelling all Lutherans from the Catholic states in his realm. Kepler, living in Catholic territory, was a refugee and Tycho granted him refuge. Kepler, without other alternatives, accepted Tycho's offer. As with Kepler's involuntary career in mathematics, this was a fateful moment that equally defined Kepler's career. Had Kepler been able to realize his preferred position, a position at Tubingen, he would have never gained access to the material that made him immortal. Tycho convinced Rudolf to allow Kepler to live in Prague under Tycho's service. Despite his dire circumstances, Kepler had done all he could to dash the deal. Tycho was the target of Kepler's personality oscillations that nearly caused their relation to fall apart. Had the relations done so, history would surely have taken a different route.

It was not charity that moved Tycho to assist Kepler, but rather Tycho's vision of his own destiny. Tycho had his own model of the universe. Tycho's universe lives in a halfway house between Ptolemy and Copernicus. Following the Ptolemaic system, earth is not a planet but a fixed body at the center of the universe around which the sun and stars revolve. Following Copernicus, the planets revolve around the sun (as the sun itself revolves about the earth). Tycho was an experienced empiricist who had assembled the best astronomical data in all of Europe. Kepler was the young, imaginative, yet untrained theorist. Tycho Brahe saw in Kepler someone who might intuit something from the data that he himself could not. Tycho's vision was that Kepler, like the sun, would illuminate Tycho's universe and Tycho's monument would forever shine on earth. On his deathbed, 18 months after Kepler came under Tycho's employment, Tycho repeated this wish to Kepler. When Tycho expired, the light went out on his universe. Kepler inherited the position of imperial mathematician that Tycho left behind and, without giving Tycho's model a thought for more than a second, pursued Copernicus' heliocentrism over a 6-year period like a salmon swimming to its spawning grounds.

In discussing the reasons that Copernicus was so reluctant to publish his works, one crucial ingredient was overlooked. Copernicus was wrong and Copernicus knew it—all the more reason to fear mockery. While correct on the big picture of heliocentrism, Copernicus was off base on the details. Circular orbits around the sun do not match observed data. The only devices Copernicus had at his disposal for matching data were the very unpleasing devices of Ptolemy. Replace the position of the sun in Ptolemy's universe with that of the earth and the moon, remove the equants, and throw in a few more epicycles and, lo and behold, you have Copernicus' inspiration for Rube Goldberg. But like the Ptolemaic system, measurements recorded in Tycho Brahe's data proved that the Copernican arrow missed the bull's eye. Kepler as guardian of Tycho Brahe's impeccable data knew this better than any man on the planet. Kepler's mission to prove Copernicanism was one of salvation, and he was compelled toward the mission by his belief that it was a path toward God. But the path would not be easy.

There is a saying by Newton, "If I have seen further than others it is because I have stood on the shoulders of giants." This saying has survived because it is a truth about the progress of science that is fitting in any age. Indeed, the saying has been wrongly attributed to Einstein when describing how he arrived at his theory of general relativity. While Einstein never said this, because Einstein stood on the shoulders of Riemann, it would be fitting if he had. Kepler too would require the shoulders of giants and he searched in an unlikely place, the opposition's camp. Kepler chose the same giant as Ptolemy when Ptolemy wished to salvage the geocentric system, Apollonius. While Apollonius proposed the wheel-on-wheel system with the specific intention of preserving the geocentric ideal, the historical irony is that his detailed study of the ellipse, for which he had no application, would be precisely what Kepler needed to save Copernicus. One other giant that Kepler chose to destroy Ptolemy was Ptolemy himself. Ptolemy introduced the concept of an equant, a point not quite at the orbit's center about which a planet moves with constant angular speed. While Ptolemy gave birth to the equant, in fact it was a premature concept that needed further incubation. Reseeded in the fertile mind of Kepler, the equant emerges as the site where the sun chooses to sit and illuminate the planets. The planets orbit the sun along an Apollonian pathway, an ellipse with the sun as its focus. With this model Kepler successfully saved Copernicus and uncovered God's plan. The data confirmed a bull's eye.

In his publication *New Astronomy* (1609), Kepler describes his journey to the ellipse. Scientifically, the text presents several accomplishments beyond a description of a planet's pathway. Kepler was also able to correctly calculate the velocity of a planet along its orbit using the following quantitative proposition. If one measures the area swept out by a line segment connecting the sun with a planet, the measurement will be proportional to the time duration of the measurement. For this proposition, Kepler once again ventured into enemy camp to recruit the shoulders of giants; both Ptolemy and Archimedes gave their support. Ptolemy's equants require the speed of the planet to change as it proceeds about its path. Kepler seized upon the idea of varying speed and incorporated it into his theory. Kepler reasoned that the engine providing the force that governed the motions of the planets was the sun and that the speed of the planets would be faster as the orbit came closer to the sun. Attributing the idea to a work of Archimedes in which Archimedes divides the area of a circle into an infinite number of triangles, Kepler performs a similar operation on the planetary orbits and arrives at his proposition. (Finally, after two millennia the work of Archimedes is available and in the hands of someone who can correctly use it.) This proposition is well known to today's student of physics as the conservation of angular momentum. The quantitative proposition that is identical to the conservation of angular momentum uses a qualitative proposition that Kepler also states in his work; a force from the sun causes the planets to orbit the sun and the force decreases with distance. Newton, standing on Kepler's shoulders, would later demonstrate the elliptical orbit using his theory of motion.

Aside from its scientific merit, Kepler's book reveals much insight into his character. It resembles a diary providing a detailed account of his scientific efforts over four noncontiguous years. At times the account is brutal. In one chapter, he presents volumes of tedious calculations with a hint toward a final resolution. The reader, exhausted by the tedium, feels Kepler's hope and hopes along with him. But no, Kepler ends the chapter with the conclusion that it was all dead wrong. The book also displays Kepler's self-deprecating humor and self criticism. In a passage that describes his foray down a wrong path, Kepler writes:

> *But then what they say in the proverb, "A hasty dog bears blind pups", happened to me . . . If I had embarked upon this path a little more thoughtfully, I might have immediately arrived at the truth of the matter. But since I was blind from desire, I did not pay attention to each and every part of chapter 39, staying instead with the first thought to offer itself . . . and thus entering into new labyrinths . . . , (New Astronomy (1993), p. 455)*

It was with great excitement that Kepler sent his yet-unpublished text, *New Astronomy*, to his university mentor, Michael Mastlin. Kepler was hoping for some assistance with its publication. Although Mastlin had assisted Kepler in publishing a previous book (that launched Kepler's career), he was hostile to Kepler's *New Astronomy* and did not lend the prestige of Tubingen to the effort. This and a dispute with Tycho Brahe's heirs over ownership of the Mars data delayed publication for 2 years. Prior to publication of *New Astronomy*, Kepler had risen to the level of scientific elite through his other works. Upon publication, *New Astronomy* seems to have fallen flat. Kepler had written a daring baroque while Europe was still listening to a lighter classical. The audience did not boo and hiss, they simply did not understand it. Nevertheless, the detail of the work, the significant calculations, the geometric figures elucidating the arguments, and the insights that can come only from genius are impressive and they bolstered Kepler's stature. If the Nobel Prize had been around back then, Kepler would have bagged it. Kepler was the acknowledged preeminent scientist in all of Europe. But times for a scientific superstar were very different from today; could one imagine a current Nobel laureate being excommunicated and forced into a position where he would have to expend an inordinate amount of time and effort to defend his mother against the charge of witchcraft?

Kepler was caught between the Catholics and the Protestants, and among his Lutheran brethren his loyalty was suspect. Suspicions began during his student years when he showed himself to be an independent thinker with his formidable debating skills. Furthermore, Kepler had spent years as the imperial mathematician in the office of Catholics. He was born a Lutheran and believed in its philosophy with strong conviction. Yet he was flexible and understood the viewpoint of others. He hoped that all Christians would find enough common ground to coexist. But most of all Kepler believed in every man's right to seek God's truth without the intercession of a religious authority, be the authority Protestant or Catholic. Kepler never had political designs and, while perhaps disagreeing with church authorities on particular points, would never make a public issue of his disagreement. The strength of his convictions did however cause him to reveal his thoughts to the wrong people at the wrong time. His detractors misconstrued his independent thinking as disloyalty.

Loyalty was critical during these years. Both sides of the religious divide were suspicious of one another and relations between the Catholic Habsburgs and the

Protestant states that they ruled were tense. Each side had its supporters and each side conscripted armies. The situation was extremely volatile and any provocation could set off hostilities; as always the ambitions of European nobility would guarantee provocation.

The Habsburg emperor Rudolf II was ineffectual and without heir. His brother, Matthias, challenged Rudolf's authority. To enhance his position, Matthias made concessions to Protestant states and signed peace treaties with adversaries. In 1608, Matthias cobbled together a Protestant army and marched upon the imperial seat in Prague. In Rudolf's haste to counter his brother, Rudolf also turned to adversaries, the Protestant Bohemians, who surprisingly furnished an army to counter Matthias. Rather than fight, the two brothers came to an agreement to split the empire.

The agreement was temporary as Rudolf showed his true intentions. Rudolf entrusted an incompetent nephew to assemble a Catholic army that would strike at both the Bohemians and Matthias, return all the possessions of the Habsburgs to Rudolf, and restore Catholic prestige. The Catholic army performed as competently as their leader. In 1611, they comically marched on Prague and went after the very Protestant army that had most recently saved Rudolf's royal behind, but they were no match for the Protestants. Once within the city walls, the Bohemian Protestant forces overwhelmed them. Seizing the moment, Matthias brought his own Protestant army to bear and the rout of Rudolf's Catholic army was complete. The Protestant armies celebrated by rioting. They entered the Catholic and Jewish quarters of Prague killing and thieving at will. Fortunately for Kepler, he lived within a safe enclave. The Protestant armies rallied around Matthias and he assumed the *de facto* powers of leader for all of the Holy Roman Empire; in a gesture of mercy, Matthias allowed Rudolf to maintain his titular role.

The instability in Prague caused Kepler to consider returning to Tubingen as a professor. He was considered the preeminent scientist in Europe and, as an alumnus, certainly believed he had more than a good chance. The only thing that matched his brilliance was the naive perspective through which he viewed the Lutheran church. Kepler looked upon the Lutheran church as a philosophical institution when in fact it had very quickly mimicked the Catholic church in becoming a political institution.

During Matthias' rule, a poisoned peace endured; Protestant and Catholics lived under a haze of mutual suspicion. Among the Lutherans, an informal body of elders maintained loyalty, acting in a similar capacity as the Catholic Inquisition. In 1611, with an eye toward returning to Tubingen, Kepler arranged a meeting with Pastor Hitzler, a respected member of the Protestant community and an alumnus of Tubingen. Suspicious of Kepler's loyalties, Pastor Hitzler put Kepler to the test by requesting that Kepler sign the Formula of Concord, a declaration of Lutheran principles. Kepler was most likely unaware that he was being offered a binary are you for us or against us option. While agreeing with most of the document, certain aspects affronted Kepler's beliefs as an independent thinker. He would not sign unconditionally. Passions of the time left no room for accommodation and the Lutheran interpretation to Kepler's refusal was, he is against us. The already difficult relations between Kepler and his fellow Lutherans had reached the breaking point. From then onward, Kepler was *persona non grata*.

In 1613, Kepler's mother was accused of witchcraft by a rather ignoble village neighbor. The accusation of witchcraft was not a trifling matter. There were common occurrences throughout Germany, and the accused, mostly women, often faced trial, torture, and a gruesome execution. The accusation against Katharina Kepler would amount to little more than public gossip over the next 2 years, but gossip in a small village is dangerous. Through a set of unfortunate circumstances spanning 5 years, the accusation came to trial in 1621. Throughout those 5 years, Kepler worked furiously at great expense to have the charges dismissed. He fired off letters reminiscent of his Spock-like debates to all authorities that might be of assistance in dismissing the charges. He once more sought the assistance of his alma mater. Certainly, the most learned and respected men of the Lutheran church could have used their influence to halt this dangerous charade, but the university had long ago turned their backs on their most famous alumni when he refused Hitzler's request to sign the Formula of Concord. He was *de facto* no longer one of them and Hitzler formalized this sentence by excommunicating Kepler in 1619. As the tensions between Catholic and Lutheran went from bad to worse, Kepler stood alone.

In 1618, upon the death of Matthias, Matthias' cousin Ferdinand II assumed the position as emperor of the Holy Roman Empire. The cold peace between the Habsburgs and the Protestant states cracked. Suspicion between the Protestants of Bohemia and their new ruler led to a revolt and a Protestant army sacked Prague. Ferdinand sought the assistance of his Spanish patrons, who obliged their beleaguered relative. In 1621, an army of mercenaries under the sponsorship of the Spaniards besieged Prague. The Protestant army proved to be no match for the Spanish-supported force; their defeat was complete and Ferdinand assumed his position in Prague firmly in charge. The war would have ended at that point, 2 years from its inception, but the political machinations across Europe would not allow peace.

Also in 1621, after years of delayed trial dates, letters, legal review, legal fees, exile, and imprisonment, Katharina Kepler stood trial for witchcraft. The trial began on January 8 and the defense's final legal brief, much of it written by Johannes Kepler, was finalized and sent to Tubingen on August 22. The prosecution also delivered its arguments and requested Cognition of Torture, a sentence designed to terrorize the accused into confession by placing instruments of torture in front of the accused and describing their application. The law faculty at Tubingen found the accused guilty and sentenced Katharina Kepler accordingly.

On September 26, with her famous son standing outside, Katharina Kepler faced the executioner, who carried out the sentence of Cognition of Torture. After listening to the executioner describe the use of the instruments of torture, Katharina responded, "They may do whatever they wish to me. Even if they pulled one vein after the other out of my body, I would have nothing to confess." Then Katharina fell to her knees and prayed. The sentence had been executed. Katharina was led back to her prison cell and several days later released. She died shortly thereafter. After his mother's trial, Kepler returned to Austria where he worked on his science for the remainder of his life.

The world around Kepler mirrored his personal crisis. The success of the Spanish to France's east had prompted an ignoble response from the French. France, a Catholic power, aided the Dutch, a Lutheran state, for the purpose of weakening the Spanish. The war that had been contained within Eastern Europe spread across Western Europe. When the Spanish made short order of the Dutch, the French enlisted Denmark, another Lutheran state, to challenge the Spanish. The war now spread north and raged along the Baltic. Denmark, like the Netherlands before it, succumbed. Nearly all Protestant lands had been subjugated and it was not hard for France to persuade the last Protestant hope, Sweden, to join in the fray. The Swedish army proved up to the task. As central Europe had endured armies sweeping across the continent from the east to the west to the north, now it would endure those battles sweeping down from north to south. With the Spaniards weakened by the Swedes, France relinquished its sideline role as instigator and joined in on the fray. After a 10-year struggle, the French defeated the Spaniards on French soil. Thirty years of fighting accompanied by famine and disease left a demoralized and reduced population across the German states. In 1646 hostilities ceased.

Kepler died in 1630. Kepler's works tilted the intellectual community toward the heliocentric system. To be sure, there were scholars whose feet were planted in Aristotle's universe, but their numbers were decreasing. Many preeminent Catholic scholars were Jesuits; the Jesuit community had its fair share of individuals who acknowledged the brilliance of Kepler's works and accepted Copernicanism. While no authority in Europe could credibly dismiss Kepler's works or the heliocentric system, that did not stop the church, which was out of touch with the rest of Europe, from trying. This great man deserved far more than the circumstances of his times permitted him. Across the Alps to the south, another great man would also be a captive of the times.

Up to the end of his life, the Catholic church unsuccessfully attempted to convert Kepler; bringing a man of Kepler's prestige into the Catholic faith would have been a coup. Only 3 years after Kepler died, the same church that had attempted to convert the man who established the most convincing arguments in favor of Copernican astronomy held an Inquisition against Galileo for writing a piece of work with the intention of convincing the world of the truth of Copernican astronomy.

Before resetting our coordinates for Galileo, there are other scientific contributions of Kepler that are worth mentioning. Kepler proposed that an attractive force is inherent in all bodies. With this belief, he is the first European to correctly attribute the rise and fall of the tides to the attractive forces of the moon. As a purely geometric exercise, Kepler also developed a method to determine the volume of a certain class of shapes known as bodies of revolution. The method is a close relative of Archimedes' techniques, perhaps inspired by Archimedes' works. It is a direct predecessor to integral calculus. This was a near miss for Kepler. He did not associate the method with the study of motion, a task that was left to Newton to complete. Kepler was also influential in optics. He transformed optical science from the science of perception to the physics of light. He was able to furnish a satisfactory account of the workings of a telescope, an account that would be of great use to Galileo. Finally, nearly two decades after publishing *New Astronomy*, Kepler formulated another law of planetary motion that relates the time required for a planet to make a complete orbit with the length of the major axis of the ellipse. Now let us teleport to Galileo.

In 1609, Galileo was professor of mathematics at the University of Padua in Florence. Florence was a thriving city under the control of the Duke of Florence, a member of the illustrious and well-connected Medici family. The Medicis were successful merchants and bankers with the church among their clientele. It is not surprising that a portion of the taxes collected by the church should find its way into the Medici's pockets; three members of the Medici family became God's personal representative on earth occupying the office of the pope. The financial support of the Medicis allowed the arts in Florence to flourish throughout the Renaissance. Galileo occupied a most prestigious scientific position in Italy. Soon his prestige would even be greater. He was the first to use the telescope, crafted from his own hands, for astronomical observations—a feat that captured the imagination of Europe. In addition to his position at Padua, he would become the chief mathematician to the court of the Duke of Florence. The Medici's support would be of great assistance to Galileo, except for one member who was more religiously disposed than others.

As with Kepler, Galileo was a formidable public debater. Unlike Kepler, Galileo was vain. Scientific debates in Italy somewhat took on the spectacle of an intellectual wrestling match. There was a lot of publicity prior to the debates, during which it was not uncommon for each side to perform an intellectual's analog of a gorilla's chest beating. Public declarations predicting that a debater would flatten his opponent were all part of the spectacle. Galileo came from a musical family and, in addition to his scientific prowess, he had a literary flair. Nobody was better equipped to debate than Galileo. In fact, he was so good that his first move in a debate would be to support his opponent's position. He did so with such enthusiasm and clarity that his opponent was left in a speechless stupor, for the opponent could not argue his own position with such skill. And then when all possible opposition arguments had been placed with the opposition totally silenced, Galileo would proceed one by one to destroy the arguments that Galileo himself had voiced. The effect was devastating and humiliating. In his scientific literature, Galileo could be equally humiliating. He had dished out more than a few insults at the expense of members of the preeminent Jesuit institution in Italy, the Collegio Romano. One recipe for producing enemies is to display a public mixture of vanity for oneself with contempt for others. Galileo followed this recipe and the Jesuits would have their chance to respond.

The first time that Galileo attracted the attention of the Inquisition was in 1616; the entire 1616 incident could have been avoided had Galileo been a bit more politically adept. And if Galileo had been more adept in his management of the 1616 incident, the later more famous event may have never occurred. Galileo produced the best telescopes in Europe and turned his telescopes toward the night skies. In 1610, Galileo was the first man to see the moons of Jupiter. He charted the position of the moons and could conclusively determine that they indeed revolve about Jupiter. Galileo, like Kepler, had favored Copernicus over Ptolemy and his observation was the first visual evidence of one celestial body revolving about another. That this was counter to Aristotle and boosted the Copernican argument did not escape Galileo, an already devout Copernican who after this experience was more fired up than ever.

Galileo documented and published his findings and was direct in using his promotion of Copernican astronomy. He corresponded with Kepler who responded most enthusiastically (this is the only issue over which the two men shared a brief correspondence). Indeed, the booster rockets behind Galileo's flight toward fame were fueled by Kepler's endorsement of Galileo's work. Unlike Kepler's work, which was accessible to a few individuals with a high proficiency in mathematics, anyone could gaze through Galileo's telescopes. The telescopes that Galileo crafted and then furnished to influential men across Europe were a sensation across the continent. Galileo enjoyed the prestige that was showered upon him.

Inspired and with his telescopes in hand, Galileo went directly to the top with his findings. In 1611, with the assistance of his benefactor, the Grand Duke Cosimo Medici, Galileo traveled to Rome for a meeting with the pope. Upon his arrival, Galileo was warmly greeted by the Jesuits of the prestigious Collegio Romano; the Jesuits were the preeminent Vatican authorities over scientific issues. It would be difficult to discern later conflicts between Galileo and the Jesuits from their gracious hosting of Galileo in 1611. The trip to Rome culminated in a successful audience with Pope Paul V. This was a coup of which Galileo could rightly be proud as the pope took little interest in scientific endeavors.

There were detractors who claimed that the Jovian moons were not real; they were only illusions created by the telescope. Galileo could well afford to ignore the detractors; he was at the top of his game. But it just was not in his character to ignore any challenger. One particularly worrisome challenger came in the form of the matriarch of the Medici family, the Grand Duchess Christina, who was concerned by the religious implications of Galileo's work. The Grand Duchess and Galileo had a cordial relationship. Galileo, as court mathematician in the service of the Medicis, had tutored her son. In 1615, the Grand Duchess was in the company of Benadetto Castelli, a monastic student of Galileo, and Doctor Boscaglia. The conversation turned to Galileo's telescopes; the Grand Duchess was eager to learn of the observations from one who had used the telescopes firsthand. What she heard displeased the Grand Duchess as she came to the conclusion that the scientific endeavors contravened the Holy Scriptures and Doctor Boscaglia, an anti-Galilean detractor, encouraged the Grand Duchess' concern.

I know of no one who has entered their student years without observing or participating in a debate between science and faith. The fighters in the debate might believe they are in the same ring but are arguing in different dimensions. While swinging furiously, neither is able to land a blow on their opponent, who is after all inaccessible to their reach. And so I know of few who have departed their student years without leaving such debates behind as fruitless endeavors. Every spectator of such debates knows that there is one set of circumstances in which there will be a knockout—when the side fighting for science enters the dimension of faith. This unfortunate fighter has foolishly exposed his body to all of the blows that the side of faith can deliver while his scientific counterpunches have no target.

In Galileo's day, particularly in Italy, a scientist's position was precarious. If he entered the debate, he had no choice but to fervently acknowledge his faith; it was best to avoid such circumstances. Yet, it is amazing that several times in his life Galileo jumped into the ring when he might have avoided it all together.

Castelli informed Gallileo of his encounter with the Grand Duchess and Galileo responded by penning his thoughts on science and faith in a letter. While the letter is addressed to Castelli, with pride Galileo encouraged Castelli to circulate it freely, and thus Galileo entered the ring. The letter was soon further circulated, reaching the hands of Galileo's detractors. Christmas Day of 1614, Father Thomas Caccini, preaching from the pulpit of the church of Santa Maria Novella, denounced "practitioners of diabolic arts." His references to Galileo's letter made his target clear.

With accusations of heresy made public but still on his feet, Galileo felt the need to defend himself. He took a two-pronged attack that a military man might call a pincer movement, but the only man squeezed was Galileo. Clueless to the damage already caused by his previous letter, Galileo penned another letter, this time to the Grand Duchess. Like the previous letter, this one presents his scientific views through the lens of faith but then goes further. Galileo denounces his denouncers, not specifically by name, but it is clear that he sees these are men as abusing their position in the church. Anyone believing that a company email is confidential between sender and receiver is in for a rude awakening. Whether Galileo was aware of the rumor grist or not, there was a certain predictability that the letter would be circulated once more among a less-than-friendly crowd. And as it circulated, editorial license was taken. The letter, or an edited version, did make its way to Rome. The second prong of Galileo's attack was even more foolhardy. Galileo set himself upon a mission to convert the pope to Copernicanism.

In 1616, Galileo undertook a second trip to Rome. The first had gone well, as evidenced by his audience with the pope. But the impetus for the pope's adoration had worn off, the novelty of Galileo's telescopes had dissipated, and feelings toward Galileo were more ambivalent. In the intervening years between visits, Galileo had dished out some rather harsh criticism of the Jesuits at Collegio Romano. The criticism had a lingering sting, especially given the favorable treatment that they had shown Galileo. While this sting was an undesirable background to Galileo's visit, the crux of Galileo's folly is that Galileo never grasped the essence of what the church was. The church was more of a political institution than a philosophical one. No philosophical argument could ever persuade the church to do something politically naive. It was a naive Galileo who, fueled by his vanity, believed that the pope could be persuaded by scientific reasoning.

Galileo literally took it to the pope in the form of a treatise that explains the tides as a result of the earth's rotational and revolutionary movements. When Galileo arrived in Rome, he passed the manuscript to a cardinal, who then passed it on to the pope, who then quite predictably never read it. In fact, given the times, in his political role the pope had few options. Since Luther, the church had been under siege on the fronts of nationalism and religious interpretation. The pope could hardly personally endorse a third front, scientific reasoning. Galileo's gambit left the pope with no choice but to address the Copernican menace and hit hard. The pope left the heavy lifting to his most educated Jesuits under the guidance of Cardinal Bellarmino, who reported their findings to the Office of the Inquisition. The Jesuit response to Galileo and Copernicus was a biblical interpretation of the structure of the universe. Citing passages in the Bible, the Jesuits proclaimed a geocentric universe and furthermore stated that those who supported the theory of Copernicus were guilty of heresy.

After 2 weeks, Cardinal Bellarmino summoned Galileo to his office. Speaking on the pope's behalf, Cardinal Bellarmino informed Galileo of the church's position. Also present in Cardinal Bellarmino's office was Father Michelangelo Seghizzi, a high-ranking official from the Office of the Inquisition. Father Seghizzi threatened to use the Office of the Inquisition to proceed against Galileo should Galileo persist in espousing Copernican astronomy. Galileo acquiesced.

One week after Galileo's visit with Cardinal Bellarmino and Father Seghizzi, the church published an official edict conveying the official church position on the structure of the universe. In no uncertain terms, Copernicus was rejected. The edict also censured Copernicus' original book, suspending further printing pending review and editing. Additionally, the church explicitly forbade continued publication of a book by Father Paolo Foscarini in which the priest attempts to reconcile biblical interpretation with Copernican astronomy. The church indicated that scientific theory had no role in theological interpretations. The printer of Father Foscarini's book was shortly afterward arrested and 2 months later the 36-year-old Father Foscarini died.

Galileo must have considered himself to be lucky. None of his works were censured and he was not specifically mentioned in the edict. But the warning had been issued; Galileo may not have understood that gravity was the cause of the tides, but he fully comprehended the gravity behind the Inquisition's demand to abandon Copernicus. Galileo left the ring but remained in the gym. For 7 years Galileo abided by his agreement with the church leadership. While Galileo did offer one more publication, *The Assayer*, in which he assails a prominent Jesuit scholar, he was mute on the subject of Copernicus. But nothing is permanent. People change office. New circumstances cause institutions to go forward and not dwell in what is no longer relevant or so it would seem.

In 1623, Pope Paul V passed away. Maffeo Barbarini assumed the papacy and took the name Pope Urban VIII. This was a great sign for Galileo. Galileo and the new pope had clicked during Galileo's first visit to Rome. The new pope agreed to an audience with Galileo and the two seemed to have renewed their friendship. In addition to these favorable signs, Kepler had published, *Introduction to Copernican Astronomy*. This text was more accessible to a wider audience than Kepler's earlier work. With this publication, the intellectual argument between Ptolemy and Copernicus shifted even further toward Copernicus, and among the church's flock there were many shifters. To Galileo, there appeared to be a new era in which his commitment of 1616 was irrelevant.

In the same year as the ascension of Pope Urban, Galileo embarked on a new project. Using his literary skills, he wrote an argument in favor of Copernican astronomy that could be followed by the layman. His approach was influenced by his years of public debate; the argument was in the form of a play in which one of the characters debates in favor of Ptolemy while another debates in favor of Copernicus. The fate of the character who debates on the side of Ptolemy, Galileo named him Simplicius, follows the real live humiliation of those who went against Galileo in his actual debates. Galileo named the play *Dialogue Concerning the Two Chief World Systems*.

Galileo first wanted to ensure that he would have the church's support for the endeavor. As was his nature, he went to the top. With this limited objective in mind, as opposed to his previous attempt at conversion, Galileo received an audience with Pope Urban. Galileo described his work and stressed that whatever view the church takes, it must demonstrate that its view was not based on ignorance and that the church understood all of the arguments for each position. Indeed, it seemed that the times were different, for Galileo successfully prevailed upon the pope to lend his authority to the project.

It took Galileo years to complete the play; during those years, Europe was in chaos. With the onset of the Thirty Years' War in 1618, wartime politics preoccupied the church. The Urban papacy exacerbated an already tense situation. The Catholic armies that were ostensibly assembled to fight off the Protestants were split into two camps, French and Spanish. The Habsburgs had the support of Spain, while the French wished to use the circumstances in the Habsburg empire to undermine Spanish influence. Rather than reconciling the two sides so that the Catholics could fight with a united army, Pope Urban came out openly in favor of the French, while at the very same time the French were undermining the Catholic cause through their alliances and support of Protestant states. As the war proceeded, it became obvious that the church no longer held its previous authority among Catholics across Europe. The church's response was to sulk and hunker down in the city states of Italy.

Upon completion of *Dialogue* in 1630, Galileo (2001) followed the protocol of censorship that all such works in Italy were subject to. The church would have preferred wider censorship authority, but it had no more control of the publication process across Europe than a modern government, no matter how nefarious, has control over a website published outside its borders. However, within Italy, the church had censorship authority guaranteeing that Italy would not be a full participant in the scientific revolution that was yet to come. Galileo dutifully sent his work to church censors for approval. The process was a long one. One could only imagine a seething Galileo repulsed by the very idea of being subject to a censorship board that was far less qualified than him on the topic of the book and yet there was nothing he could do but play the game. And he did play it well. Galileo managed to stack the board of censors with individuals favorably disposed toward him. There were passages that the censors found offensive and Galileo edited these passages until the reviewers found them acceptable. After a little over a year of back and forth, the censors gave their approval and the play *Dialogue* was ready for publication in the fall of 1631.

This was no ordinary play in any sense of "ordinary." It was a literary play written by a mathematician on a subject that resided within a scientific realm over which religious authorities laid claim. It was long, over 500 pages. Once *Dialogue* had cleared church censorship, the typeset on Mr. Gutenberg's printing machine had to be readied, and the pages had to be printed and bound. It was not until February of 1632, 10 years after Galileo conceived of the project, that *Dialogue* was available to the public. The book was well received almost everywhere, the exception being certain offices in the church, including that of the papacy. There, it was viewed as another volley against church authority, equally threatening, if not more so, than the Protestant army that had sacked Rome nearly a century before. There was one dissimilarity between the two attacks; the Pope was defenseless against the previous attackers but not against the current attacker.

In September 1632, the church suspended publication of *Dialogue*. Furthermore, Galileo was summoned to the Office of the Inquisition. This was unlike Galileo's previous 1616 meeting with an officer of the Inquisition where a warning was issued. This was an official proceeding. Claiming ill health, Galileo requested that he respond to the Inquisition's inquiry from Tuscany, a request that was denied. In the presence of an official from the Inquisition's Florentine office, Galileo subjected himself to a panel of physicians who reported on the state of Galileo's health. The panel's report of Galileo's infirmities confirmed that a journey to Rome would endanger his life. The church ignored the report and once more demanded that Galileo respond to the Inquisition's summons by traveling to Rome. At the age of 69, Galileo made the journey to Rome. The actual condition of Gaileo's health is unknown. Some biographers claim that Galileo was indeed in bad shape, while others claim that Galileo's claims of ill health were a ruse. Galileo did survive the trip and outlived his progeny to the ripe age of 80.

Galileo reached Rome on February 13, 1633. Later in the spring, he stood before the Inquisition on the charge of not faithfully executing his commitment of 1616. Despite receiving papal approval before commencing with the project and despite having subjected himself to the church's censorship process, it was a fact that Galileo did ignore his 1616 commitment to cease espousing Copernican astronomy. Whether or not Galileo learned the lesson that the past can always strike back and finally understood that the church was first and foremost a political entity is unknown. But understand it or not, Galileo would feel the painful blow of church authority. On June 22, the church passed its sentence. The Inquisition found Galileo guilty of heresy for defending Copernican astronomy.

Public humiliation was a normal element of church justice and Galileo's sentencing was no exception. Feigning mercy, the sentence of the Inquisition agreed to grant absolution to Galileo, so that it would still be possible to access heaven for eternity, provided that Galileo publicly renounce his misdeeds. The renunciation had been prepared and Bruno certainly would have empathetically encouraged Galileo to read it, better than a spike through the tongue ending in a crispy departure at the stake. As Galileo read the renunciation, the members of the Inquisition as well as an assembly of witnesses listened:

I, Galileo, son of the late Vincenzio Galilei, Florentine, aged 70 years, arraigned personally before this tribunal, and kneeling before You, Most Eminent and Reverend Lord Cardinals, Inquisitors-General against heretical depravity throughout the Christian commonwealth, having before my eyes and touching with my hands the Holy Gospels, swear that I have always believed, I believe now, and with God's help I will in future believe all that is held, preached, and taught by the Holy Catholic and Apostolic church. But whereas-after having been admonished by his Holy Office entirely to abandon the false opinion that the Sun is the center of the world and immovable, and that the Earth is not the center of the same and that it moves, and that I must not hold, defend, nor teach in any manner whatever, either orally or in writing, the said false doctrine, and after it had been notified to me that the said doctrine was contrary to Holy Writ-I wrote and caused to be printed in a book in which I treat of the already condemned doctrine, and adduce arguments of much efficacy in its favor, without arriving at any solution: I have been judged vehemently suspected of heresy, that is, of having held and believed that the Sun is the center of the world and immovable, and that the Earth is not the center and moves. Therefore, wishing to remove from the minds of your Eminences and all faithful Christians this vehement suspicion justly conceived against me, I abjure with a sincere heart and unfeigned faith, I curse and detest the said errors and heresies, and generally all and every error and sect contrary to the Holy Catholic church. And I swear that for the future I will never again say nor assert in speaking or writing such things as may bring upon similar suspicion; and if I know any heretic, or person suspected of heresy, I will denounce him to this Holy Office, or to the Inquisitor or Ordinary of the place where I may be. I also swear and promise to adopt and observe entirely all the penances which have been or may be imposed on me by this Holy Office. And if I contravene any of these said promises, protests, or oaths (which God forbid!), I submit myself to all the pains and penalties imposed and promulgated by the Sacred Canons and other Decrees, general and particular, against such offenders. So help me God and these His Holy Gospels, which I touch with my own hands. I, the said Galileo Galilei, have abjured, sworn, promised, and bound myself as above; and in witness of the truth, with my own hand have subscribed the present document of my abjuration, and have recited it word by word in Rome at the convent of Minerva, this 22nd day of June 1633. I Galileo Galilei, have abjured as above, with my own hand. (Sobel (1999), p. 275)

After his public humiliation, Galileo was placed under house arrest for the remainder of his life. As a temporary measure, the church arranged for Galileo's internment in the city of Siena under the stewardship of Archbishop Piccolomini. While this is the same city in which the infamous Cardinal Borgia held his orgies, there would be no celebrations for Galileo. Nevertheless, Archbishop Piccolomini was an admirer of Galileo and sought to heal his famous prisoner's broken spirit. Archbishop Piccolomini arranged for contact with scholars across Europe on a wide range of topics that studiously avoided the Copernican debate. The response of the broader European community may have provided a measure of comfort to the frail scientist.

Through its minions, the church got wind of Galileo's favorable internment in Siena and hastened his move to a more permanent location. A house was built near San Matteo, a monastery in the hills of Tuscany. Within the monastery, Galileo's three illegitimate daughters had lived as nuns. They were all born to the same mother, Galileo's lover whom he never married. Throughout Galileo's ordeals, his most beloved daughter, Marie Celeste, comforted and consoled him. There are many surviving letters that attest to a loving relationship. Galileo was imprisoned in his new location and a Church-appointed overseer restricted his social contacts. Galileo confronted one more ordeal in his last years, the death of Marie Celeste in 1634. Then in 1642, still under house arrest, Galileo passed away. He literally reunited with his beloved daughter in the grave: A later exhumation discovered the presence of Marie Celeste.

Galileo left an impressive scientific legacy. His greatest contribution was in the study of motion. The legend of Galileo's dropping of two weights from the Tower of Pisa is an unconfirmed story. But it is drawn from factual accounts of other experiments that Galileo did perform, for Galileo was very much an experimentalist. The most significant of his experiments led to a mathematical description of free-falling bodies.

Galileo designed low-friction roller coaster–like tracks upon which he would roll spheres. Using a novel timing mechanism, he timed the descent of the spheres from a standstill position to the bottom of the track. The timing mechanism was similar to an hourglass, but rather than using sand, Galileo used water. A water source was connected by tubing to a pan below. He started the flow of water by opening a clamp on the tubing at the beginning of the sphere's descent and then closed the clamp at the end of the sphere's descent. The water would collect itself in the pan and then Galileo would weigh time, for the weight of the water was proportional to the time of the descent.

Through this experiment, Galileo discovered that the distance that a body descends is proportional to the square of the time of the descent and independent of the body weight. Galileo mathematically described the motion of an object as having a constant acceleration that is independent of the object's weight. With this description, Galileo rediscovered Oresme's work of two and a half centuries earlier. Galileo recognized that the pathway of an object falling in accordance with this description is a parabola, a conic section studied by the same Apollonius who furnished Kepler with his ellipse.

There were two ramifications of this experiment: one toward the past and one toward the future. Looking to the past, this was a refutation of Aristotle, who predicted that the descent of a heavy object would be faster than the descent of a light object. Looking toward the future, this was a step toward the development of calculus and the laws of motion; it established a concrete example of a specific motion along with a law for the description of that motion. A concrete example is valuable to scientists for two reasons. First, the example may be generalized into a broader theory. Second, it furnishes a test case for a more general theory; the generalized theory must produce the results of the concrete example or it is rejected.

Einstein was particularly partial to Galileo, calling Galileo the father of modern science. In defending the Copernican system, Galileo proposed the first notions of relativity. Within the Aristotelian universe, the earth is fixed, and points in space can be clearly identified. However, the Copernican system sets the earth adrift. Galileo recognized that it is not possible to fix a point of reference within a system that is itself adrift. Accordingly, measurements of motion are not fixed but are relative with respect to the motion of the observer. Any laws of motion must account for relative measurements. Einstein's theory of general relativity expands upon Galileo's ideas.

One wonders if the church would have prosecuted Galileo had it foreseen the damaging consequences to its own cause. The church's 1616 bout with Galileo was already damaging enough. Throughout Europe, scientists had their own telescopes, many of them gifts from Galileo, and they could judge for themselves whether the Jovian moons were images or reality. With Kepler's explanation of the workings of a telescope, public opinion fell behind Galileo. The publication of Kepler's work, *Epitome of Copernican Astronomy*, further increased the distance between the church and public opinion. The prosecution of Galileo over *Dialogue* reinforced the trend. While the church banned the printing of *Dialogue* from the time of Galileo's trial, the ban was only enforceable in Italy. Elsewhere throughout Europe, *Dialogue* became a sensation. Galileo's prosecution was the church's last battle for the control of the individual's intellect. The church had already lost this war but would not concede.

The humiliation of Galileo was seen for what it was, bullying out of frustration, and the European scientific community tossed aside both Aristotle and the church. After a millennium, European scientists were freed from church dogma. Institutes of higher learning were no longer church-sponsored theological seminaries; they transformed into centers of independent inquiry. In this setting, a mere 23 years passed between Galileo's death and Newton's discovery of calculus.

Greatness springs from many sources. Newton was born into humble circumstances in the village of Woolsthorpe-by-Costerworth. As if to presage his greatness, Newton was born in the very same year as Galileo's death, but Woolsthorpe-by-Costerworth was an unlikely place for Galileo to pass his baton. Woolsthorpe-by-Costerworth was a village where young men went to church and followed in the path of their fathers, often uneducated. Like Kepler, Newton's father passed away when Newton was a boy. His mother remarried and, also like Kepler, Isaac was sent to live with his grandparents. Newton attended the local elementary school and later the King's School. At school, Newton was a loner. He was for the most part an unnoticed and uncommunicative child. During recess, one can imagine his inner thoughts intensely focused on a subject of curiosity while his classmates were playing games. And in class, one can imagine Newton's mind far removed from the subjects at hand that he would find trifling to the point of unbearable boredom. He was a misunderstood child who developed a capacity for turning inward, a skill that assisted him in his dogged pursuit of research. However, he did not develop the social skills that would later be necessary for dealing with the broader community.

Fortunately, there was one educator who understood Newton a bit more than others. Prior to completing high school, Newton's family removed him from school so that he could oversee the family farm, as was expected of him. In a short time, it became apparent to Newton's family that he was completely ill suited to this endeavor. Henry Stokes, the headmaster at the Kings School, perceived that Newton's intellectual capacity was quite strong, although no one could imagine his brilliance, and persuaded Newton's mother to allow Newton to finish his studies and enter college.

In 1661, Newton began his university training at Cambridge under the tutelage of Isaac Barrow, holder of the prestigious Lucasian Chair of Mathematics. For the first time, Newton was exposed to a circle of men who could stimulate his intellect and direct his growth. At Barrow's request, Newton reviewed Euclid's *The Elements*. Then Newton immersed himself in the most current mathematical texts and kept up with the most current mathematical issues. Among the works that Newton found most exciting was Rene Descartes' text on analytic geometry. Rene Descartes had successfully combined algebraic concepts, as developed by the Arab world while Europe experienced the Dark Ages, with the geometry of the ancient Greeks. He synthesized the two through the creation of a coordinate axis system that we now refer to as Cartesian coordinates and proposed the general concept of a function. Geometric objects become numeric entities by their association with points on the coordinates. Equations with quantities expressed as variables represent geometric shapes. Analysis of the geometry is possible through algebraic manipulation of the variables. The conic sections of Apollonius and in particular the ellipse and the parabola are investigated in Descartes' work.

Aside from books, Newton was exposed to scientific shop talk. European mathematicians had worked out specific cases of differential and integral calculus; Isaac

Barrow was a significant contributor. Through publications and letters, these advances circulated to European universities where they were topics of discussion. The greatest discourse revolved around the works of Kepler and Galileo. While these two men had accurately described trajectories of motion, one for planetary orbit and one for a falling body, there was a desire to find a general theory of motion that would account for Kepler's and Galileo's discoveries. Established intellectuals from all over Europe committed themselves to this endeavor. Into the mix stepped a novice, the young Isaac Newton, who was ready for the moment.

In 1664, a particularly nasty spell of the bubonic plague gripped England hitting densely populated areas. Within a year, one of six Londoners had succumbed. In 1665, as a precautionary measure, Cambridge University shut its doors and did not reopen until 1667. At the age of 22 Isaac Newton returned to his hometown and in isolation engendered the most productive burst of scientific discovery in history. In optics he developed a new theory of light. In mechanics he laid the groundwork for the laws of motion and made progress toward a theory of gravity and planetary motion. These were the beginnings of an effort that he would complete two decades later. Newton's most remarkable achievement came in mathematics. Newton jump-started the field of power series with his discovery of the binomial theorem. But the real treasure was an area that Newton named the theory of fluxions. Prior to Newton, the mathematical tools required to model natural phenomena were feeble. Scientists were chiseling away at the nature's secrets with a pocket knife. With his theory of fluxions, Newton invented an earth mover that would later be called calculus.

Upon his return to Cambridge, Newton shared his discoveries with Isaac Barrow. One can only imagine the impression that Newton's work made upon Barrow. Past men had achieved greatness for lifetime contributions that amounted to a smidgen of what Newton had intuited in isolation during the plague-induced closure of Cambridge. Barrow was overwhelmed. Although he garnered significant prestige in academic circles, he understood that Newton was a phenomenon. He knew that Newton was far more worthy of the Lucasian Chair that he held. Of his own accord, Barrow resigned the position and passed it on to Newton.

Unfortunately for Newton and mathematics during his time, there were those in the scientific community who were unlike Isaac Barrow and were not gentlemen. Unfortunately for science, Newton did not have the tools to cope with such men. During the plague years, Newton had produced an enormous quantity of research; as noted above, his research on motion and calculus was only a portion. One of the offspring of Newton's research was the reflecting telescope, a new design of telescope built by Newton that was far more powerful than its predecessor. As Galileo had discovered, the telescope captures the imagination of the public more than abstract mathematical theory. The success of Newton's telescope caused a stir and influential men prevailed upon Newton to publish something that would explain its workings. Newton responded with a publication of his new theory of light.

Newton's theory was as novel in his day and age as was Aristarchus' heliocentric theory in his. And like Aristarchus' bold theory, Newton's publication met with an overwhelmingly hostile reaction. In Newton's day, there was not a satisfactory explanation for color. Newton, ever the tinkerer, had placed a prism across a ray of light that he allowed to enter his room through a pinhole in a blind. He observed the ray of white light separate into streams of color. The thought occurred to Newton to attempt to restore the light to its original color by inserting a second prism in just the right fashion across the streams of color emerging from the first prism. Newton found that he could indeed recover white light. His conclusion, now accepted, was that white light is a cluster of colored light. Newton reported his findings and his conclusion in his first ever publication.

Preposterous, cried more than one esteemed member of the scientific elite. It went against all intuition that the mixing of many colors could result in white. This was as obvious to anyone who mixed paint as the geocentrism of Aristarchus' contemporaries. Newton found himself under siege. A leading figure in the assault was Robert Hooke (1635–1703). Hooke was a nasty, cantankerous know-it-all set on proving that he was the greatest. Hooke had his own theory concerning light and colors that was in conflict with Newton's theory. Today, few know and nobody cares about Hooke's theory because it is wrong and irrelevant. But in Newton's time, Hooke's prestige could not be ignored. Newton found himself defending his ideas in an exhausting and continuous battle. Newton fired off many letters in response to Hooke's salvos in his direction. The whole experience disturbed Newton and his response must have been reflexive; Newton turned inward. Not wishing to be mired in controversy, Newton did not return to the task of publishing his other discoveries for decades. He sparingly doled out his ideas to a small circle of friends and let it be known that he did not wish his notes to be published prior to his death. Then he all but retreated from science for a period of more than a decade. Throughout the 1670s and early 1680s, Newton, a man endowed with one of the most brilliant scientific minds in all of history, engaged himself in the pursuit of alchemy.

The appearance of comets was a mystery to the men of the Middle Ages. The appearance of motion in the heavens, where God resides in perfection, was at once disturbing and inspiring. Astrologers would forecast doom or fortunes at the sight of a comet. The post-Renaissance appearance of a comet in 1681 observed by a young Englishman named Edmond Halley was equally mysterious. An astrologer's forecast of a significant event would have been dead accurate. In 1684, Halley's wish to understand comets led him to Newton. Halley challenged Newton to present a description of a comet's path; this was the prod that Newton needed to reconnect with the work that he started during the days of the plague. Newton solved the comet and indeed the paths of all bodies orbiting the sun using his method of fluxions. With the mathematical tool, Newton revealed Kepler's ellipse.

This is the place where our story should end. Calculus has been invented and the ellipse revealed. The historical trail from Aristarchus to Newton has been closed and, as the ellipse, it returns to its starting point, a heliocentric system. In between, the trail passes through Euclid, Apollonius, Ptolemy, Copernicus, Kepler, and Galileo and follows the mathematical achievements that allow science to uncover the nature of our universe. The trail illuminates the worst as well as the best of man: intolerance, intimidation, injustice, intelligence, imagination, and grit. The trail tells a story of intellectual triumph and makes the case for freedom of expression. It is a perfect ending to a dramatic story. But history does not allow us to end the story here. One more twist is left in the development of calculus. And of course we cannot abandon Archimedes without bringing him into the fold of developments.

Newton came to believe that his notes were inadvertently circulated to a German philosopher, Leibniz. Leibniz was born in the German town of Leipzig in 1646. His father was a professor at Leipzig University. Leibniz, like Newton and Kepler, became fatherless at a young age, but the similarity in upbringings stops there. Leibniz' family nourished his education from the outset. He was not a loner but wherever he went was the toast of the school. A brilliant future was foreseen from his early days. Leibniz lived up to the expectations and was quickly accepted into the elite intellectual circles of Europe. Although Leibniz matriculated with a degree in law, at the age of 20 he became interested in mathematics. The facility he had with men of great talent was of service, for one of Europe's most preeminent mathematicians, Christian Huygens, tutored Leibniz long before Leibniz had a reputation of his own. Within a short time, Leibniz surpassed his teacher.

Leibniz was a grand visionary with a keen mind for abstraction and theory. The historical record shows that Leibniz independently discovered calculus in 1675 and published his first works on calculus in 1684. Although Newton discovered calculus 10 years before Leibniz, Leibniz' work was not superfluous. His notation is superior and it is Leibniz' notation that has been adopted as an international standard. What was Leibniz' eureka moment? As with Newton, Leibniz was aware of initial efforts in the direction of calculus. Using his ability for generalization, it is conceivable that Leibniz was able to distill the initial efforts of his contemporaries into theory and then expand upon the theory; perhaps this is how it happened. But Leibniz himself posits a suggestion of what occurred. Leibniz once stated, "He who understands Archimedes and Apollonius will admire less the achievements of the foremost men of latter times." Calculus could easily have been spawned out of the mix of the trained and theoretical mind of Leibniz with the pioneering work of Archimedes. So, after a 2000-year hiatus, indeed the road to calculus may have passed through Archimedes.

It is fitting to end our story with Leibniz. The best known of his sayings is perhaps the least understood, "We live in the best of all possible worlds." Leibniz was not blind to the cruelty of the world around him, and quite the contrary, he had a very firm grasp of the world and knew that cruelty was very much a part of the human experience. His view is that the most negative aspects of the flawed human character are inherent in any possible world. It is within this constraint that Leibniz made his observation. So did mankind have to pass along such a tortuous and at times depraved path toward one of its most outstanding achievements? Perhaps so. Given our flawed character, we took the only path we possibly could. And are we doomed to set a similarly depraved course toward our future or can we overcome our negative traits—Let us move onto mathematics.

2.2   NUMBERS

This book presents the mathematics required to derive Kepler's ellipse using Newton's calculus. The presentation passes through geometry, algebra, trigonometry, and finally calculus. But mathematics begins with numbers and so we begin with an investigation of numbers. If one imagines numbers as points on a number line, there is no obvious distinction among the points. And yet, the properties of the distinct points and the way we express them is very different. In this section, differences between these seemingly similar points are investigated. The differences result in the classification of real numbers into integers, rational numbers, and irrational numbers. Examples motivate the classification of numbers. Finally, this section explores the impact that different numbers have on two related systems, artistic designs and the collision of an asteroid with the earth.

2.2.1   Integers, Rational Numbers, and Irrational Numbers

2.2.1.1 Integers   The concept of a whole number, 1, 2, 3, and so on, is so intuitive that no explanation will be given. One is able to use whole numbers to solve a variety of problems. Consider, for example, payment for a basket of goods. One simply multiplies the unit price of each item by the quantity of purchase for each item to find the payment for each good, and then one sums across all the goods. This can be accomplished using whole numbers.

Let us complicate the situation and see that restricting ourselves to whole numbers causes difficulties. Suppose that two individuals, Mr. G and Mr. K, are trading baskets of goods and that they will settle the difference with a cash payment, assuming that there is an agreed-upon market price for all goods. Let us consider the value of the swap prior to cash settlement from the perspective of each party and then determine how to settle the cash payment. Each individual calculates the benefit of the swap by subtracting the value that he gives from the value that he receives. The calculation appears as follows:

$$V_G = B_G - B_K \qquad V_K = B_K - B_G$$

where

- V_G is the value of the swap to Mr. G
- V_K is the value of the swap to Mr. K
- B_G is the value of the basket that Mr. G receives
- B_K is the value of the basket that Mr. K receives.
- All values are given in currency units that are henceforth denoted by CUs

Unless the values of the baskets are identical, one of these individuals will receive less than he gives and the value of the swap will not be a whole number. Set, for example, B_G to 10 CUs and B_K to 15 CUs. Then for Mr. K, the value of the swap is 5 CUs, but for Mr. G, there is no whole number that can express the value of the swap. The problem requires numbers that are less than zero. So we couple the concept of a negative number with that of the whole numbers to arrive at integers. In the problem above, for Mr. G, the value of the swap is given by the integer value of -5 CUs. The settlement payment that accompanies the swap is then the value 5 CUs from Mr. K to Mr. G so that the value of the complete transaction to both individuals is identically zero. This is a fair transaction.

2.2.1.2 Rational Numbers and Conversion   With a slight complication of the problem, we see that it is necessary to expand beyond integers to determine meaningful quantities. Consider that Mr. G wishes to obtain 100 L of Lowenbrau beer and he has arranged to swap a fair quantity of Chianti. Let us imagine that one can purchase 3 L of Lowenbrau for 1 CU and sell 2 L of Chianti for 1 CU and these values indicate a fair market price. The problem is to find the quantity of Chianti that Mr. G must swap.

Since the market value for 3 L of Lowenbrau is identical to the market value for 2 L of Chianti, for every 3 L of Lowenbrau that Mr. G wishes to obtain, he must provide 2 L of Chianti. There are several ways to arrive at a solution. One way is to separate the 100 L of Lowenbrau into individual pitchers of 3 L and then associate 2 L of Chianti with each pitcher of Lowenbrau. Mathematically, the number of liters of Chianti that Mr. G must sell is expressed as follows:

$$\text{Pitchers of Loewenbrau} = 100 \div 3$$

$$\text{Liters of Chianti} = \text{pitchers of Loewenbrau} \times 2$$

$$= (100 \div 3) \times 2$$

Immediately, we can see that neither the pitchers of beer nor the equivalent liters of Chianti is integer valued. There are a bit more than 33 but a bit less than 34 pitchers of beer. We must enlarge our set of numbers to include rational numbers, those numbers that can be expressed as fractions with an integer in the numerator and an integer in the denominator. With the assistance of rational numbers and the ability to perform arithmetic with them, the answer becomes $\frac{200}{3} = 66\frac{2}{3}$ L of Chianti that Mr. G must furnish.

Within the word *rational* lies its basis, ratio. A rational number is a ratio between two integers and thinking in this manner can yield some insight. In the above problem, one can arrive at a ratio of the quantity of Chianti to the quantity of Lowenbrau that equates their market values; the ratio is $\frac{2}{3}$. Notice that this ratio is independent of units; 2 L of Chianti is equal in value to 3 L of Lowenbrau and 2 barrels of Chianti is equal in value to 3 barrels of Lowenbrau. The quantity of Chianti that is equal in value to any given quantity of Lowenbrau is obtained by multiplying the given quantity of Chianti by $\frac{2}{3}$.

The problem above is a conversion problem and in this case the conversion of liters of Chianti to an equivalent value of liters of Lowenbrau. Rational numbers frequently arise in problems of conversion as a ratio between values must be expressed. We present a few more examples, focusing on those involving motion. Indeed, the examples with motion will later be generalized in our development of calculus.

Example 2.1   CONVERSION OF SPEED TO DISTANCE

Suppose that in Galileo's experiments he set his track so that the sphere rolls downhill and then flattens out for quite a distance. Once it reaches the flat section, the sphere rolls at a constant speed. Let us say that Galileo measures that the sphere rolls a distance of 7 m in 2 s. Assuming there is sufficient track, how far will the sphere roll in 30 s?

Solution $30 \, s \times \frac{7}{2} \, m/s = 105 \, m$

While the example may appear simple, with an understanding of this example, you are halfway to the fundamental theorem of calculus (the remainder of the book takes you all the way there). So some remarks are in order:

- As a precursor to the problem, we note that we were able to calculate the speed, $\frac{7}{2}$ m/s, from a measurement of time and distance. Differential calculus generalizes this to more complex motions; given a particle's distance as a function of time, find its speed.

- The problem performs the inverse; once the speed is known, calculate the distance. Integral calculus generalizes a method for accomplishing this for more complex motions.

- The fundamental theorem of calculus is nothing more than a realization of the relationship between speed and distance. This is why understanding this example establishes a good basis for understanding the fundamental theorem of calculus.

- Note the inclusion of units in the calculation. Units cancel one another, so that the unit seconds (s) appearing with 30 cancels with the unit seconds (s) in the denominator of the expression for the speed. There is no cancellation of the length unit, meters, so it correctly remains in the answer.

- Also, notice that the nature of the ratio $\frac{7}{2}$ is very different from the ratio $\frac{2}{3}$ that was found to equate Lowenbrau with Chianti. Whereas $\frac{2}{3}$ in the beverage problem is a universal constant independent of units, $\frac{7}{2}$ in the speed problem is not. It very much depends upon the units. Indeed, if we express the speed in miles per hour, feet per second, or furlongs per score, we will get a different constant. This observation leads to our next example.

Example 2.2   CHANGE-OF-SPEED MEASUREMENTS

Let us take the above example but determine how far the ball travels in 2 min.

Solution   The solution is to multiply speed by time, but the time units associated with each measurement must be identical. Since the time is given in minutes, let us convert the speed from meters per second to meters per minute. This is accomplished by multiplying with another ratio expressing the number of seconds per minute, $\frac{60}{1}$, or just 60:

$$\frac{7}{2} \, m/s \times 60 \, s/min = 210 \, m/min$$

With the time units all set in minutes, it is possible to determine the distance:

$$2 \, min \times 210 \, m/min = 420 \, m$$

This example generalizes to a formula known as the chain rule in differential calculus and a principle known as change of variables in integral calculus. It is important to understand the concept of converting from one unit to another and how ratios are used to perform the conversion.

Example 2.3 WEIGHING TIME

This example follows the spirit of Galileo's measurements. Recall that Galileo did not measure time with a clock but weighed time by measuring the amount of water that flows into a pan from an hourglass-type device. Let us assume that in the above problem the sphere travels along the flat part of the track for a distance of 15 m and the amount of water collected during that time weighs 2 kg. Using the speed $\frac{7}{2}$ m/s, find the flow rate of the water in kilograms per second.

Solution

$$\tfrac{2}{15} \text{ kg/s} \times \tfrac{7}{2} \text{ m/s} = \tfrac{7}{15} \text{ kg/s}$$

Example 2.4 CONVERTING WEIGHT TO SPEED

We can pose the above example from a different perspective. Suppose that it is known that the measuring device flows at a rate of $\frac{7}{15}$ kg/s and, as above, the amount of water collected after the sphere goes 15 m is 2 kg. What is the speed of the sphere?

Solution

$$\tfrac{15}{2} \text{ m/kg} \times \tfrac{7}{15} \text{ kg/s} = \tfrac{7}{2} \text{ m/s}$$

Note the similarity between the examples. Conversion from one unit to another is performed by a multiplication of ratios. The labeling of units greatly assists with getting the ratios correct.

2.2.1.3 Irrational Numbers This section demonstrates that the rational numbers do not fill out the number line. Consider the following problem: given the side of a square, find the diagonal of that square. This is in fact a conversion problem. Similarity of squares shows that there is a constant ratio between the diagonal and the side (Figure 2.1). If one can find the ratio of the diagonal to the side, then that ratio can be used to convert a given length for the side of any arbitrary square to a length for the diagonal of the arbitrary square. Let us attempt to do this using rational numbers.

Let D be the length of the diagonal and S be the length of the side. To find a rational number that represents the ratio D/S, one must solve the following equation:

$$qD = pS \tag{2.1}$$

where both p and q are nonzero integers. Then the ratio D/S is the rational number p/q. How do we know if we can or cannot equate an integral number of diagonals, D, with an integral number of sides, S? The ancient Greeks addressed this problem and we follow their path. Using the Pythagorean theorem (presented in Chapter 3), it

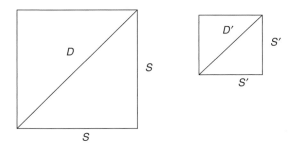

$$D/S = D'/S'$$

Figure 2.1   Similiraty of squares.

is possible to determine D for a given value of S:

$$S^2 + S^2 = D^2$$

$$2S^2 = D^2$$

$$D = \sqrt{2}S$$

Therefore, the ratio of D to S is $D/S = \sqrt{2}$. Our original aim is to constrain the numbers p and q in equation (2.1) to rational numbers. So let us proceed and see if this is possible:

$$qD = pS$$

$$\frac{p}{q} = \frac{D}{S} = \sqrt{2}$$

It is possible to express p/q in its most reduced form. This means that both presumed integer numbers p and q cannot be even:

$$p = \sqrt{2}q$$

$$p^2 = 2q^2$$

The right side of the above equality, $2q^2$, is an even number so the left side, p^2, must also be even. Since the square of an even number is even, the value p must be even. Furthermore, since p is even, there is an integer, r, with $p = 2r$. Substituting for p, we have the following:

$$4r^2 = 2q^2$$

$$2r^2 = q^2$$

Using the final equation, $2r^2 = q^2$, it is seen that q must be even by applying the identical argument that showed p is even; q^2 is even and so q is also even. It appears that both p and q are even. But since p/q is in its reduced form, both numbers cannot be even. There is something amiss, and let us try to identify why this inconsistency arises.

The equations are all constructed correctly. There is nothing wrong with assuming that p/q is in reduced form because it is always possible to reduce rational fractions. The only other cause of the inconsistency is our attempt to find a rational number that represents $\sqrt{2}$. The fact that this attempt leads to an inconsistency demonstrates that the attempt itself is folly and indeed $\sqrt{2}$ is not a rational number.

The objective of this section is complete. We have found a number that is not rational and, in doing so, demonstrated that the rational numbers do not fill out the number line.

2.2.2   The Size of the Irrational Numbers

Let us take a pragmatic perspective toward mathematics. From this perspective, numbers must be able to quantify specific properties; such as the area of a wall so that someone knows how much paint to buy, the time it takes to drive to the mall so that one knows when to leave for an appointment, or the weight of a Roman boat so that the necessary leverage for lifting it out of the water is known.

Any measuring device has discrete limitations that restrict the device's measurements to the rational numbers. This is because associated with all measuring devices is a smallest unit of measurement that perfectly divides a standard unit of measurement. For example, a metric stick used to measure length may be divided into millimeters, one thousandth of a meter, and all measurements are accurate within 1 mm. One cannot measure a length of $\sqrt{2}$ m because $\sqrt{2}$ is not a rational number and hence cannot be an integral multiple of millimeters. The fact that we are unable to solve equation (2.1) for rational p and q means that no matter how fine we take our smallest unit of measurement, nanometers, for example (10^{-9} m), we will not be able to precisely measure an object of length $\sqrt{2}$ m using the measuring stick.

At first glance, from our pragmatic perspective this appears very disconcerting. There are numbers that we cannot measure. Perhaps $\sqrt{2}$ is a fluke. Perhaps there are not too many irrational numbers, so it is not necessary to be too concerned about them. So let us ponder whether or not irrational numbers are commonplace.

The following concrete problem is posed to address the issue. Let us first restrict the problem to the interval from 0 to 1. The length of this interval is 1. The irrational numbers within the interval from 0 to 1 are a subset and perhaps it is possible to find their length. If that length turns out to be really small, perhaps it is not necessary to worry about the irrational numbers. Alternatively, if that length turns out to be significant, we better learn how to deal with irrational numbers.

Finding the length of the irrational numbers is not a trivial problem; it requires substantial theoretical firepower. Nevertheless, we find the length substituting intuition for firm theory when necessary.

Let us begin by dividing the interval from 0 to 1 into two sets, the set of rational numbers, which is denoted by S_r, and the set of irrational numbers, which is denoted by S_i. Every number is either rational or irrational, a number can be expressed as a ratio of integers or it cannot, so every number in the interval is in either S_r or S_i, and there cannot be a number in both sets. For example, $\frac{3}{8}$ is in the set S_r while $\sqrt{2}$ is in the set S_i. The length of the interval from 0 to 1 is just 1. This length must be the same as the sum of the lengths of S_r and S_i. The following equation expresses this equality:

$$L(S_r) + L(S_i) = 1$$

where

- $L(S_r)$is the length of the set of rational numbers in the interval 0 to 1
- $L(S_i)$is the length of the set of irrational numbers in the interval 0 to 1

It turns out that finding $L(S_r)$ is easier than finding the $L(S_i)$. So let us concentrate on this problem. Afterward, $L(S_i)$ is determined from the above expression, $L(S_i) = 1 - L(S_r)$.

The length $L(S_r)$ is determined by summing the lengths of every point within S_i. Is it possible to do this? Yes, provided that it is possible to identify each term of the sum: the first term, second term, third term, and so forth. Alternatively, if it is not possible to enumerate the terms in a sum, it is not possible to perform the sum. The property of being able to enumerate elements in a set is called countability. Specifically, a set is countable if each member of that set can be associated with a unique positive integer. Below are examples of countable sets in which the association is made explicit.

Example 2.5   THE ALPHABET

The alphabet is a countable set. Associate A with 1, B with 2, C with 3, and so on. Generalizing this example, any set with a finite number of elements is a countable set.

Example 2.6   A FINITE SET WITH NEGATIVE NUMBERS

Consider all the integers between the numbers -4.1 and 5.1. Even though there are negative numbers, it is possible to associate a positive integer with each value. The following table presents an association:

Whole number	1	2	3	4	5	6	7	8	9	10
Original value	-4	-3	-2	-1	0	1	2	3	4	5

Notice that this is not a unique association. The next table presents another association:

Whole number	1	2	3	4	5	6	7	8	9	10
Original value	0	1	-1	2	-2	3	-3	4	-4	5

One can describe the second association with the following formula:

$$f(n) = \begin{cases} 2n & \text{for all positive } n \\ -2n + 1 & \text{for all other } n \end{cases}$$

Example 2.7   INTEGERS, AN INFINITE SET

The integers form a countable set. Any integer could be inserted in the second table of the preceding example by extending the table. The formula in the preceding example presents an association over all integers.

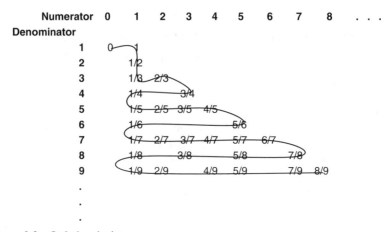

Figure 2.2 Ordering the integers.

It turns out that the rational numbers between 0 and 1 form a countable set. In fact, the entire set of rational numbers is countable, but we focus on those between 0 and 1. Figure 2.2 illustrates the construction of the association. In the figure, the horizontal numbers at the top represent the numerator of a rational number, while the vertical numbers represent the denominator. A rational number is given a place in the matrix; for example, the rational number $\frac{1}{4}$ occupies the second column and fourth row. While the Figure 2.2 only displays a finite set of numerators and denominators, the actual matrix goes on indefinitely in each direction. In this manner, all the rational numbers are placed in the matrix. Notice that the numbers are only entered in the matrix in their reduced form. Otherwise, the matrix position is left blank. For example, there is no entry for $\frac{2}{4}$.

To demonstrate countability, it is necessary to enumerate the entries. The squiggly curve through the numbers show how this is done. The table below is created by following the squiggly curve for the first 10 entries:

Whole number	1	2	3	4	5	6	7	8	9	10
Original value	0	1	$\frac{1}{2}$	$\frac{1}{3}$	$\frac{2}{3}$	$\frac{3}{4}$	$\frac{1}{4}$	$\frac{1}{5}$	$\frac{2}{5}$	$\frac{3}{5}$

We have constructed the necessary association of rational numbers, between 0 and 1, with the positive integers to demonstrate that they are countable.

Having demonstrated that it is possible to enumerate the rational numbers between 0 and 1, it is possible to determine $L(S_r)$. The length of a countable set is the sum of the lengths of its elements. For $L(S_r)$, this is expressed in the following equation:

$$L(S_r) = L(R_1) + L(R_2) + L(R_3) + L(R_4) + \cdots \tag{2.2}$$

where

- R_j is the jth rational point
- the sum goes on indefinitely

Each of our rational numbers is a point with no length; $L(R_j) = 0$ for every positive integer j. Substituting 0 for each $L(R_j)$ in equation (2.2) shows that the length of our set S_r is a countably infinite sum of zeroes, which itself is just zero.

We have found that $L(S_r) = 0$, which means that $L(S_i) = 1$ [recall $L(S_i) = 1 - L(S_r)$] and the set of irrational numbers is much larger and certainly more commonplace than the set of rational numbers. The size of the rational numbers in comparison with that of the irrationals is so miniscule that the rationals are insignificant. Let us place this notion in a different context. If there was a truly random process that picked a number between 0 and 1, then after any number of attempts the probability of the process picking a single rational number is zero. The hope that $\sqrt{2}$ is a fluke is completely dashed. The opposite is true, the rational numbers are flukes, while the irrationals abound.

We close this section with one remark. There is the question of why not apply the procedure of adding up the lengths of the individual points to the irrational numbers. Following the above process, the length of irrational numbers is zero as well because each individual irrational point has length zero. But the irrational numbers are not countable; there are too many of them so that enumeration is impossible. It is not possible to follow the above process and directly sum an uncountable quantity of terms; a direct sum can only be performed if each term can be enumerated.

2.2.3   Suitability of Rationals and the Decimal System

From the preceding section, one can deduce that most quantities, that is, the area of a wall so that someone knows how much paint to buy, or the time it takes to drive to the mall so that one knows when to leave for an appointment, or the weight of a Roman boat so that the necessary leverage for lifting it out of the water is known, have irrational measures. From a pragmatic perspective, irrational numbers cannot be ignored. We must come to terms with them. How do we proceed?

The answer comes from our common sense, which says that it is not necessary to have such accurate measurements. If it is possible to find the area of a wall to within a square meter, we know how many cans of paint to buy. If the time required to get to an appointment is known within 10 min, we can plan just fine. If Archimedes knows the weight of a Roman ship to within 50 kg, he can engineer his mechanical lifters. Each situation has a tolerance band. For some, such as the manufacturing of microchips, the tolerance band may be severe, but it is not zero. So long as it is possible to meet any nonzero tolerance band, the rational numbers can do any job—not bad for a group of insignificant numbers.

Let us bring some rigor to the notion of tolerance band with an example. Suppose we need to cut a strand of gold to within $\left(\frac{1}{10}\right)$ th of a millimeter $\left[\left(\frac{1}{10,000}\right) \text{ th of a meter}\right]$ of $\sqrt{2}$ m. How do we proceed?

To find a solution, note that all decimal values with a finite number of terms are rational numbers. For example, the number 0.5698 is $\frac{5698}{10,000}$. Our approach to finding a length within the tolerance band is to use the decimal system by finding a solution one digit at a time. The method is self-explanatory.

Let us go for the ones digit.

$1^2 < 2 < 2^2$, which is the same as $1 < \sqrt{2} < 2$
So the first digit is 1. Let us go for the one-tenths digit.

$1.4^2 < 2 < 1.5^2$, which is the same as $1.4 < \sqrt{2} < 1.5$
So the one-tenths digit is 4. Let us go for the one-hundredths digit.

$1.41^2 < 2 < 1.42^2$, which is the same as $1.41 < \sqrt{2} < 1.42$
So the one-hundredths digit is 1. Let us go for the one-thousandths digit.

$1.414^2 < 2 < 1.415^2$, which is the same as $1.414 < \sqrt{2} < 1.415$
So the one-thousandths digit is 4. Let us go for the one ten-thousandths digit.

$1.4142^2 < 2 < 1.4143^2$, which is the same as $1.4142 < \sqrt{2} < 1.4143$

The required tolerance has been established. As long as there is an instrument that can measure to within one-tenth of a millimeter of accuracy, the instrument should measure between 1.4142 and 1.4143 m. Taken between these measurements, the cut is within the tolerance band. If greater accuracy is required, it is possible to continue down this path and get more digits. So, indeed, using the rational numbers, it is possible to get as close to $\sqrt{2}$ as desired.

In this regard, there is nothing unique about $\sqrt{2}$. Given any irrational number, it is possible to find a rational number within any tolerance band using the above process; go one digit at a time bounding the number above and below until the number is bounded as tightly as is required. Chapter 6 demonstrates that the methods of calculus generalize this process; using calculus one determines an unknown quantity by approaching it using quantities that can be determined.

Having satisfactorily addressed our practical concerns, let us turn to the more abstract idea of accurately representing an irrational number. The Greeks were quite concerned by this and took to geometric representations. For example, $\sqrt{2}$ would be represented as the diagonal of a square with sides of unit length. The modern-day solution follows a more pragmatic approach. We simply write an expression that describes the number's significant property and do not give it any more thought. For example, $\sqrt{2}$, $5^{1.3}$, π, and $\cos(\pi/7)$ are all modern-day representations of irrational numbers, that are, the square root of 2, 5 raised to a power of 1.3, the ratio of the circumference of a circle to the diameter of the circle, and the cosine of the angle $\pi/7$, respectively. Any further specification is unnecessary.

We close with some remarks of interest:

- This method of approaching the unknown through the known is a theme of this book. Archimedes uses the approach to approximate the value of pi and determine volumes of several odd shapes. As noted above, the method is central to calculus.

- The process used to approximate an irrational number shows one of the advantages of a decimal system. The record indicates that the Mesopotamians were the first to introduce a decimal-like system. The difference between our modern system and theirs is that the base for their system was 60 as opposed to our system, which is base 10. The ancient Chinese system was a decimal base-10 system like the modern system.

- As a testimony to the capacity of rational numbers to do the job, all of Kepler's data were in the form of rational numbers; Tycho Brahe could only make rational measurements. And yet, with this slightly imperfect data, Kepler was able to find the correct orbit of Mars.

- The above process for finding a rational approximation of an irrational number is not an efficient process. This process finds one digit at a time. In the case of approximating the square root of a number, there are more efficient methods that allow one to get more than a single digit with each approximation.

2.2.4   Rational and Irrational Outcomes

This book pursues the ellipse. Occasionally, during the pursuit, something of interest comes into view and out of temptation we follow it. This section is the first sojourn off the path of the ellipse.

There is a tendency to think of a number as only providing a quantity and that two numbers that are close are very similar. This section questions the notion through an analysis of an asteroid collision. Before taking on the problem of the asteroid collision, an artistic design is analyzed. The design follows a simple mathematical formula and the outcome of the design may change dramatically as a parameter in the formula changes. The design yields insight into asteroid collisions.

2.2.4.1 *Mathematical Art*   Look at the patterns shown in Figure 2.3.

These are examples of mathematical art in which an artistic waif who is nevertheless mathematically savvy can create beautiful patterns. In this section, these patterns are analyzed.

Let us imagine a circular table with nine pegs set equally about the edge of the table and label the pegs counterclockwise from 0 to 8. Attach a string to peg 0, moving clockwise skip two pegs, and attach the string to the third peg over. From

Figure 2.3   Mathematical artwork.

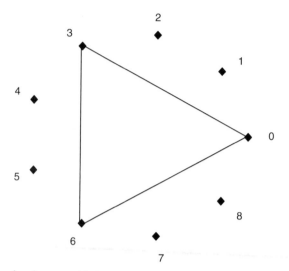

Figure 2.4   A cycle, nine pegs skip two.

that peg, again moving clockwise, skip two pegs and attach the string to the third one over. Continue this pattern. What happens?

Eventually the string returns to peg 0. Why is this so? Notice that each peg leads to a unique peg along the path and there is only one peg that can precede it. For example, the path always proceeds from peg 3 to peg 6 and there is no other peg that precedes peg 6. Since there are only nine pegs, after a while, you have to reconnect to a peg that has already been connected. Let the first instance of a reconnection be at peg j. If j is not 0, then there is more than one peg that leads to j, which cannot happen so the first instance of reconnection must be at peg 0 and indeed the string returns to peg 0. We call the path from peg 0 back to peg 0 a cycle.

Once the string has returned to peg 0, the pattern repeats. The cycle takes the path 0–3–6–0. Figure 2.4 depicts the cycle. In the remainder of this section, this pattern is referred to as the first pattern.

Notice that the string never connects to pegs 1, 2, 4, and 5.

Let us see what happens if there is a slight change in the construction, instead of skipping over two pegs between connections, skip over three pegs and attach to the fourth one over. Figure 2.5 illustrates the result. The cycle assumes the path 0–4–8–3–7–2–6–1–5–0. In the remainder of this section, this pattern is referred to as the second pattern.

There is a visual difference between the two patterns; the first pattern is much more plain and the second far more striking. Is there a quantitative explanation for the qualitative difference?

There are a number of quantities of interest: the number of circuits in a cycle, the number of legs that have been made, and a concept for length. A circuit occurs each time the path passes 0 whether or not it reconnects to peg 0. A leg is a segment between two connected pegs. In the first pattern, there is one circuit and there are three legs. In the second pattern, there are four circuits (the path skips

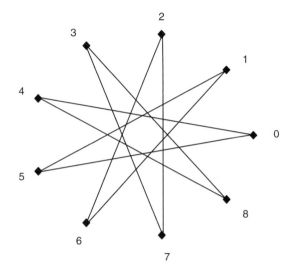

Figure 2.5   A cycle, nine pegs skip three.

over peg 0 three times and then returns to peg 0 on the fourth go around) and nine legs.

   The length of a portion of a path is the number of pegs from the starting peg to the final peg of the portion. The length of any path is the sum of the lengths of the portions. Accordingly, the leg length is the number of pegs from one attachment to the next, and in the first pattern the leg length is 3, while in the second pattern it is 4. The cycle length is obtained by adding up each of the leg lengths within the cycle. Since the leg lengths are all identical, the cycle length is the product of the leg length and the number of legs. In the first example, the cycle length is $3 \times 3 = 9$. In the second example, the cycle length is $4 \times 9 = 36$.

   Is there another way to get the cycle length? Consistent with our previous definition, take the circuit length to be the number of pegs around one circuit; this is just the total number of pegs on the table. In both of the above patterns, the circuit length is 9. Then the cycle length is the sum of all the circuit lengths in the cycle. Since the circuit lengths are all identical, the cycle length is the product of the circuit length and the number of circuits in one cycle. In the first example, the cycle length is $9 \times 1 = 9$. In the second example, the cycle length is $9 \times 4 = 36$.

   We have calculated the cycle length in two ways:

$$\text{Cycle length} = \text{leg length} \times \text{number of legs in a cycle}$$

$$\text{Cycle length} = \text{circuit length} \times \text{number of circuits in a cycle}$$

Equating the two calculations of cycle length yields the following equation:

$$\text{Leg length} \times \text{number of legs in a cycle} = \text{circuit length}$$

$$\times \text{number of circuits in a cycle} \quad (2.3)$$

Consider the general case in which there are a total of N pegs on the circular table and an attachment occurs every Mth peg; that is, the circuit length is N and the leg length is M. As in the above patterns, because there are a finite number of pegs, the string must return to peg 0 and form a cycle. Let q be the number of circuits in a cycle and p be the number of legs. Rewriting equation (2.3) using M, N, q, and p results in the following equation:

$$Mp = Nq \tag{2.4}$$

In equation (2.4) and the remainder of this section, M, N, p, and q are all positive integers.

Notice the similarity between equations (2.4) and (2.1). Equation (2.1) was established to determine if an integral multiple of the length of the side of a square is equal to an integral multiple of the length of the square's diameter. (This was not the case.) Similarly, equation (2.4) finds a pair, p and q, in which an integral multiple of the leg length equals an integral multiple of the circuit length. In this equation, there is always a solution; set $p = N$ and $q = M$.

When a positive solution is found, the path has gone exactly through p legs and q circuits, and the path has returned to peg 0. One cycle must consist of the smallest positive values of p and q that satisfy equation (2.4), the least number of legs that corresponds with a perfect multiple of circuits.

Is the solution, $p = N$ and $q = M$, the smallest solution over positive integers? We return to our patterns hoping to find an answer. For the first pattern, $M = 3$ and $N = 9$:

$$3p = 9q$$

This equation can be rewritten as

$$\frac{3}{9} = \frac{q}{p}$$

By placing the fraction on the right into its reduced form, it is possible to find the smallest values of p and q:

$$\frac{1}{3} = \frac{q}{p}$$

The solution is that $p = 3$, three legs, and $q = 1$, one circuit. For the second pattern, we have the following:

$$4p = 9q$$

$$\frac{4}{9} = \frac{q}{p}$$

Since $\frac{4}{9}$ is in reduced form, $p = 9$, nine legs, and $q = 4$, four circuits, are the smallest solutions. The above arguments are generalized into the following algorithm for

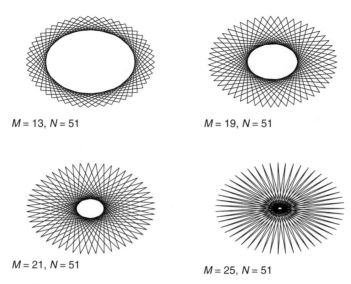

$M = 13, N = 51$ $M = 19, N = 51$

$M = 21, N = 51$

$M = 25, N = 51$

Figure 2.6   Patterns.

finding the number of circuits and number of legs in a cycle:

Step 1. Create the fraction M/N.

Step 2. Reduce the fraction and call it q/p.

Step 3. The number of circuits is q and the number of legs is p. The cycle length is $p \times q$.

The patterns shown in Figure 2.6 are created for the indicated values of M and N.

We close this section with some remarks:

- As long as the circuit length is a prime number, then the fraction M/N is already in reduced form and a cycle will connect to every peg.

- In all the examples, the leg length is chosen as less than half of the circuit length. This is sufficient to determine the design of all the patterns using the symmetry of the system. For example, the pattern created by the values 5 for the connection length and 9 for the circuit is a reflective symmetry of the pattern created by the values 4 for the leg length and 9 for the circuit length. This happens because the connection points of the system with $M = 5$, $N = 9$ are identical to the connection points of a system in which $M = 4$, $N = 9$, but the pegs have been labeled clockwise as opposed to counterclockwise.

- It is possible to consider leg lengths with any integer value, not just positive integers. Negative integers would indicate moving around in a clockwise, as opposed to counterclockwise, manner.

- In general, the patterns with high cycle length yield complex figures, while the patterns with low cycle length are simple. The type of pattern that is more visually pleasing (simple or complex) is a matter of individual taste. But the mathematically astute can create whichever is desired.

2.2.4.2 Incoming Asteroid

The designs in the previous section are governed by the reduced form of the rational number M/N. Two designs that are governed by numbers that are very close can have very different outcomes. For example, think of two designs, D_1 and D_2, with 100 pegs, and $N = 100$ for both systems. Let $M_1 = 50$ and $M_2 = 49$. Design D_1 yields a simple design; there are only two legs and one circuit in the cycle. The path initiates at peg 0, proceeds to peg 50, and then returns to peg 0. However, since $\frac{49}{100}$ is in reduced form, there are 100 legs and 49 circuits in D_2; all the pegs are connected through a starlike design of 49 circuits. The numbers $\frac{50}{100}$ and $\frac{49}{100}$ are reasonably close, yet the designs are very different. (If these two numbers do not seem close enough, try $\frac{500}{1000}$ and $\frac{499}{1000}$.) The study of motion yields similar phenomena. Motion that is governed by a single or a small set of parameters may differ substantially by changing a parameter ever so slightly. This section presents an example that generalizes the results of the previous section.

The setup is the following. Allow an object to move counterclockwise around a circle at constant speed; assume that distance is measured in a system where the circumference has length 1. After a fixed time interval, record its position. Continue recording the position over the same fixed time interval indefinitely. The position is identified by the length of the arc initiating from the object's initial position to its current position in a counterclockwise direction. Because the speed of the object remains constant and the time interval between recordings is fixed, the arc length between two recordings is a constant length. Let us denote this length by A and denote the recordings by R_j, where j represents the recording after the jth time interval. We will examine the behavior of the record of positions.

The following expressions establish the records R_j:

$$R_1 = A \qquad R_2 = 2A \bmod 1 \qquad R_3 = 3A \bmod 1$$

In general, the following relations hold:

$$R_j = jA \bmod 1 = (R_{j-1} + A) \bmod 1 \qquad (2.5)$$

The function mod is taken to maintain the recording between 0 and 1; once the object has completed a full revolution, the measurement is reset to zero.

What can be said about the behavior of the record? Let us begin by associating this system with that of the previous section. Assume that the arc length between recordings is $\left(\frac{4}{9}\right)$th the entire circumference, that is, A is $\frac{4}{9}$. One can divide the circle into nine arcs of length $\frac{1}{9}$, and the end points of these arcs are identical to the position of the pegs in the second pattern of the previous section. Then the object visits each end point in exactly the same order as the second pattern of the previous section (see Figure 2.5). A complete cycle is given by $R_0 = 0$, $R_1 = \frac{4}{9}$, $R_2 = \frac{8}{9}$, $R_3 = \frac{3}{9}$, $R_4 = \frac{7}{9}$, $R_5 = \frac{2}{9}$, $R_6 = \frac{6}{9}$, $R_7 = \frac{1}{9}$, $R_8 = \frac{5}{9}$, and $R_9 = 0$. Compare this with the cycle of pattern 2 of the previous section; recall that cycle is 0–4–8–3–7–2–6–1–5–0. The only difference is that the length of the circle in this section is 1, whereas the length of the circuit in the second pattern is 9. This difference causes a rescaling of all lengths by a factor of 9; the leg lengths, 4, are rescaled to give the arc lengths, $\frac{4}{9}$.

Can the results of the previous section be generalized? Let us assume the same notation in which M, the leg length, has the interpretation of arc length and N, the circuit length, is the circumference length. Reexamine equation (2.4): $Mp = Nq$. Integer solutions for the values of p and q yield a similar conclusion to what was found in the previous section. On the recording R_p, the object has gone around the circle exactly q times. Therefore, the object has returned to its initial position. A cycle occurs when p and q are the smallest nonzero integer solutions. Restating the result in terms of the value A, we note that the circumference length N is 1 and the arc length M is given by A:

$$Ap = q$$
$$A = \frac{q}{p} \tag{2.6}$$

The conclusion is that if A is a rational number, the object cycles. If the rational fraction q/p is in its reduced form, then on the recording R_p a cycle is complete and the object has revolved about the circle exactly q times.

What happens if A is not a rational number? In this case, the behavior is quite different. Equation (2.4) and its equivalent form, equation (2.6), have no integer solutions. It can never be that the recording R_p ends on the initial position after q revolutions for any nonzero, integer values of p and q. The observations continue indefinitely without ever repeating themselves.

In the case of irrational A, more can be said about the way that the observations are distributed about the circle. The circle is a symmetric object and the observations should be spaced out over the circle symmetrically; they should not clump in any one region more than any other. The following property specifies the notion of a symmetric distribution of the observations.

Density Property. *The observations come arbitrarily close to any arbitrary point on the circle.*

The property is so named because a set of points that come arbitrarily close to a larger set is said to be dense within the larger set. For example, since any real number can be approximated by a rational number within any given tolerance band, the rational numbers are dense within the real numbers. The density statement for the recordings in this section similarly means that for any given tolerance band around any point on the circle some point on the record of a system governed by an irrational value of A falls within the tolerance band.

A formal proof of the density property can be found in *Geometric Methods in the Theory of Ordinary Differential Equations* (1983) by V. I. Arnold. The idea behind the proof is to use symmetry to show that a property that is true in an interval of the circle can be extended to the entire circle and is accordingly true for the entire circle. The density statement states that the object comes arbitrarily close to any point. On the contrary, suppose that there is a point from which the object remains at a fixed distance; there is a gap in the circle and there are no recordings in that gap. One can extend this gap to fill up the entire circle, so the property that there are no

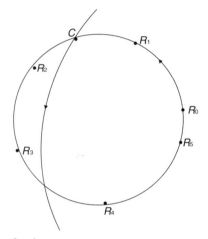

Figure 2.7   Intersection of pathways.

recordings holds for the entire circle, which is patently false. The only way to avoid this circumstance is by rejecting the supposition that a gap exists.

What are the ramifications for a planet that rotates in a circular orbit with an asteroid periodically penetrating and exiting the circle. The situation is depicted in Figure 2.7. Although there are two points of intersection between the pathway of the asteroid and the planet, the analysis for each is identical, so we consider only one. The analysis assumes that the asteroid's trajectory and the planet's trajectory are fixed so that the asteroid indefinitely returns to the same exact point in a fixed time.

At some initial time, the asteroid is located on the planet's circular pathway at the point C, while the planet is at the point R_0. The next time the asteroid returns to point C, perhaps hundreds of years later, the planet is at the point R_1. Let the arc length between R_0 and R_1 be given by A. Each successive time that the asteroid returns to C, the planet's position has shifted by an arc of length A. Indeed, the relations of equation (2.5) hold.

Will the asteroid and the planet ever collide? Let us assume that A is rational. Then the planet naturally executes a cycle and returns to its initial position. If a point of the cycle is close enough to C to cause a collision, then a collision occurs; otherwise the planet completes its cycle and repeats it indefinitely with no collision.

Alternatively, if the arc length A is irrational, then by the density property, the planet will at some time come close enough to point C to cause a collision. In Section 2.2.2, it was seen that an arbitrary real number is most likely irrational. In all likelihood, a collision occurs.

We close this chapter with some remarks:

- It is possible to strengthen the density statement as follows:

Ergodic Property. *The length of any arc is equal to the proportion of obser-vations that lie within the arc.* Suppose that one wants to measure the length of an arbitrary arc with some room for error. It is possible to collect enough recordings, so that the proportion of recordings that lie within the arc indicates the length of the arc; the difference between the actual length and the proportion is smaller than the margin of error. As the requirement for accuracy increases (decreasing tolerance band), the number of recordings necessary to guarantee that one is within the prescribed error also increases.

- The ergodic statement is an expression of the uniformity with which the recordings are distributed. For all arcs of identical length, provided the number of recordings is sufficiently large, the number of observations within each arc is nearly the same.

- The ergodic statement is rather remarkable. Note that any arc contains all the irrational points within its end points and is accordingly an uncountable set of points. Yet using the ergodic property, one can approximate the length of an arc (an uncountable set) by determining the proportion of recordings within the arc length over a finite record of observations, and the approximation approaches the actual length as the number of observations (countable) increases. Similarly, a rational value can approximate an irrational value using a finite number of digits and the approximation approaches the actual irrational value as the number of digits increases. As noted in the discussion of approximating irrational values, integral calculus follows a similar process.

- The example illustrates a difficulty of approximating an irrational number with a rational one. Qualitative differences emerge from slight quantitative differences. In one case recordings cycle; in the other case they do not. Problems with a computer's inability to represent an irrational number lead to inaccuracies that over the long run become consequential. If one wants to be close to the true trajectory for a given time, there is a rational number that will remain close to the trajectory of the irrational one for that length of time, but not indefinitely.

- Honoring the spirit of self-criticism, I note that this book is all about the ellipse, and in particular its objective is to show how planets orbit the sun in an ellipse. In the example, the orbit is circular, which is a bit of a throwback. This criticism could be addressed with a bit of effort. One could map the actual elliptic orbit onto a phantom circular orbit that would amount to the same thing. There is no need to go into this level of detail as it would obscure the main point of this section, a demonstration that quantitative and qualitative differences can result from approximations of irrational numbers using rational numbers.

CHAPTER 3

THE SPACE: GEOMETRY

The starting point of our pathway toward calculus and the ellipse is with Euclid. In this we are not alone. Euclid initiates the mathematical education of many mathematicians. Concerning the man, whose influence is matched with few other historical figures, little is known. Folklore ascribes two quips that if accurate indicate that he had quite a wit. According to one story, a student challenged Euclid to demonstrate any value of his teachings. Euclid responded by requesting of his slave, "Give him three pence since he must make gain out of what he learns." In another story, the King, Ptolemy, asks Euclid if there is an easy way toward the understanding of *The Elements*. Euclid replied something to the effect that there is no special road to geometry even for a king.

Euclid's legacy is beyond what he or his colleagues could ever have imagined. *The Elements* (Euclid, 2002) has been translated into more languages than any other book, save the Bible. It ranks among history's bestseller list and is still available at bookstores throughout the world. Indeed, if Euclid could capture royalties from book sales, he would be able to boast about many pence that have come his way as a result of this book and his impudent student would be unduly impressed. Beyond its popularity, as a standard of mathematics education over a span of 2000 years in both Arab and European cultures, the book has shaped mathematical and scientific methodology. Axiomatic deduction is the standard of modern mathematical discourse and this stems from the success of *The Elements*. Isaac Barrow, Newton's mentor, instructed Newton to become thoroughly familiar with *The Elements* as a first step in his mathematics education. Inscribed above the doorway of the library through which Leibniz passed as a boy are the words "Let all who pass through study Euclid." Whether King Ptolemy passed through the doorway is unknown, but certainly both Newton and Leibniz did.

By the time of Euclid, the Greeks had made much mathematical progress. Axiomatic deduction had been central in the school of Plato. Plato's student Eudoxus was preeminent in this endeavor. The axiomatic method begins with statements, called axioms, which appeal to common sense as self-evident. From the axioms, one deduces new statements, known as theorems. Once a formal deduction of the theorem is established, the theorem becomes a part of the body of knowledge of mathematics and may then be used to generate further theorems. In his work *The Elements*, Euclid formalizes axiomatic deduction by providing known proofs of existing theorems, establishes theorems from known conjectures by supplying the proof, and adds

---

The Ellipse: A Historical and Mathematical Journey   by Arthur Mazer
Copyright © 2010 by John Wiley & Sons, Inc.

original theorems along with their proofs. The dividing line between what was known prior to Euclid and Euclid's original contributions is unclear.

Euclid follows the axiomatic–deductive method with uncanny skill. The critical step is the first one, establishing the axioms. The axioms germinate either directly or indirectly all the theorems. As such they must be sufficiently robust to produce an interesting body of theorems but sufficiently narrow so as not to overstretch the notion of what is self-evident. Other required features of the axioms are that they are independent—no axiom can be deduced from the remaining axioms—and consistent—the axioms cannot lead to statements that are contradictory. Euclid's first order of business is to place forward his axioms and in this he demonstrates brilliance. Euclid chooses five axioms from which he derives results that number into the hundreds. The five axioms are given below:

1. A straight line can be drawn joining any two points.

2. Any straight-line segment can be extended indefinitely in a straight line.

3. Given any line segment, a circle can be drawn having the segment as radius and one end point as center.

4. All right angles are congruent.

5. If two lines are drawn that intersect a third in such a way that the sum of the inner angles on one side is less than two right angles, then the two lines inevitably must intersect each other on that side if extended far enough.

Euclid's choice of axioms reflects the geometric underpinnings that dominate the Greek approach to mathematics. In this regard, numbers are considered as lengths or areas of geometric objects. Accordingly, all of what Euclid would demonstrate must be accomplished through the construction of geometric objects and the only allowable instruments for their construction are a straight edge and a compass. The axioms must describe allowable geometric figures that the first three axioms address.

As an example to illustrate the geometric perspective that the Greeks adopted, the Greek demonstration that $(x + y)^2 = x^2 + 2xy + y^2$ looks like the one shown in Figure 3.1.

Note the largest square of length $x + y$ is composed of two squares, one of length x and one of length y, and two rectangles, each with sides of lengths x and y. The equality states that the area of the largest square is equal to the area of its composite

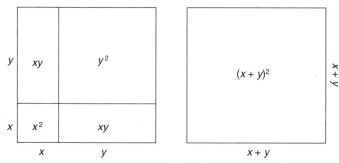

Figure 3.1   Geometric demonstration: $(x + y)^2 = x^2 + 2xy + y^2$.

parts, a large square, two rectangles of equal size, and a small square. To assert this requires the construction of each square using a compass and a straight edge. The above axioms allow for this construction.

The axioms have undergone tremendous scrutiny over their lifetime. There have been efforts to eliminate some as redundant or unnecessary. Every effort to alter the axioms has ended in failure, a testimony to Euclid's brilliance.

The final axiom does warrant some remarks. The axiom is known as the parallel postulate as it determines the conditions under which two lines are parallel. This has been the subject of the most controversy and the target of most efforts to alter the axioms. Several very talented mathematicians have attempted to prove its redundancy. Following the works of Islamic mathematicians al-Haytham (915–1039), Oman Khayyam (1048–1123), and Nasir al-Din al-Tusi (1201–1274), Girolamo Saccheri (1667–1733), a Jesuit, imagined what would happen if one were to eliminate the fifth axiom. How would this change the geometry of space? Which of *The Elements'* theorems would remain, and what new theorems would result? Saccheri's questions were largely rhetorical. He believed that it was impossible to violate the fifth axiom; if one were to attempt to describe a geometry that violates the parallel postulate, a statement that is in contradiction of the remaining axioms would surface. This would in effect be a proof of the parallel postulate, demonstrating its redundancy and reducing its status to a theorem. Saccheri committed himself to this endeavor. As Ponce de Leon searched in vain for the fountain of youth, so did Saccheri search for a contradiction. Saccheri believed his work to be a failure when, in fact, the mathematical exploration charted new territory that would have much broader implications than Ponce de Leon's excursion.

Johann Lambert (1728–1777) followed up on Saccheri's work with a more open perspective. He looked not for a contradiction but for an actual example of a geometric system in which there is no fifth axiom and he succeeded. Lambert's geometric system is a curved space that is akin to the space–time continuum underlying Einstein's general theory of relativity. The road from Saccheri to Einstein is another historic journey that we do not pursue. Instead we remain firmly planted in Euclid's world. Euclid's geometry appeals to our common sense as developed by our everyday experiences. Furthermore, the calculus of Newton and Leibniz is grounded on Euclidean geometry and Euclidean space is where the ellipse resides.

Topics in this chapter include a discussion of dimension, the Pythagorean theorem, a theorem of Cavalieri (1598–1647) and its application toward finding volumes of different shapes, and Archimedes' approach toward menstruation. The chapter takes two excursions, one exploring the notion of a fractal dimension and another into equal-area maps.

3.1   EUCLIDEAN SPACE, DIMENSION AND RESCALING

3.1.1   Euclidean Space and Objects

Greek geometers studied objects (squares, circles, triangles, and so on) and considered space as a boundless envelope that contains objects. A point can contain no objects and is zero dimensional. The shortest path between two points gives a line segment that can be extended endlessly into a line; the line is a one-dimensional space. Given two intersecting lines, the space generated by taking all the lines that are parallel to one while intersecting the other is a two-dimensional plane. From the two-dimensional plane and another line that intersects the plane in a single point, three-dimensional space may be generated by taking all the planes that are parallel to the initial plane while intersecting the additional line in a single point. The Greeks considered objects to be of one, two, or three dimensions.

3.1.2 Euclidean Space in Higher Dimensions

The modern notion of Euclidean space is a generalization of three-dimensional Euclidean space into higher dimensions. While the concept of dimensions beyond 3 may seem foreign, in fact it is necessary to consider higher dimensions to describe many physical systems, even very simple ones. We give a description of the motion for a single rigid body as an example. Consider a dimension as an attribute associated with the rigid body's motion. The motion has many attributes, such as location, velocity, orientation, and change in orientation; a description of the body's motion requires a space that is an extension of physical space.

The starting point is to locate the body's position while at rest. The location of an object indicates the position of the object's center of mass and is accordingly a single point. An object at rest has no degrees of freedom; its center of mass is fixed and the dimension of the space necessary to describe its position is 0; we can describe its position with a single point. If we allow the object to move in a line, say from left to right, the object can move along one degree of freedom; accordingly, the dimension of the space in which the object can move is 1. Providing an additional degree of freedom so that the object can move from front to back, as well as left to right, the dimension of the space is 2. Finally, allowing the object to move in any direction, left to right, front to back, and up and down, the object has three degrees of freedom and the object can move in three dimensions. At most three numbers suffice to describe the location of an object at a given time. If we wish to designate the object's location by time, a fourth number is needed, namely, the time. Accordingly, the dimension of the space where the object is described is 4.

Ascribing more quantitative properties to the object requires additional dimensions, because each additional property requires another number. For example, if we wish to give not only the object's location but also its velocity, an additional three dimensions are required. The additional dimensions each provide a speed in a prescribed direction. There is a speed in the left-to-right direction, another speed in the front-to-back direction, and finally a speed in the up-and-down direction. The velocity is the composite of all the speeds. Now there are seven numbers required to describe the object: three for position, three for velocity, and one for time. Mathematically, the space required to describe the object is seven dimensional.

Beyond the location and velocity, one might wish to describe the object's orientation. Consider, for example, that the object is a rigid hat that has been tossed in the air. At various times the hat may be flipped upside down or its bill may be pointed in a particular direction. In general, the hat or any object may rotate about its center of mass while the center of mass moves. To describe the orientation requires a description of the object's rotation about its center of mass. It turns out that three more dimensions are required to specify the rotation. And if one wishes to describe the velocity of the rotation, an additional three dimensions are required. Altogether, 13 numbers are required to specify the motion of the object; mathematically, its motion requires a 13-dimensional space. This book ends with a demonstration that the orbit of a planet around the sun is elliptic. This demonstration requires four dimensions, two for position in a plane and two for planar velocity; orientation is not considered.

All the above dimensions apply to a single rigid body. For each additional rigid body under consideration, another 12 dimensions must be included (3 for position, 3 for velocity, 3 for orientation, 3 for velocity of rotation). If one wishes to ascribe other properties to the body (that is, temperature, mass), more dimensions are necessary.

Hopefully enough fuss about dimensionality has been made to provide a convincing argument that pursuing geometric structures in dimensions above 3 is worthwhile. Geometric concepts, such as distance between points, perpendicular intersection of lines, hyperplanes (planes of dimension higher than 2), and measurement, can be generalized from standard three-dimensional space to higher dimensions. This chapter is firmly planted in three dimensions (except for an excursion into fractals). In Chapter 4, concepts from this chapter are generalized to higher dimensions.

3.1.3   Unit Measurements and Measures of Objects

A primary interest of Greek geometry is to determine the measure of a given object. The starting point is to define a unit measurement for every dimensional space: unit length, unit area, and unit volume. The measurement of a given object, if it is well defined, is then the greatest number, whole or otherwise, of objects with unit measurement that the initial object can contain.

For example, in one dimension one designates a selected line segment as having unit length. Then the length of another arbitrary line segment is the maximum number of unit lines or fractions thereof that fit within the arbitrary segment. Similarly, in two dimensions, one designates a square with unit sides as having unit area. Then the area of a given two-dimensional object is the maximum number of unit squares or fractions thereof that fit within the given object. Finally, in three dimensions, one designates a cube with unit sides as having unit volume. Then the volume of a given three-dimensional object is the maximum number of unit cubes or fractions thereof that fit within the given object.

A rather obvious property concerning these measurements is that they are all additive. That is, the measure of an object that is composed by bringing two or more objects together is the sum of the measures of each object within the composition. This property is very useful; it allows one to determine the measure of a complex object by decomposing the complex object into simpler pieces. One can then determine the measure of the simpler objects, recompose the complex object, and take the measure of the complex object as the sum of the measures of the simpler objects.

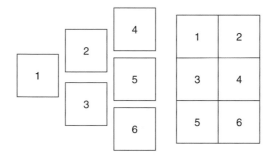

Figure 3.2   The whole equals the sum of the parts.

A simple example is given in Figure 3.2, a rectangle of length 3 and width 2. Using the definition, the maximum number of unit squares that fit into the rectangle is 6; the area of the rectangle is accordingly 6. Alternatively, one can decompose the rectangle into six squares of unit area and sum up the areas of each of these squares to arrive once again at an area for the rectangle of 6.

Remark

- Our modern concept of measurement has evolved from the Greek concept given above. Objects that are currently considered are more abstract, such as the set of all irrational numbers between 0 and 1.
- There are alternative approaches to determine the measure. For example, instead of taking the sum of units that an object can contain, one could take the smallest number of units that contain the object. If the object has a well-defined area or volume, the two approaches yield the same result.

3.1.4   Rescaling, Measurement, and Dimension

The above concepts work well when the objects considered are simple, for example, if a one-dimensional object can be cut from a line or a two-dimensional object can be carved from a plane. But what happens when the object is a curve, such as the circumference of a circle, or a curved surface, such as the surface of a sphere? In such cases, how do we even know the dimension of the object?

This section finds a relation between measurement and dimension through rescaling. The notion of rescaling used in this book may differ from that in other sources. In this book, rescaling is nothing more than rescaling the unit quantity; the rescaled unit length is a multiple of the original unit length. It is possible to determine the measurement of an object in rescaled units knowing its measurement in original units. The relationship depends upon the dimension of the object. We proceed with an example.

Example 3.1   RESCALING OF A LINE, RECTANGLE, AND CUBE

Suppose there are three objects, a line segment, a rectangle, and a cube, each with measurements in yards given, respectively, by L, A, and V.

Consider the units to be measured in feet. For the line, each yard is equivalent to $3 = 3^1$ ft; the length in feet is $3 = 3^1$ times the length in yards:

$$\text{Length in feet} = L \times 3\,\text{ft}$$

where L is the length in yards. For the rectangle, each square yard is equivalent to $9 = 3^2\,\text{ft}^2$; the area of the rectangle in square feet is $9 = 3^2$ times the area in square yards:

$$\text{Area in square feet} = A \times 3^2\,\text{ft}^2$$

where A is the area in square yards. Finally, for the cube, each cubic yard is equivalent to $27 = 3^3\,\text{ft}^3$; the volume of the cube in cubic feet is $27 = 3^3$ times the volume in cubic yards:

$$\text{Volume in cubic feet} = V \times 3^3\,\text{ft}^3$$

where V is the volume in cubic yards. In each case, the measurement in rescaled units becomes

$$\widetilde{M} = M \times s^d \tag{3.1}$$

where

- $\widetilde{M}$ is the rescaled measurement
- M is the original measurement
- s is the scaling factor (in the above cases, $s = 3$)
- and d is the dimension of the object

One can take equation (3.1) as the definition for dimension. Using this definition, if one can work out the relationship between measurement and rescaling, then the dimension of the object can be found.

Below, we apply the definition to find the dimension of a circle's circumference and then find the dimension of a fractal object. To examine the case of a circle, the method of exhaustion as described by Eudoxus is used. (Eudoxus was a student of Plato, and Archimedes credits Eudoxus with developing the method of exhaustion.) The method of exhaustion allows one to approximate complicated shapes by simpler shapes that lie within the complicated shape and, through refinement, improves the approximation to any arbitrary degree of accuracy. One way to apply this is to the case of a circle as described below and illustrated in Figure 3.3.

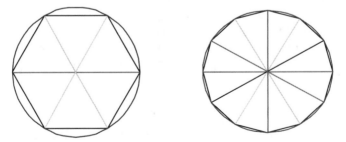

Figure 3.3   Six- and 12-sided approximations of the circle.

Start by approximating the circle with six inscribed triangles. The object formed by the bases of all triangles approximates the circumference of the circle and, accordingly, summing their lengths results in an approximation for the measure of the circumference. To get a better approximation, double the number of triangles to 12 and sum the lengths of the bases. Continue improving the approximation by doubling the number of triangles and summing the lengths of the bases of these triangles. Measurements are made using a standard unit of length.

Once we accept that the circumference may be approximated by the bases of the triangles, which are line segments, it is apparent that the circumference is one dimensional. Nevertheless, let us apply the definition as given by equation (3.1). In equations, the approximations are as follows:

$$M_1 = B_{1,1} + B_{1,2} + \cdots + B_{1,6} = 6B_1$$

$$M_2 = B_{2,1} + B_{2,2} + \cdots + B_{2,12} = 12B_2$$

$$M_n = B_{n,1} + B_{n,2} + \cdots + B_{n,k} = kB_n$$

where

- M_n is the measure of the nth approximation in original units
- $B_{n,1}, B_{n,2}, \ldots$ are the lengths of the bases of the inscribed triangles in the nth approximation
- k is the number of triangles, $k = 6 \times 2^{(n-1)}$
- the common length of all the bases for the nth approximation is given by B_n

Let us rescale units. Let the initial unit measurement be s times the new measurement. Then in the new units, the length of the base is given as follows:

$$\text{Rescaled length of a base} = \widetilde{B} = B_n s$$

Using the above equations, in new units the approximations of the circumference become $\widetilde{M}_n = M_n s$, where $\widetilde{M}_n$ indicates the measurement in rescaled units. The dimension of each approximation is 1 because in the rescaling equation (3.1) the power associated with the rescaling factor is 1. Since the approximations approach the measurement of the actual circumference, the dimension of the circumference is also 1.

Remarks

- Eudoxus' method of exhaustion is found in Euclid. This is a significant first step toward the integral calculus, but Eudoxus does not take the next step of applying the statement and making an actual measurement. Archimedes is the first who works out the actual measurements by finding the limiting case for the approximations.
- We were less than rigorous in the above treatment. Specifically, we did not demonstrate convergence of the approximations. In employing the method of exhaustion, Archimedes demonstrates convergence by bounding the object of interest between two series of approximations, one that is smaller than the

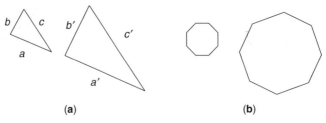

Figure 3.4   Resizing.

object of interest and one that is larger. He then demonstrates that the smaller and larger approximations converge to the same value. This argument is used in Chapter 6, where calculus is presented.

Associated with rescaling is the notion of resizing. Resizing of an object is the creation of a geometrically similar object whose size is a common multiple of the original object along each dimension. Figure 3.4 illustrates the concept of resizing applied to a triangle and an octagon. A resizing factor that plays the same role as the rescaling factor in equation (3.1) may be obtained by finding the ratio between any similar line segments. For example, the resizing factor s in resizing the triangle from the initial Figure 3.4a to the resized Figure 3.4b is obtained by taking any of the ratios as follows:

$$s = \frac{a'}{a} = \frac{b'}{b} = \frac{c'}{c}$$

The relationship between resizing and rescaling is as follows. For any resizing and rescaling factors of the same values, the measurement of the resized figure in original units is the same as the measurement of the original figure in rescaled units. The relationship equation (3.1) with s taken as the resizing factor holds. Because of the similarity between resizing and rescaling, many texts do not distinguish between the two. Indeed, resizing is not a common term.

3.1.5   Koch's Snowflake, a Fractal Object

This section investigates an object with fractal dimension. The section is another excursion and as all excursions the material is tangential to the remaining material in the book.

In equation (3.1), which defines the dimension of an object, there is a priori no reason that the dimension must be integer valued. In this section, an object with fractal dimension is considered. Neither Euclid nor any of the other ancient Greeks ever considered such a possibility. Karl Weierstrass (1815–1897) was the first to propose a fractal object, although he did not consider its dimensionality. Weierstrass' construction is artificial; however, the concept now finds application for describing geometries found in nature. Examples are river deltas, surfaces of highly porous medium, and systems displaying chaotic motion. The example presented in this section is like Weierstrass' example artificial, but it illustrates the possibility of fractal dimensions. Its creator is Helge von Koch (1870–1924) and it is known as Koch's snowflake.

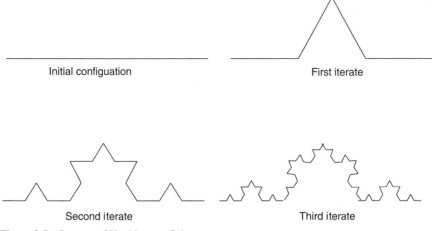

| Initial configuation | First iterate |
| Second iterate | Third iterate |

Figure 3.5   Iterates of Koch's snowflake.

Koch's snowflake is produced iteratively. Figure 3.5 illustrates the initial configuration and the first three iterations. Note that the initial configuration is simply a line between the points 0 and 1. The first iterate is obtained by placing an equilateral triangle over the middle third of the original line segment and then removing the base of the equilateral triangle. Successive iterations are formed in the same fashion; on top of the middle third of every line segment, one places an equilateral triangle and then removes the base. Koch's snowflake is the limiting set of the iterations.

Let us try to get the dimension of the snowflake. As a first attempt, assume that the snowflake's dimension is 1 and find its length. Summing up all the line segments in an iterate, the reader may verify that the length of the nth iterate is $\left(\frac{4}{3}\right)^n$, where $n = 0$ gives the initial configuration in Figure 3.5. Since the lengths of the iterates increase without bound, when considered as a one-dimensional object, the snowflake has infinite length. However, the snowflake itself is bounded within the unit square and perhaps ascribing a different dimension would cause a bounded measurement for this object.

An approach toward finding the snowflake's dimension is to apply the definition as proposed in equation (3.1). We do so using a resizing factor of 3; one initial unit is the equivalent of three rescaled units. Applying equation (3.1) with s set to 3 results in the following equation:

$$\widetilde{M} = M \times 3^d \qquad (3.2)$$

where

- $\widetilde{M}$ is the measurement in new units
- M is the original measurement

The upper object in Figure 3.6 shows the second iterate in the snowflake's construction. In recognition of the limitations of the drawing, the author requests that readers exercise their imagination and consider the object as the complete snowflake viewed

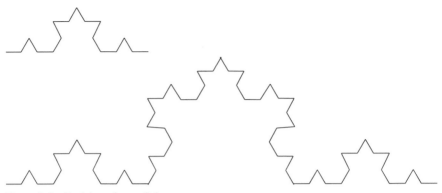

Figure 3.6   Resizing of snowflake.

with poor resolution so that all the details are not visible. The lower object, also to be considered the complete snowflake, is a resizing of the upper object by a factor of 3. Note that upon resizing the lower object contains four replicas of the original snowflake, so that the measure of the entire lower snowflake is four times the measure of the upper snowflake. This observation leads to the following equation:

$$\widetilde{M} = 4M \tag{3.3}$$

Equating equations (3.2) and (3.3) and noting that $s = 3$ yield the following:

$$M \times 3^d = 4M$$
$$3^d = 4$$

Taking the logarithm of both sides reveals the dimension

$$d \log 3 = \log 4$$
$$d = \frac{\log 4}{\log 3} \approx 1.26186$$

Using the relation between resizing and dimension establishes the dimension of Koch's snowflake somewhere between 1.2 and 1.3. With this dimension, one could prescribe the unit measure of one to the original snowflake, and the measure of any composition of snowflakes or partial snowflakes is determined with respect to the standard unit snowflake.

3.2 MEASUREMENTS OF VARIOUS OBJECTS

In this section, we determine the length, area, or volume of various geometric objects. Formulas for different shapes are determined. Shapes include the triangle, circle, cone, and sphere.

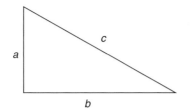

Figure 3.7   Right triangle of Pythagorean theorem.

3.2.1 Pythagorean Theorem, Length of the Hypotenuse

The Pythagorean theorem is named after the Greek mathematician Pythagoras (circa 580–500 B.C.E.). The theorem was known to the Mesopotamians long before Pythagoras' time. Ancient Chinese and Indian mathematicians also independently discovered the theorem.

The theorem relates the sides of a right triangle to the length of its hypotenuse. We have already seen that its application to finding the diagonal of a square led the Pythagoreans to the discovery of irrational numbers. In the next chapter, the theorem is useful for finding the distance between points in a Cartesian coordinate system.

The theorem states that the sum of the squares of the sides of a right triangle is equal to the square of the triangle's hypotenuse. The equation is the following (Figure 3.7):

$$a^2 + b^2 = c^2$$

where

- a is the length of the triangle's base
- b is the length of the triangle's height
- c is the length of the triangle's hypotenuse

There are several ways to prove this result. We demonstrate two.

For the first approach, consider Figure 3.8. The figure is constructed from four replicas of an arbitrary right triangle. The triangles are arranged to form a large square with sides of length $a + b$. Inscribed within the outer square is a smaller square with sides of length c. There are two ways to find the area of the outer square, each giving its own formula. Equating the two formulas and simplifying yield the Pythagorean theorem.

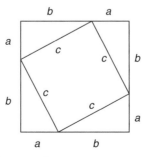

Figure 3.8   Illustration of Pythagorean theorem.

The first way to find the area of the outer square is to simply square the outer square's sides. This gives the following equation:

$$\text{Area} = (a + b)^2 = a^2 + 2ab + b^2$$

The second way to find the area of the outer square is to sum the area of the triangles and the inner square. The area of one triangle is $\frac{1}{2}ab$, so the area of all four triangles is $\frac{4}{2}ab$ or equivalently $2ab$. Also, note that the area of the inner square is c^2. This results in the second equation for the area of the outer square:

$$\text{Area} = 2ab + c^2$$

Equating the two formulas for the area of the outer square and simplifying yield the result:

$$a^2 + 2ab + b^2 = 2ab + c^2$$
$$a^2 + b^2 = c^2$$

Another approach to the proof of the Pythagorean theorem is of interest because it applies the concepts of rescaling and dimension from the previous section. Consider Figure 3.9. The figure consists of a right triangle that is split into two triangles in a manner so that all three triangles are geometrically similar.

The area of the largest triangle is equal to the sum of the areas of the smaller triangles. As all the triangles are geometrically similar, the areas of the smaller triangles are obtained by the use of equation (3.1) with the correct resizing factor. In each case, the dimension d is 2. Also, in each case, the resizing factor is given by the ratio of the hypotenuse of the triangle of interest to the hypotenuse of the largest triangle. For the left triangle $s = a/c$, whereas for the right triangle $s' = b/c$. Applying equation (3.1) to the left triangle gives the following:

$$A_L = A \left(\frac{a}{c}\right)^2$$

where

- A_L is the area of the left triangle
- A is the area of the largest triangle

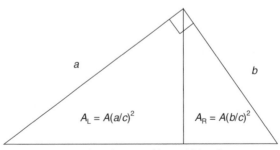

$A_L = A(a/c)^2$ $A_R = A(b/c)^2$

c = hypotenuse of largest triangle

Figure 3.9   Pythagorean theorem: another proof.

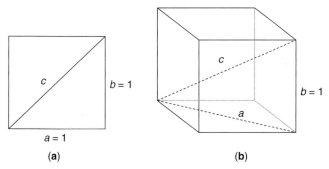

Figure 3.10   Diagonals of a square and a cube.

Similarly, applying equation (3.1) to the right triangle gives the following:

$$A_R = A \left(\frac{b}{c} \right)^2$$

where A_R is the area of the right triangle.

Summing the areas of the left and right triangles to get the area of the largest triangle and then simplifying yield the following result:

$$A \left(\frac{a}{c} \right)^2 + A \left(\frac{b}{c} \right)^2 = A$$

$$\left(\frac{a}{c} \right)^2 + \left(\frac{b}{c} \right)^2 = 1$$

$$a^2 + b^2 = c^2$$

Example 3.2   DIAGONAL OF A SQUARE

Find the diagonal of the unit square. Referring to Figure 3.10a, $a = b = 1$:

$$a^2 + b^2 = c^2$$

$$1^2 + 1^2 = 2$$

$$c = \sqrt{2}$$

Example 3.3   DIAGONAL OF A CUBE

Find the diagonal of the unit cube. Referring to Figure 3.10b, note that a is the diagonal of the unit square. From the previous example, we see that $a = \sqrt{2}$. Also, $b = 1$:

$$a^2 + b^2 = c^2$$

$$2 + 1^2 = 3$$

$$c = \sqrt{3}$$

3.2.2 Cavalieri's Theorem in Two Dimensions

Cavalieri (1598–1647), a disciple of Galileo, thought considerably about measurements of different shapes. His best known result bears his name, Cavalieri's theorem. Interestingly enough, two applications of Cavalieri's theorem are known to have occurred long before Cavalieri. Zu Gengzhi (429–500), following the works of Liu Hui (circa 250) in China, used Cavalieri's theorem to derive the formula for the volume of a sphere. Archimedes used the theorem to demonstrate the relation between the volume of a sphere and the sphere's surface area. Later in this chapter, we replicate Archimedes' result.

Cavalieri presents his theorem in the form of a construction. He constructs two-dimensional objects between two initial parallel lines. Then he asserts that if the two objects' cross-sectional lengths are equal for cross sections along every parallel between the initial parallels, the two objects have equal area. Figure 3.11 illustrates the concept. In this figure, two horizontal lines bound two figures from above and below and the objects' cross sections are displayed along an arbitrary parallel. Cavalieri's theorem states that if all such cross sections have equal length the areas of the objects are equal.

Cavalieri proposes the argument illustrated by Figure 3.12 as proof of his theorem. Two objects having the property that their cross sections along any given parallel are of the same length are said to satisfy the Cavalieri property. Any such objects can be superimposed so that their intersection has positive area. Superimpose object G upon object K in such a manner that the area of the intersection is maximum. Remove the intersection from the superimposed objects noting that the removed areas are the same. If there is any difference in the areas of G and K, there must be a difference in the areas of their respective remaining objects. Label these objects G' and K', with G' the remains of G and K' the remains of K. The objects G' and K' also satisfy the Cavalieri property. Once again superimpose the objects G' and K' and remove the intersection. Cavalieri claims that continuing the process would eventually deplete both objects G and K; since there would be nothing left, the objects could not differ in area.

Cavalieri's theorem is an example of dead-on instinct with a somewhat flawed (though clever) proof. His claim that repeatedly superimposing and removing intersections eventually depletes the objects is not demonstrated and not in general true.

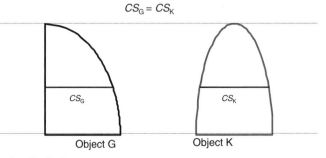

Figure 3.11   Cavalieri's theorem.

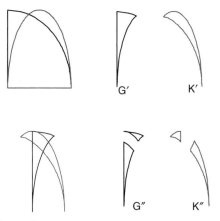

Figure 3.12   Cavalieri's proof.

With additional conditions on the shapes, the argument is true; however, a rigorous proof of the theorem would have to await calculus.

3.2.3 Cavalieri's Theorem, Archimedes Weighs In

The symbol for justice is a balance scale held by a blindfolded woman. The balance scale is a metaphor for carefully weighing each side of a legal case. Archimedes was once involved in a legal case; in his case, the balance scale was not a metaphor but was the actual instrument with which Archimedes weighed justice. The problem was to determine if a crown commissioned by the king was constructed of gold as required or if the artisan used a gold alloy. Archimedes knew that the volume of an alloy would be larger than the volume of an equivalent weight of gold. He also knew that the buoyancy of the alloy with larger volume would be greater than the buoyancy of the gold with smaller volume.

With this information Archimedes solved the case. He placed the crown on one side of the scale and placed an equal weight of gold on the other, and the scale balanced. Next, with both the crown and the gold remaining on their respective sides of the scale, he immersed the scale in water. Within the water, the side of the scale with the crown raised to a higher level than the side with the gold, demonstrating that the crown was more buoyant and thus of larger volume than the equivalent weight in gold. This was proof positive that the crown was an alloy and this is how Archimedes weighed justice.

Archimedes was a tinkerer whose experiments assisted in his mathematical endeavors. To find the volumes and centers of masses of objects, he performed thought experiments with a scale. Following Archimedes, let us perform a thought experiment on Cavalieri's theorem.

Suppose two planar shapes satisfying the Cavalieri property are cut out from a sheet of material that is of uniform thickness. We will show that they do have the same weight and hence must have the same area.

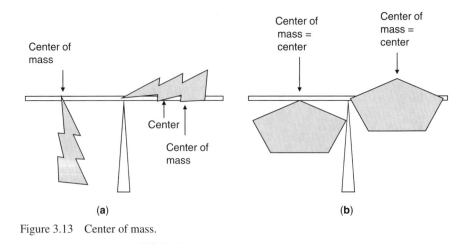

Figure 3.13　Center of mass.

The scale that we consider is a balance scale; there is a central pivot and weights can be suspended at different positions from the pivot. For an object that is stretched out across one arm of the scale, the center of mass of the object is the single point where one would suspend the object in order to induce the same effect upon the scale as the originally placed object (Figure 3.13a). For an object that is symmetrically distributed about its center, the center of mass and the center are one and the same (Figure 3.13b).

Let two planar shapes G and K be placed on the left and the right sides of the scale so that their corresponding axis is aligned with the corresponding left and right arms (Figure 3.14a). Also, ensure that their starting distances and ending distances from the pivot are the same; this can be done because their axes are of the same length. Select any distance to the left and the right of the pivot; because of the Cavalieri property, the cross-sectional length of each shape at the set distance is identical and these points are balanced. Because every point is balanced, the two objects are balanced.

We have not yet shown that the weights are equal; this only happens if the scale is in balance when each object is suspended from a single point and the points (one for each object) are equidistant from the pivot. Let us take exact replicas of G and K and place them on the scale reversing the direction of the axis as shown in Figure 3.14b. The left side now has G and its replica, while the right side has K and its replica. The placement of the replicas maintains the Cavalieri property for the ensemble, so the ensemble is in balance.

Also, for each side the ensemble is symmetric about its center. As a result of the symmetry, the center of mass for both objects is at the object's midpoint. Remove each ensemble and then suspend them at the center of mass for their corresponding arms; nothing changes, so the scale remains in balance (see Figure 3.14c). Note that the centers of mass are equidistant from the pivot. Therefore, the ensembles are of equal weight. As the ensemble is twice the weight of the original shapes, the shapes are also of equal weight.

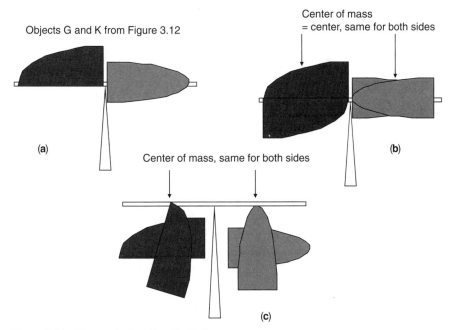

Figure 3.14   Demonstrating Cavalieri's theorem.

Remark. This demonstration of Cavalieri's theorem is not a proof in the sense of Euclid. Indeed, it is quite removed from Euclid's initial axioms. Nevertheless, it is complete provided that one accepts the behavior of the scale as used in the theorem. There is no written record of Archimedes proving Cavalieri's theorem, although as previously noted, he uses it to demonstrate the relationship between the volume of a sphere and the sphere's surface area.

3.2.4 Simple Applications of Cavalieri's Theorem

In this section, we apply Cavalieri's theorem to simple objects, parallelograms and triangles. Of course, it is possible to find areas for these objects without resorting to Cavalieri's theorem. But this section presents introductory material for following sections where Cavalieri's theorem is used to find formulas concerning areas and volumes of more complicated objects.

The theorem asserts the equivalence of areas of rectangles with sides a and b and parallelograms with height a and base b (Figure 3.15). For both the rectangle (Figure 3.15a) and the parallelogram (Figure 3.15b), the length of the cross section along any parallel is identically b. Knowing the area of the rectangle is ab, we conclude that the area of the parallelogram is also ab.

The cross sections in object C above are also of constant length, b. This object has equal area to both the rectangle and the parallelogram.

Cavalieri's theorem also asserts that any triangles of equal height and base have the same area. In Figure 3.16, two triangles, G and K, are drawn between two

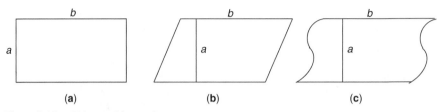

Figure 3.15   Objects with equal area.

parallels; the base of each triangle is of the same length, b. We demonstrate that the cross-sectional lengths along any arbitrary parallel are the same. Select a parallel that is a distance d from the uppermost parallel. Each portion of the triangle that lies above the selected parallel forms a triangle that is geometrically similar to its respective original triangle. Designate the length of the base of the smaller triangle by m. By similarity, the ratios b/h and m/d are equal. Equating these ratios and solving for m yields the length of the selected cross section:

$$\frac{m}{d} = \frac{b}{h}$$

$$m = \frac{b}{h}d$$

Since h, d, and b are equal for both triangles, the lengths of the cross sections along the parallel are equal. This is true for every parallel between the initial parallels. By Cavalieri's theorem, the two triangles are of equal area.

Using the above result, we can assert that the formula for the area of a single triangle with a given height and base is in fact quite general and applies to any triangle with the same height and base. The simplest triangle to work with is a right triangle (Figure 3.17). A right triangle is half of a rectangle with the same height and base, so its area is half the area of the associated rectangle, $A = \frac{1}{2}hb$.

3.2.5   The Circle

This section presents the formulas for the circumference and area of a circle using rescaling arguments. Cavalieri's theorem is then applied to find a relation between

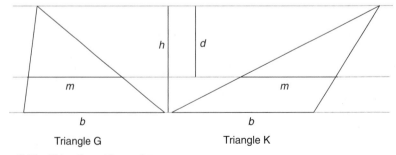

Triangle G   Triangle K

Figure 3.16   Triangles with equal area.

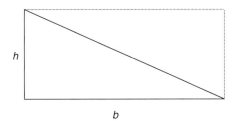

Figure 3.17 A right triangle as half a rectangle.

circumference and area. The result is stated in terms of the constant π, which is approximated in Section 4.3.2.

Consider a circle with radius 1. The circumference has a length that we call a constant ρ. Next resize the circle to have a radius of r. The resizing factor s in equation (3.1) is the ratio of the resized radius to the original radius, $s = r/1 = r$. Denote the length of the circumference of the unit circle by λ. Using equation (3.1) to calculate the length of the resized circumference with $\widetilde{M} = c$, $M = \lambda$, $s = r$, and $d = 1$ yields the following result:

$$\widetilde{M} = M \times s^d \qquad c = \lambda r \tag{3.4}$$

Accordingly, if the circumference of the unit circle can be found, then it is possible to find the circumference of any circle. In Section 4.3.2, the circumference of the unit circle is approximated using the method of Archimedes.

A similar resizing argument shows that the area of a circle is proportional to the square of its radius. Consider once again a circle of radius 1 and resize the circle to have radius r. As above, $s = r$. Denote the area of the original circle by a constant α. The dimension of the circle $d = 2$. Using equation (3.1) to calculate the area of the resized circle with $\widetilde{M} = A$, $M = \alpha$, $s = r$, and $d = 2$ yields the result $A = \alpha r^2$.

To find the area of a circle in terms of its circumference, we cut it, unravel it, and determine the area of the resulting shape. Given a circle of radius r, cut the circle from its circumference to its center as shown in Figure 3.18. Next unravel the circle so that the circumference lies along a horizontal line. What object results? The circle is composed of a family of smaller circles all centered at the same point with

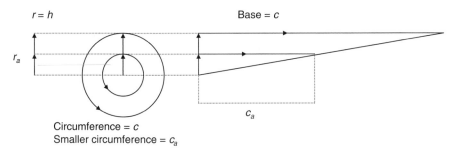

Figure 3.18 Unraveling the circle into a triangle.

radii ranging from 0 to r. Consider the circle with radius $r_a < r$ and circumference $c_a < c$. The circumference is unfolded onto a line that is parallel to the unfolded circumference of the original circle. Unfolding every circle in this manner results in the triangle of Figure 3.18. The area of the circle is the same as the area of the triangle, $A = \frac{1}{2}hb$. In this case, the height h is given by the radius r and the base b is given by the length of the circumference. Using equation (3.4) for the length of the circumference, the area is given by the following formula:

$$A = \tfrac{1}{2}\lambda r^2 \tag{3.5}$$

Define another constant, π, as $\frac{1}{2}\lambda$. Then equations (3.4) and (3.5) are the familiar formulas for the circumference and area of a circle:

$$c = 2\pi r \qquad A = \pi r^2$$

Remark. Archimedes mentions this construction in his work on spheres.

3.2.6   Surface Area of the Cone

Combining the technique used to find the area of a circle with the Pythagorean theorem results in the area of the surface of a cone. Let the height and the radius of the base of the cone be given (Figure 3.19). The length from the base to the surface is given by the Pythagorean theorem, $L = \sqrt{r^2 + h^2}$. Cut a line from the base to the top of the cone and then unfold each of the circles that is parallel to the base in a fashion similar to what was done when finding the area of a circle. The result is the triangle shown in Figure 3.19. The base of the triangle is the circumference of the cone's base circle; this has length $c = 2\pi r$. The height of the triangle is given by L. The area of

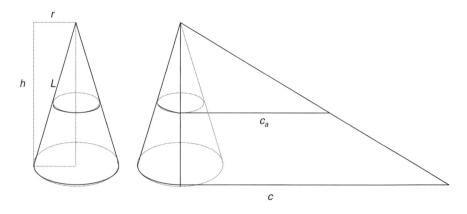

Circumference of base circle = c
Smaller circumference = c_a

Figure 3.19   Unraveling the cone.

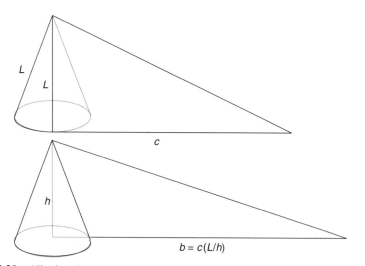

Figure 3.20 Aligning the triangle with the central axis.

the surface of the cone is given by the area of the triangle:

$$A = \tfrac{1}{2} 2\pi r L = \pi r \sqrt{r^2 + h^2}$$

Before continuing to the next shape, we consider the possibility of peeling the circles off the cone and placing them along the vertical axis of length h while maintaining the area of the cone. A similar operation is used to determine the surface area of a sphere. Here, it is presented on the cone as a warm-up exercise. One certainly can cut out every circle and append to the axis at its corresponding height and then unwrap the circles on the vertical line. The area of the corresponding triangle is $\pi r h$. This is smaller than the actual area. What has happened and can we make a correction?

The vertical axis is shorter than the slanted line of the original cut. Plucking the circles off the slanted line and clipping them onto the vertical axis requires that we squeeze them together. To maintain the original area, it is necessary to compensate for the squeezing in one direction by stretching in the other. This is accomplished by multiplying each circle by the ratio L/h; the circles are squeezed by the ratio h/L so that they can fit onto the shorter line, and so we stretch them by the ratio L/h to maintain the area (Figure 3.20).

Note that the stretching factor is the ratio of the slope length to the axial length.

3.2.7 Cavalieri's Theorem a Stronger Version in Three Dimensions

Cavalieri presents his theorem in both two and three dimensions. The three-dimensional analog to the Cavalieri property of Section 3.2.2 is given by changing the parallel lines of that section to parallel planes and changing the cross-sectional lines

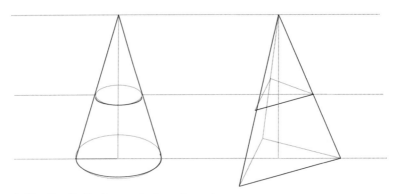

Figure 3.21   Cavalieri's theorem in three dimensions.

of that section to cross-sectional planes with equivalent areas. Figure 3.21 illustrates the construction.

The conclusion for the three-dimensional case is that objects satisfying the Cavalieri property have the same volume.

The two-dimensional proof that Cavalieri proposed can be adopted for three dimensions. Establishing that iterative elimination of intersecting sets depletes both objects is even more daunting in the three-dimensional case than the two-dimensional case. On the other hand, the balancing scale argument for the three-dimensional case follows that of the two-dimensional case with minor adjustments.

A useful result that is a bit stronger than the Cavalieri theorem is the following. Imagine two objects G and K that satisfy the following property:

$$A_{hG} = sA_{hK} \tag{3.6}$$

where

- A_{hG} is the area of the cross section within object G at height h
- A_{hK} is the area of the cross section within object K
- s is a constant value for every height

The feature to note is that the constant s applies to each height, so the ratio of areas is constant across the entire object. Then the following relation holds:

$$V_G = sV_K \tag{3.7}$$

where

- V_G is the volume of object G
- V_K is the volume of object K

For rational values of s, the strengthened Cavalieri result can be demonstrated as follows. First, it may be assumed that all the cross sections of object K are squares (if not, an object of equal volume could be constructed using square cross sections). Since s is a rational number, $s = p/q$ with both p and q integers. Split the object K into q pieces as illustrated in Figure 3.22. In the figure, the cross-sectional area of

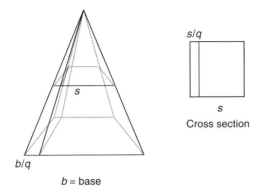

Figure 3.22   Demonstrating Cavalieri's theorem.

each piece is identical; each piece's cross-sectional area is A_K/q. Then by Cavalieri's theorem, each piece has the same volume, V_K/q. Construct an object, $\widetilde{K}$, using p replicates of any of the pieces. Then the volume of $\widetilde{K}$ is $p(V_K/q)$ (Figure 3.22):

$$V_{\widetilde{K}} = p\frac{V_K}{q} = sV_K$$

Also, by construction, G and $\widetilde{K}$ have cross sections of equal area that by the Cavaliei theorem results in equal volumes:

$$V_G = V_{\widetilde{K}}$$

Substituting for the value of $V_{\widetilde{K}}$ yields the result:

$$V_G = V_{\widetilde{K}} = sV_K$$

Remark. The result when s is irrational requires a bit, but not much, more sophistication. It will not be demonstrated in detail here. The idea uses approximations over rational numbers in a similar way that $\sqrt{2}$ was approximated by rational numbers in Section 3.2. Establish a small tolerance band and take a number r that is a rational approximation of s lying within the tolerance band. Replace s with r and construct $\widetilde{K}$. It can be demonstrated that the difference between V_G and $V_{\widetilde{K}}$ goes to zero as the tolerance band is narrowed. The same holds for the difference between $V_{\widetilde{K}}$ and sV_K. Allowing the tolerance band to approach zero yields the result.

3.2.8   Generalized Pyramids

In this section, we determine the volumes of generalized pyramids. Generalized pyramids are all objects constructed of cross-sectional areas that are equal to the areas of the cross sections of a standard pyramid; the standard pyramid has rectangular cross sections. The identifying feature of a pyramid is that every cross section is a resizing of the base and the resizing factor is given by the square of the ratio of the distance of the cross section from the top of the pyramid to the height of the pyramid. We take some time to give an example using a standard pyramid.

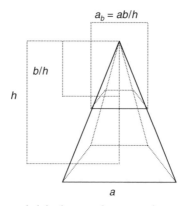

Figure 3.23   Relation between height, base, and cross section.

In Figure 3.23, the height of the pyramid is given by h and the base is a square with sides of length a and area $A = a^2$. Each cross section is also a square. The sides of the cross section at distance b from the top are given by the following:

$$a_b = a\frac{b}{h}$$

where a_b is the side of the square on the cross section that is at a distance b from the top.

The area of the cross section, A_b, is given by the following:

$$A_b = a_b^2 = a^2 \left(\frac{b}{h}\right)^2 = A \left(\frac{b}{h}\right)^2 \qquad (3.8)$$

Equation (3.8) presents the property of a generalized pyramid; the area of each cross section must be equal to the area of its base times the square of the ratio of the cross section's height to the height of the object. The cross sections need not be geometrically similar, although in the two examples we present, the cross sections are in fact similar. Two examples of generalized pyramids are given in Figure 3.21 and include a cone (circular cross sections) and a three-sided pyramid (triangular cross sections).

By Cavalieri's theorem, a formula for the standard pyramid applies to the generalized pyramid because both have identical cross-sectional areas as given by equation (3.8). Because the areas of all the cross sections depend solely upon the area of the base and the height of the generalized pyramid, there must be a formula relating the volume of a generalized pyramid to the height and area of the base. We seek this formula.

It is possible to construct a cube using six equally sized pyramids. Figure 3.24 illustrates the construction. Figure 3.24a shows three of the six pyramids (the left, top, and front pyramids), while Figure 3.24b shows the cube composed of six pyramids. Designate the length of an edge of the cube by a. Then the volume of the cube is a^3, the area of its base A is a^2, and the height h is $a/2$ (equivalently $a = 2h$).

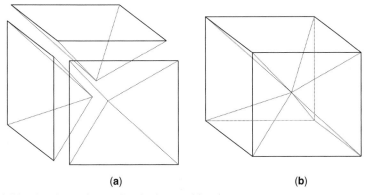

(a)   (b)

Figure 3.24   A cube as six equally sized pyramids.

Equating the volume of the cube with that of the six pyramids of equal volume yields the following identity:

$$6V_P = V_C$$

$$V_P = \frac{V_C}{6}$$

$$V_P = \frac{a^3}{6} = \frac{aa^2}{6} = \frac{hA}{3} \qquad (3.9)$$

where

- V_P is the volume of the pyramid
- V_C is the volume of the cube
- $2h$ is substituted for a value of a in the final equality
- A is substituted for the area of the base, a^2, in the final equality

A formula has been proposed for the volume of a pyramid, but the formula is obtained by looking at a pyramid with a specified geometry. The ratio of the height to the square root of the area of the base is set at $\frac{1}{2}$, $h/\sqrt{A} = \frac{1}{2}$, restricting the use of the formula to those pyramids that satisfy the ratio. This restriction is overcome using the strengthened version of Cavalieri's theorem. We next present the result for a general pyramid.

Let a general pyramid with height h and base area A be given. Construct another pyramid with the same height h and area $A' = 4h^2$. Note that by the way A' is chosen, $h/\sqrt{A} = \frac{1}{2}$. Set $s = A/A'$ and the relation equation (3.6) holds, allowing one to apply the generalized Cavalieri formula, equation (3.7). Using equations (3.6) and (3.7) results in the following formula for the generalized pyramid:

$$V_A = sV_{A'} = s\frac{hA'}{3} = \frac{hA}{3} \qquad (3.10)$$

3.2.9 The Sphere as a Generalized Pyramid

In his work *On the Sphere and Cylinder*, Archimedes asserts the following:

> *Judging from the fact that any circle is equal to a triangle with base equal to the circumference and height equal to the radius of the circle, I apprehended that, in like manner, any sphere is equal to a solid cone with base equal to the surface of the sphere and height equal to the radius.* (Heath (1921), p. 35)

Archimedes recognizes the sphere as a generalized pyramid. He draws an analogy between the unfolding of the circle as demonstrated in Section 3.2.5 (see Figure 3.18) and the unfolding of the sphere to obtain the pyramid (a solid cone). The trick is to note that the sphere is made up of an infinite number of spherical shells, all with a common center. To construct the generalized pyramid, create cross sections of equal area to each shell and stack the cross sections on an axis of height equal to the radius of the original sphere (Figure 3.25). In Figure 3.25, the center of the sphere becomes the apex of the pyramid marked by the letter C. The inner shell becomes the triangular cross-section at $h = r$ and the outer surface of the sphere becomes the triangular base of the pyramid.

Using a rescaling argument, we demonstrate that this construction produces a generalized pyramid. The argument must show that the relation given by equation (3.8) holds. Referring to Figure 3.25, note that the radius of the sphere, r, gives the height of the pyramid, h. Consider the shell with radius b; this shell corresponds to a cross section with a distance b from the top of the pyramid. Also, this shell is a resizing of the outermost shell, and indeed all shells are geometrically similar. The resizing factor is b/r. Using equation (3.1) with A, the area of the outermost shell, A_b, the area of the shell with radius b, and noting that the dimension of the shells is 2, we obtain the following:

$$A_b = A \left(\frac{b}{r} \right)^2$$

This is the precise relation, equation (3.8), that must be demonstrated. Because the sphere is a generalized pyramid, equation (3.10) holds, $V = rA/3$.

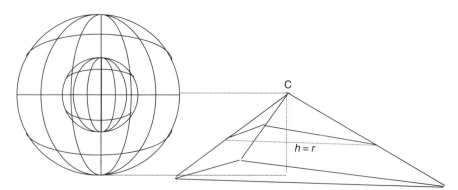

Figure 3.25   A sphere as a pyramid.

It is common to denote the surface area of the sphere by S and the formula is frequently given as follows:

$$V = \frac{rS}{3} \tag{3.11}$$

3.2.10 The Surface Area and Volume of the Sphere

Up to this point, we have followed Archimedes. Archimedes next finds a formula for the volume of a sphere and uses equation (3.11) to find the surface area. Archimedes' proof is the method that is used in modern calculus and is presented in Chapter 6. In this section, we proceed along a different route; we find the surface area and use equation (3.11) to determine the volume.

The approach is similar to the unfolding of the circle and the cone in Sections 3.2.5 and 3.2.6, respectively. Start with a sphere of radius r. Note that the sphere is a collection of circles of latitude. We could cut across a longitudinal section of the sphere and unwind each circle of latitude into a line segment. The result is depicted in Figure 3.26, where the cut is made along the central meridian. A similar process on the cone yields a simple shape, a triangle, and we were able to determine the area of the triangle. The process here yields something like an extremely curved shoe horn and finding its area is a bit more complicated.

Recall that at the end of our discussion on the surface area of the cone we unlatched the straightened circles from the cut and reattached them to their corresponding positions on the cone's axis. Recall that in doing so each circle is squeezed into a smaller area, and to find the initial area, it is necessary to apply a stretching correction that counteracts the squeezing. The same procedure is applied to the sphere. The first step is to figure out the length of each unfolded circle at each point on the sphere's polar axis. The next step is to calculate the stretching ratio. The final step is to calculate the area of the resulting figure.

Let the position of a point on the axis be denoted by z with $z = 0$ the axis center. Each value of z denotes a plane perpendicular to the axis that contains one and only one circle of latitude. We must determine the length of each of these circles. Referring to Figure 3.27, the radius of the circle at position z, denoted by r_z, follows from the Pythagorean theorem:

$$r_z^2 + z^2 = r^2$$

$$r_z = \sqrt{r^2 - z^2}$$

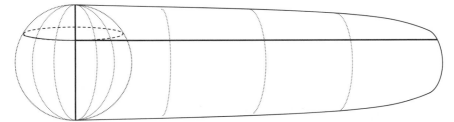

Figure 3.26   Unraveling the surface of a sphere.

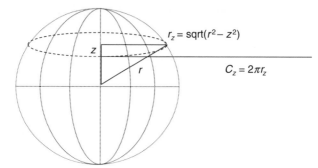

Figure 3.27   Length of a given latitude.

The circumference of the circle at z, c_z, is then given by the following equation:

$$c_z = 2\pi \sqrt{r^2 - z^2}$$

Now that the length of each circumference is known, the next step is to find the correct stretching ratio to compensate for the squeezing that is necessary to fit the circles from the circle of longitude onto the shorter polar axis. We make some observations. Around the equator there is some squeezing but not much. In fact, at the equator, the circle of longitude is parallel to the sphere's polar axis, so the equator does not get squeezed or stretched. The stretching factor close to the equator should accordingly be close to 1. Around the poles, there is a lot of squeezing. The circles around the poles are squeezed to fit them onto the polar axis and the stretching factor near the poles should be very large. From this we note that, unlike the case of the cone, the stretching factor depends upon the circle's latitude. But the cone is still of use; just apply a different cone at each point along the original cut.

For a point on the cut, the cone of interest is the one that is tangent to the sphere (Figure 3.28). Recall from Section 3.2.6 (see Figure 3.20) that the stretching ratio is the ratio of the slope length to the axial length. Referring to Figure 3.28, the stretching factor at height z, s_z, is given by the following equation:

$$s_z = \frac{L}{h}$$

Also, Figure 3.28 illustrates that the triangle formed by the slope of the cone, the radius r_z, and the cone's axis is similar to the triangle formed by the radius of the sphere, r, the radius r_z, and the line segment along the axis to the point z. The result is the stretching ratio s_z:

$$s_z = \frac{r}{r_z}$$

Note that the stretching factor matches the initial observations. At the equator, the stretching factor is 1. As we move toward the poles, the stretching factor becomes increasingly larger.

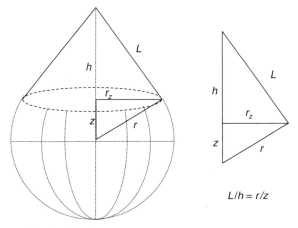

Figure 3.28   Geometry of latitude cone.

The lines of length c_z that lie across the cut are plucked from the cut, fixed to the longitudinal axis, and stretched into lines of length $c_z s_z$:

$$c_z s_z = 2\pi r_z \frac{r}{r_z} = 2\pi r$$

A wonderful thing has happened. The lines that we attach to the longitudinal axis are all of the same length. The original shape of Figure 3.26, which appears impossible to measure, has been transformed into a rectangle of equal area as in Figure 3.29. The length of the rectangle is the length of the longitudinal axis, $2r$. The width of the rectangle is $2\pi r$. The area of the rectangle, which is the same as the surface area of the sphere, is the product of its length and height:

$$S = 4\pi r^2$$

Combining the result of the surface area of a circle with equation (3.11), the volume of the sphere is given by the following equation:

$$V = \frac{4}{3}\pi r^3$$

Figure 3.29   Aligning the unraveled sphere with the central axis.

In every method used to find the volume of the sphere, a seemingly great stroke of luck occurs. For Archimedes, the great stroke of luck is that the volume of the sphere is expressible in terms of cones and cylinders. For the Chinese mathematician and astronomer Zu Gengzhi (480–525), the great stroke of luck is that an object whose volume he needs to calculate turns out to be a generalized pyramid with an elementary base shape; calculation of the area for the base shape is trivial. In our method, the great stroke of luck is that the mapping of the sphere onto the longitudinal axis results in a very simple figure, the rectangle. A priori one cannot expect such luck and yet it occurs. There is more going on than we realize. The luck occurs because of the symmetry of the sphere. Often, there is an obscure simplification that yields a solution to a seemingly intractable problem. Great mathematicians have an instinct for recognizing problems that may be simplified and uncovering the simplification. Often a method yields further results than originally envisioned. We examine this in the next section.

3.2.11 Equal-Area Maps, Another Excursion

Once again we embark upon an excursion and as always the material is tangential to the remainder of the book.

The history of cartography folds into the history of the studies of planetary motion beginning with Ptolemy. In addition to being the authority on planetary motion, Ptolemy achieved equal distinction in the area of cartography. Among many other subjects, Ptolemy's work *Geographia* presents the modern latitudinal and longitudinal coordinate system for the earth. This work competes with Euclid's *Elements* as one of the most popular and influential works in history.

Ptolemy presents several transformations of the sphere onto a flat surface; these furnish different ways to map the earth's surface. His development of transformations of the sphere is motivated by his investigations in astronomy and cartography. As an astronomer, he wished to map the celestial sphere and as a cartographer he wished to map the earth's surface.

Ptolemy was a prolific scientist. As noted in Chapter 2, he produced significant works in astronomy, geography, and mathematics. One of his investigations in mathematics was an attempt to prove the redundancy of Euclid's fifth axiom. Great mathematicians throughout history attempted to prove that the axiom is a consequence of the other axioms. Ptolemy was among those who tried and failed. The failed attempts were caused by man's limited perspective; space looks Euclidean and the essence of space is captured by the first four axioms. Why should not the fifth be a consequence of the first four?

If the fifth axiom were truly independent from the remaining axioms, one could replace it with its opposite statement and a consistent set of theorems would result. If it were not independent of the remaining axioms, when replacing the fifth axiom with its opposite, the resulting system would not be self-consistent; theorems that contradict one another would result. The mathematician Johann Lambert put the fifth axiom to this test. Lambert never found a contradiction.

Others, notably al-Haytham (915–1039), Omar Khayyam (1048–1122), Nasir al-Din al-Tusi (1201–1274), and Giovanni Saccheri (1667–1733), had followed a similar track and, not finding a contradiction, believed that they failed. But Lambert was bolder than his predecessors. He asserted that the fifth axiom is indeed independent of the other axioms. As a result, he tossed the fifth axiom and with it tossed Euclidean geometry. In place of Euclidean geometry, Lambert discovered a geometry with other properties: lines that are not straight, spaces where no lines are parallel, and bounded spaces with infinite measure. Lambert's work was ignored. It appeared as a mathematical construction with no real consequences for certainly it was believed that we live in a Euclidean universe. In fact, this was the first known step toward describing the non-Euclidean space that is the stage for the universe of Albert Einstein's relativity theory. Physical reality is the non-Euclidean space that Lambert first describes, while Euclidean space is an imagined idealization based upon our limited perception.

Lambert most likely viewed his work in the same vein as others, an interesting mathematical construction. Lambert lived at a time when Newton was an icon and Newton was very attached to the Euclidean world. Lambert enters our story in two different ways. He is the first to write down Newton's laws of motion and gravitation using Leibniz' notation. Later in this book, we will write Newton's laws as Lambert had written them for the first time. The other way in which Lambert enters our story is through his discovery of equal-area maps. An equal-area map has the feature that equal areas of the earth's surface are seen as equal areas on the map. The transformation that is used in the previous section presents a method for creating an equal-area map, the surface of the sphere is transformed to a rectangle of equal area, and the area of every region of the sphere's surface is respected. The discoverer of this transformation is Lambert. The map of the earth using this transformation is shown in Figure 3.30.

Note that with the exception of the equator the stretching factor distorts all the latitudes. The region around the equator does not get too badly distorted, but as one moves toward the poles, the distortion becomes significant. The countries further north are squeezed in and stretched out; they appear as if they had been put on the rack. For various reasons, aesthetics among them, one might wish to make adjustments to Lambert's map while preserving the equal-area property. Suppose, for example, you are a Londoner. Distortions far away from London are perfectly fine, but the region around London should have minimal distortion. How can this be accomplished?

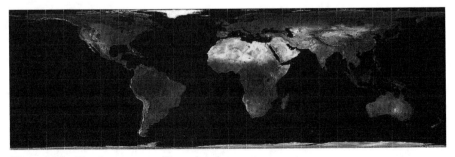

Figure 3.30   Equal-area map calibrated to the equator.

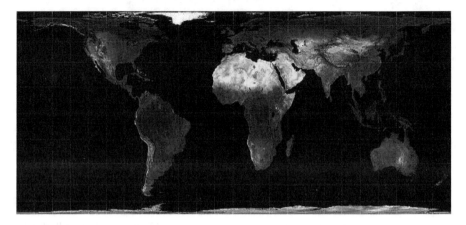

Figure 3.31 Equal-area map calibrated to London.

Recall that the stretching factor is given by the equation $s_z = r/r_z = r/(\sqrt{r^2 - z^2})$.

Set the scale of the map in units so that the radius is 1. Then the stretching factor becomes $s_z = 1/(\sqrt{1 - z^2})$.

The point z denotes the distance to the equatorial plane. Let us determine the value for London. To the nearest degree, the parallel passing through London is at $52°$, which has an associated z value of 0.788. Approximating s_z for London yields the value 1.624. In the east–west direction, we can squeeze the entire map by a factor of 1.624 so that the parallel around London is restored to its original scale and then stretch the map in the north–south direction by a factor of 1.624 to maintain an equal-area map. From the perspective of the Londoner, the distortion introduced by the initial equal-area transformation has been redressed, as the map of Figure 3.31 illustrates. With this correction, north-to-south as well as east-to-west distances are reasonably accurate around the 52nd parallel. Further away from the 52nd parallel, the representation is not as accurate and both north–south and east–west directions have been distorted. Indeed, in the map of Figure 3.31, it appears as though the continent of Africa has been put on a diet. Note that, by symmetry, in addressing the 52nd parallel in the northern hemisphere, we have also addressed the 52nd parallel in the southern hemisphere. Using this process, one can calibrate the map to any desired parallel.

Remarks

- The cartographer Gerardus Mercator (1512–1594) is the inventor of the most commonly used map, the Mercator projection. Rather than preserving equal area, the Mercator map preserves angles by applying the stretching factor not only to the east–west direction but also along the north–south direction. Mercator was another victim of the Counter Reformation. He was jailed for 7 years on suspicion of harboring Protestant sympathies.

- There are many more ways to create equal-area maps. The method used in this section can be generalized; further area-preserving transformations of the rectangle can be developed and applied to the rectangle of the previous section. The image does not have to be rectangular. Common shapes are also circles, two sets of circles, triangles, or ellipse-like objects.

- Creating a flat map requires sacrificing some properties of the original sphere. While equal-area maps preserve area, they distort shape and distance. Depending upon usage, other types of maps may be preferable.

- How is it that seemingly bizarre behavior arises in non-Euclidean geometry? One source of what might be considered aberrant behavior is a generalized concept of a line. Given a space, a line segment between two points is the shortest pathway within the space between the two points. Let us take the surface of a sphere as our entire space. It is known that the shortest pathway between two points is along the great circle passing through those points (great circles are intersections of the sphere and planes that pass through the center of the sphere). Then the lines on the space are given by great circles. The equator is a line in this space and the longitudes are line segments. All longitudes intersect the equator at 90° angles, and yet the longitudes intersect at the poles. In this way, the parallel postulate is violated and the sphere's surface, when considered a space unto itself, is non-Euclidean. Theorems from Euclidean geometry do not apply. For example, the angles of a spherical triangle do not in general add up to 180°. Consider a triangle formed by the equator and two different lines of longitude within the northern hemisphere. Each angle at the equator is 90°, so these angles alone sum to 180°. The triangle has another angle at the pole whose measure is given by the difference in the longitudes of the triangle's sides. The sum of the angles of this triangle are greater than 180°. Given that the geometry of the sphere differs from that of the Euclidean plane, it is no wonder that flap maps distort the world's geometry.

CHAPTER 4

THE LANGUAGE: ALGEBRA

Greek geometric mathematics has its limitations; it is limited to problems that are expressible in terms of geometric objects. Until the limitation was addressed, it would be difficult to develop mathematics beyond Archimedes and Apollonius. The specific areas of weakness that needed attention were a cumbersome number system that did not contain zero and the lack of symbolic notation along with standard processes for manipulating the symbols. Progress in these areas that culminate in modern algebraic symbolism is traced to many sources such as Greece, India, Arabia, and Europe. (In the Far East, independent approaches were taken.) There is no well-delineated story line illuminating the progress. While there are some signposts, they are planted in controversy and may be misleading. Perhaps this is because the story itself does not unfold uniformly along one track. Instead, repeatedly across cultures and eras, different threads are spun and then abandoned without being integrated into a coherent system, only to be reconstituted at a later time. What is known is that the signposts point to Rene Descartes, who tied each end of the story line together by popularizing a coordinate system that proposed Greek geometric shapes as algebraic expressions. We follow the signposts to Descartes.

A leading figure in the story is Abu Jafar Muhammed ibn Musa al-Khwarizmi (780–850). Al-Khwarizmi was a scholar in the House of Wisdom, the university established by the Abassid Caliph Harun-al Rashid in Baghdad. Al-Khwarizmi memorializes himself in two books. The first one, *On Calculation with Hindu Numerals*, as its title suggests, introduces the Hindu numerical system along with arithmetic operations using Hindu numerals. This book was a standard mathematics text throughout the height of Islamic power. It was one of the Arabic texts that was translated into Latin, bringing the Hindu system to the Europeans. Through this text, the Hindu numeric system became the modern international standard.

In another book, *Al-jabr wa'l muqabalah*, al-Khwarizmi describes methods for solving linear and quadratic equations through operations on the equation's expressions followed in the case of quadratic equations by geometric constructs. While the geometric constructs had been known to the Greeks, a systematic presentation of operations to simplify equations was not known from the Greek literature. The two operations that al-Khwarizmi considers are translated as "completion" and "balancing." Completion is the operation of adding a positive value to both sides of an equality in order to eliminate negative values. For example, $x^2 - 5x = 3$ becomes $x^2 = 5x + 3$ through completion. Balancing is the operation of eliminating common terms from both sides. For example, $x^2 + 2x = 7x + 2$ becomes $x^2 = 5x + 2$ by eliminating the common term, $2x$, from both sides.

Al-Khwarizmi demonstrates both limitations and sophistication with this approach. A limitation is that, lacking symbolic notation, al-Khwarizmi communicates mathematical expressions in prose. The following is a problem and solution from his book:

> *What is the square which when taken with ten of its roots will give a sum total of thirty nine? Now the roots in the problem before us are ten. Therefore take five, which multiplied by itself gives twenty five, an amount you add to thirty nine to give sixty four. Having taken the square root of this which is eight, subtract from this half the roots, five leaving three. The number three represents one root of this square, which itself, of course is nine. Nine therefore gives the square.*

Today we would simply write the problem and solution as algebraic expressions:

$$x^2 + 10x = 39$$

$$x = \left(\sqrt{\left(\tfrac{10}{2}\right)^2 + 39} \right) - \tfrac{10}{2} = 8 - 5 = 3$$

Therefore, $x^2 = 9$.

Needless to say, prose is a much more cumbersome language for the development and application of algebra than modern symbolic language.

A point of interest that some may consider a limitation is that al-Khwarizmi ultimately relies upon geometric constructions as known to the Greeks to solve quadratic equations. These constructions do not yield the unified solution currently taught in schools throughout the world, and they are unable to find negative solutions. The example above is solved through the geometric construction illustrated in Figure 4.1.

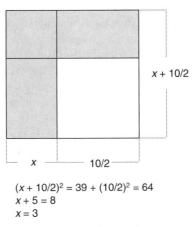

$(x + 10/2)^2 = 39 + (10/2)^2 = 64$
$x + 5 = 8$
$x = 3$

Figure 4.1   Geometric solution to the quadratic equation.

The shaded region has an area of $x^2 + 10x$, which is equal to 39. The solution described by al-Khwarizmi is to find the area of the outer square, $\left(\frac{10}{2}\right)^2 + 39$, take its square root to find the length of the larger square's side, 8, and then noting that the length of the larger square is five units larger than the side x that is to be determined, subtract 5 from 8 to get the solution, 3.

The solution is so Greek that it masks two critical contributions. The first contribution is a classification scheme whereby each quadratic equation that al-Khwarizmi investigates is categorized into one of six cases. An equally significant contribution is the use of completion and balancing to reduce any quadratic equation to one of six cases. This formalizes two algebraic operations that al-Khwarizmi states in prose but would later be applied to symbols and taught to students throughout the world.

For his efforts, the etymology of two words can be traced to al-Khwarizmi. The first word is *algebra*. This follows from al-Khwarizmi's choice to name one of his operations completion, which in Arabic is *al jabr*. The other word, *algorithm*, came about through a translational misunderstanding of al-Khwarizmi's work *On Calculation with Hindu Numerals*. The first Latin translation of the original Arabic version is *Algoritmi de Numero Indorum*, where *algoritmi* is the Latinized version of al-Khwarizmi's name. A common perception was that *algoritmi* referred to calculation methodologies. The perception took hold and today the word algorithm refers to a specified sequence of steps taken to arrive at a solution to a quantitative problem. In retrospect, the translational misunderstanding is perfectly justified for al-Khwarizmi's approach toward the solution of quadratic equations is the quintessential algorithm.

Al-Khwarizmi's work spawned efforts to introduce algebraic principles into broader applications of mathematics. Whereas the Greeks perceived numbers as measurements of geometric objects, the Arab world took a broader view. This allowed Arab mathematicians to address problems without a direct geometric interpretation using algebraic as opposed to geometric techniques. For example, Arab mathematicians determine the sum of powers, $1^p + 2^p + 3^p + \cdots + n^p$. Notable in this endeavor are al-Karaji (953–1029) and al-Haytham (965–1039). The latter, al-Haytham, is considered the greatest medieval physicist with his most important contribution in optics.

Arabs made limited headway toward the development of symbolic mathematics and always confronted the limitations of prose as a vehicle for expressing and communicating mathematics. The modern symbolism used to express mathematical concepts has evolved over a long time frame and is still expanding as the expansion of mathematical knowledge requires an increasing set of symbolic expressions. The solution to the cubic equation and the mathematical brouhaha surrounding it illustrate that the shift from prose to symbolism came about slowly.

In sixteenth-century Italy, mathematical duels were staged events with all the theatrics one would expect from the culture that gave birth to opera. Public dueling afforded the victor's prestige, possibly accompanied by teaching opportunities. One form of dueling was to exchange problem sets, the sets would be circulated to participants as well as a wider audience of experts who would act as witnesses. In 1535, Tartaglia (1500–1557) found himself preparing for a duel with a certain Mr. Fior. The latter had at his disposal a solution to the cubic equation $x^3 + ax = b$ for a and b

positive numbers. Mr. Fior received the solution from its original discoverer, Scipione del Ferro (1465–1526). Tartaglia set about finding the solution and succeeded. Fior submitted 30 questions, one of which is as follows:

> *Two men together gain 1000 ducats. The gain of the first is the cube root of the gain of the second. What is the gain of each?*

Of note is that since al-Khwarizmi little headway had been made in the realm of notation. Prose was still the language of mathematics. In modern-day parlance, the problem would be posed as solve $x^3 + x = 1000$.

At any rate, Tartaglia swept through Fior's problems with ease while Fior was stumped by Tartaglia's. Tartaglia earned a bit of notoriety and the duel came to the attention of Geronimo Cardano (1501–1576). After years of Cardano's cajoling, in 1539 Tartaglia revealed the solution to Cardano, while in return, Tartaglia elicited a promised from Cardano to keep the solution a secret so that Tartaglia could publish it in a future work. In 1545, Cardano published the solution in his work *Ars Magna*. There is a bit of controversy among historians. Some believe that upon seeing the same solution in an unpublished work predating Tartaglia, that of del Ferro, Cardano no longer felt obliged to maintain his promise to Tartaglia. Others believe Cardano never intended to keep his promise. How Cardano discovered the unpublished works of del Ferro is unknown, but just as Cardano stalked Tartaglia, it is conceivable that he followed the tracks of Fior, which led him to del Ferro's den.

Cardano's publication was well received, enhancing his already considerable stature. Tartaglia was furious with Cardano for breaking his commitment to secrecy and unloaded a canister of insults with the objective of sullying Cardano's reputation. Cardano did not respond but was represented by his very able student Ferrari. Ferrari shot a volley of plagiarism at Tartaglia, somehow linking Tartaglia's solution with that of del Ferro. The charge of plagiarism elicited a challenge. In 1547 Tartaglia, hoping to engage in a duel with Cardano, sent off a list of 31 questions, but Cardano would have none of it. Cardano was, after all, the more famous of the two and there was no reason to respond to an inferior. As a consolation, Cardano offered his student, Ferrari. Two months later Ferrari responded in kind with a list of his own. Tartaglia then responded with a set of solutions to Ferrari's questions. Ferrari ripped them up claiming only five were correct and then 3 months later responded to Tartaglia's questions. It is all a disputation with piercing invective in which everybody was miserable.

Under somewhat clouded conditions, Tartaglia agreed to a public debate between himself and Ferrari. It seems that Tartaglia had recently returned to his native town, Brescia, where he was offered a position as a lecturer. He was pressed into the debate to uphold the honor of Brescia. A public debate was not Tartaglia's preferred venue for a duel. Tartaglia had a speech impediment that was the result of a French soldier piercing his jaw with a sword in his youth. Hence, he was known by the nickname Tartaglia, the stammerer, instead of his real name, Nicolo Fontana. On top of this handicap, the debate would be on Cardano's home turf where Cardano could stack the audience. Tartaglia took a beating from which he never recovered. The only account of the debate is Tartaglia's; the account claims that a hostile crowd prevented Tartaglia from speaking so he left the debate early. Tartaglia never recovered from the humiliation and was forced out of Brescia. Ferrari, on the other hand, was showered with praise.

The literature on the debate is split among those sympathetic to Tartaglia and others claiming Tartaglia was in fact soundly defeated by Ferrari. Those in the latter camp point to the work in *Ars Magna* that Cardano attributes to Ferrari. There Ferrari presents solutions to cases of the cubic beyond the cases that Tartaglia addressed and presents a solution to the quartic polynomial. Historians on the side of Ferrari claim that Tartaglia never mastered the broader works of Ferrari and that was his undoing. We will not pursue what can only be speculation. But a trip over *Ars Magna* reveals the state of mathematical development in the sixteenth century.

In *Ars Magna*, Cardano, like al-Khwarizmi, categorizes solutions into cases and then demonstrates the solution of each case. Also, *Ars Magna*, like *Al-jabr Wa'l Muqabalah*, is rife with the geometric flavor of the Greeks. Diagrams that could have come from Euclid's hand abound. Finally, prose dominates the book. The following is Cardano's description of a solution to one case taken from a translation of his book (Chapter 12):

> *When the cube of one-third the coefficient of x is not greater than the square of one-half the constant of the equation, subtract the former from the latter and add the square root of the remainder to one-half the constant of the equation and, again, subtract it from the same half, and you will have, as was said, a binomium and its apotome, the sum of the cube roots of which constitutes the values of x.*

It is as if Cardano is al-Khwarizmi's clone. Cardano addresses a problem that is certainly a step up in degree of difficulty, but the methods and communication have not changed between the original and the clone that follows 700 years later. One cannot expect significant development in mathematics when a common language for mathematics in the form of algebraic symbols is not yet in use.

Kepler continued the tradition of prosaic and geometric mathematics and, like Archimedes nearly two millennium before his time, had close encounters with calculus. The first encounter is laid out in Kepler's description of the conservation of angular momentum of a planet's orbit. Kepler's geometric construction echoes Archimedes' results of his investigations of spiral motions. (In fact, Kepler credits Archimedes with inspiring his approach.) The next encounter between Kepler and calculus is when Kepler describes a method for determining the volume of shapes that are symmetric about an axis. This method is eerily similar to the method of Archimedes' proof of the volume of a sphere. Even the greatest among men are prone to err; Kepler was no exception. A century before Kepler, German mathematicians initiated a movement toward algebraic symbolism that was known as coss. Kepler dismissed coss as a meaningless abstraction, choosing instead the geometric language of Archimedes and like Archimedes was a victim of its limitations.

Why did Kepler dismiss coss? One can only speculate that the symbolic expressions did not contain the geometric insight that inspires mathematicians. For this reason, Kepler viewed coss as limiting, and at that stage of development Kepler was correct.

Rene Descartes (1596–1650) was a drifter. He was born in France, educated at a Jesuit College, then studied law at Poitier university. Afterward, his life resembled Europe during the Thirty Years' War, uncentered and in flux with no apparent purpose to its movements. The resemblance is not merely a metaphor. Descartes entered military service on various sides of the fragmented continent. He was in the service of the Netherlands, the first Protestant nation to challenge Catholic Spain's supremacy on the continent. Then he switched sides offering his services to Duke Maximilian I of the Holy Roman Empire, Spain's surrogate. Then he switched sides once more offering his services to France, Spain's nemesis. Apart from his military career, during the years 1620–1628, there are reported Descartes sightings in Bohemia, Germany, Holland, Hungary, Italy, and France. Perhaps a weary Descartes then decided to settle down, choosing Holland for the next 20 years. Nevertheless, his idea of settling down differs from the typical community man. He moved addresses between Amsterdam, Dordecht, Deventer, Egmond, Endegeest, Franeker, Leiden, Santpoort, and Utrecht, seemingly changing residences at random. Descartes died in Sweden while in the service of Sweden's Protestant queen. Befitting his life as a drifter, due to his Catholic heritage, Descartes was buried in a graveyard inhabited by residents who never had a home: unbaptized victims of infant mortality. The wanderer who lived most of his life in foreign lands did not form the social connections that center the lives of most men. (Although it is noteworthy that Descartes contracted the illness that eventually killed him while selflessly attending to a sick associate.) Yet it was this wanderer who connected coss with geometry, uniting the disciplines in the field of analytic geometry. From the time of Descartes' works onward, algebraic symbolism replaced prose as the language of mathematical expressions and geometry was etched into Cartesian coordinates. Preparations for a breakthrough in mathematics were complete and calculus would follow only 14 years after Descartes' death.

In this chapter, Cartesian coordinates are presented along with an investigation of several algebraic expressions for geometric objects. Using the newly found power of analytic geometry, some unfinished business from Chapter 3 is attended to: Archimedes' evaluation of π and an investigation of the four-dimensional sphere. The latter is an excursion. The chapter then moves on to cover the method of induction introduced by al-Karaji and put to use by al-Haytham. Before closing with some properties of the ellipse, the chapter develops linear algebra and matrix notation in two dimensions. The chapter assumes that the reader is familiar with the concept of a function and uses the concept throughout.

4.1 CARTESIAN COORDINATES AND TRANSLATION OF THE AXES

Cartesian coordinates provide visualization of a function or a geometric object in one, two, or three dimensions. While visualization is not possible, they serve the same role of coordinating algebra and geometry in higher dimensions. The Cartesian coordinates of a point in n dimensions are denoted by n numbers in parentheses $(x_1, x_2, x_3, \ldots, x_n)$. In one, two, and three dimensions, it is common to designate the point by x, (x, y), and (x, y, z), respectively. There is some ambiguity in the notation

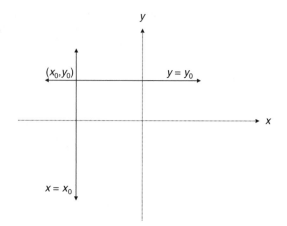

Figure 4.2 Intersection of two lines at a point.

as sometimes the letters x, y, and z represent fixed numbers and sometimes they represent variables in an equation. To reduce the ambiguity, whenever a specific point is being considered, a subscript accompanies the letter. For example, the symbols (x_0, y_0) or (x_A, y_A) denote specific points.

In this section, two examples of the use of Cartesian coordinates to illuminate commonalities between algebraic and geometric concepts are presented.

4.1.1 Intersections of Geometric Objects as Solutions to Equations

The correct interpretation of a point (x_0, y_0) on the coordinate plane is that x_0 represents a distance along the x axis and y_0 represents a distance along the y axis. There is another interpretation. Consider the following two equations and their plots on a Cartesian plane (Figure 4.2):

$$x = x_0 \qquad y = y_0$$

The first is the equation for a vertical line in which the variable x is a constant. The second is the equation for a horizontal line in which y is a constant. The two lines intersect at the point given by (x_0, y_0). Accordingly, the point (x_0, y_0) can be considered the intersection of two lines.

In three dimensions, the point (x_0, y_0, z_0) can be interpreted as representing the intersection of three equations:

$$x = x_0 \qquad y = y_0 \qquad z = z_0$$

Each of these three equations forms a plane. Two planes intersect in a line and the third plane intersects the line at a point; the intersection of three transverse planes is a point (Figure 4.3). The point is identified by the coordinates (x_0, y_0, z_0).

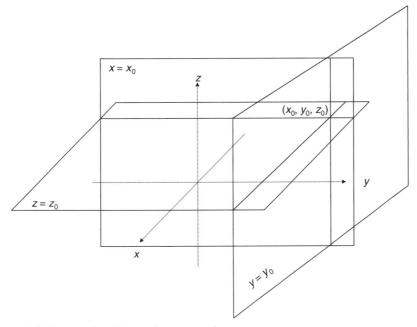

Figure 4.3   Intersection of three planes at a point.

A generalization of this perspective affords insight into a fundamental principle for solving equations. In two dimensions, the equation $f(x, y) = 0$ forms a curve. If a problem requires the specification of a point, two curves and hence two equations are required; two properly constructed curves have a point of intersection. In three dimensions, the equation $f(x, y, z) = 0$ forms a surface. If a problem requires the specification of a point, three surfaces and hence three equations are required; three properly constructed surfaces have a point of intersection. There is a higher dimensional generalization. In n dimensions, n hypersurfaces and hence n equations are required to specify a point.

This principle is an example of both an algebraic and a geometric approach toward the solution of equations. Solution points can be considered as intersections of geometric objects or solutions to algebraic equations. The Cartesian coordinate system unites these perspectives in a single framework. Later in this chapter, there are several examples in which it is necessary to determine the coordinates of points. The principle of the preceding paragraph informs us of the number of equations that are necessary to determine the coordinates.

4.1.2 Translation of Axis and Object

For clarity, this section restricts itself to two dimensions and points are given by their x and y values.

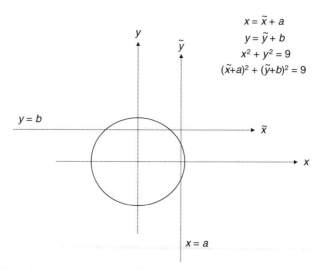

Figure 4.4   Equations in translated coordinates.

As previously noted, a curve in two-dimensional space is given by the functional relation

$$f(x, y) = 0 \tag{4.1}$$

As a concrete example, think of $x^2 + y^2 - 9 = 0$, which gives the circle of radius 3. The problem at hand is to express the functional relationship in terms of a coordinate system that is translated as indicated in Figure 4.4.

Let points in original coordinates be denoted by (x, y) and points in translated coordinates be denoted by $(\tilde{x}, \tilde{y})$. As indicated in Figure 4.4, when $\tilde{x} = 0$, $x = a$. Similarly, when $\tilde{y} = 0$, $y = b$. In general, we have the following equalities:

$$x = \tilde{x} + a \qquad y = \tilde{y} + b \tag{4.2}$$

Substituting the equalities from equation (4.2) into the equation for the curve, equation (4.1), results in the equation for the curve in translated coordinates, $f(\tilde{x} + a, \tilde{y} + b) = 0$.

Example 4.1

Express the circle $x^2 + y^2 - 9 = 0$ in a coordinate system centered at $(1, 2)$.

Solution $(\tilde{x} + 1)^2 + (\tilde{y} + 2)^2 - 9 = 0$.

Translation of the coordinates has a complementary operation, translating the geometric object. Translating the coordinates by a specified distance in a given direction yields the equivalent result of translating the object by the same distance in the opposite direction. In the example, the coordinates are moved to the point $(1, 2)$. This has the equivalent effect of moving the circle to a new location centered at a new

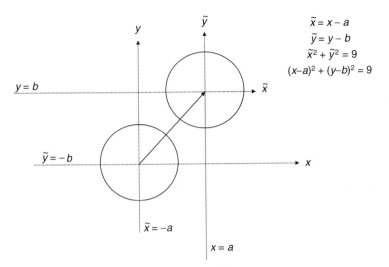

$$\tilde{x} = x - a$$
$$\tilde{y} = y - b$$
$$\tilde{x}^2 + \tilde{y}^2 = 9$$
$$(x-a)^2 + (y-b)^2 = 9$$

Figure 4.5   Translation of center from $(0, 0)$ to (a, b)

point in the opposite direction $(-1, -2)$. Accordingly, to translate a figure from a given position to a new position, subtract the values that determine the shift from the values of x and y (Figure 4.5). The formula for translating the object by the quantity (a, b) while maintaining the same coordinate axes is $f(x - a, y - b) = 0$.

Example 4.2

Shift the circle with radius 3 centered at the origin to a circle with radius 3 centered at $(-1, -2)$.

Solution $(x - (-1))^2 + (y - (-2))^2 - 9 = 0$
$$(x + 1)^2 + (y + 2)^2 - 9 = 0$$

Remarks

- This problem demonstrates the power of unifying geometric and algebraic concepts in a single framework. The geometric figure is constructed as an algebraic object. The geometry of translation is then expressed algebraically. Finally, geometric translation is algebraically introduced into the equations, yielding a new algebraic expression. All along none of the geometric intuition is lost; indeed the geometry guides the algebra.

- The example in two dimensions is generalized to n dimensions as follows. Let $f(x_1, x_2, x_3, \ldots, x_n) = 0$ be a hypersurface in n dimensions. If we wish to translate the axes so that the origin is aligned with the points $(a_1, a_2, a_3, \ldots, a_n)$, then the surface is expressed as $f(\tilde{x}_1 + a_1, \tilde{x}_2 + a_2, \tilde{x}_3 + a_3, \ldots, \tilde{x}_n + a_n) = 0$. If we wish to translate the hypersurface while maintaining the original axes, the formula is $f(x_1 - a_1, x_2 - a_2, x_3 - a_3, \ldots, x_n - a_n) = 0$.

4.2   POLYNOMIALS

An nth-order real-valued polynomial is a function of the following form:

$$f(x) = a_n x^n + a_{n-1} x^{n-1} + a_{n-1} x^{n-2} + \cdots + a_1 x + a_0 \qquad (4.3)$$

where

- n is a positive integer
- for each j, a_j is a real-valued coefficient (we only consider real numbers)

A common problem is the location of the roots of a polynomial. That is, given a polynomial $f(x)$, find all values of x such that $f(x) = 0$. While this problem does indeed find applications, the applications do not offer much insight into the solutions. It is best to view the problem as a puzzle unto itself. Historically, this has been the approach to the problem of finding roots; it is a puzzle unto itself. Surprisingly, this narrow perspective has yielded very powerful results that have broad application beyond the location of roots. For example, the concept of an imaginary number stems from the search for roots of polynomials; indeed the roots of the imaginary number lie in Cardano's work *Ars Magna*. Further analysis of complex functions demonstrates that a whole class of functions may be approximated to any level of accuracy by a polynomial function. We have seen something similar before; any real number may be approximated to any level of accuracy using only integers. This same property now exists on a functional level with the polynomials being the set of functions from which others may be approximated.

Let us see how powerful this is. Consider the evolution in time of the distance between two moving objects, perhaps the earth and Mars. In general, the distance may be a very complicated function of time and it may not be possible to precisely express the function. However, given any tolerance band around the function, it is possible to construct a polynomial that remains within the tolerance band (Figure 4.6). While polynomials may not leave much of a first impression, they indeed are much more powerful than they appear.

In the equation of a polynomial, when $n = 1$, the graph of the polynomial is a line. In the case that $n = 2$, the graph is a parabola. We shall look at lines and parabolas both algebraically and geometrically.

In subsequent results, it is sometimes necessary to equate polynomials that are obtained by different processes. Two polynomials are equal only if their coefficients, a_j, are identical for each j. Accordingly, the polynomials are set equal by equating their coefficients.

4.2.1   Lines

A line is defined by two points. If we place these two points onto a Cartesian plane, a natural ratio is apparent: their difference along the y axis over their difference along the x axis (Figure 4.7). For reasons that are apparent from the figure, the ratio is known as the slope. The line is defined by a continuous extension in such a manner that this ratio is preserved for any two points on the line; the slope is a constant. A

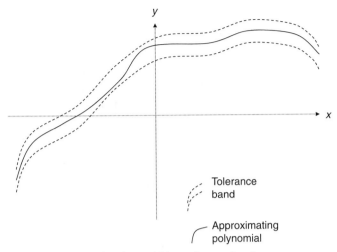

Figure 4.6   Polynomial approximations within a tolerance band.

consequence of this property is that all right triangles with their hypotenuse aligned along the line must be similar, as illustrated by the similar triangles in Figure 4.7.

We seek an algebraic expression for the line. Begin by selecting a coordinate system so that the origin, the point $(0, 0)$, is on the line. This can be accomplished by translating the origin of the original axis vertically (upward or downward) so that the translated origin lies on the line. Assume that the slope is given by the letter m and let $(\tilde{x}, \tilde{y})$ be another point on the line as described in shifted coordinates. Using the fact that the ratio is constant, we have the following:

$$\frac{\tilde{y} - 0}{\tilde{x} - 0} = m$$

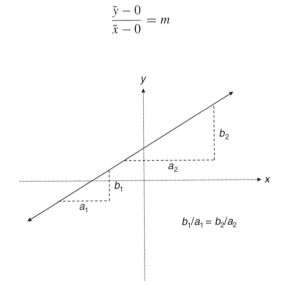

Figure 4.7   Similar triangles along a line.

$$\frac{\tilde{y}}{\tilde{x}} = m$$

$$\tilde{y} = m\tilde{x} \tag{4.4}$$

This establishes the equation of a line passing through the origin. To get the equation of the line in original coordinates, perform a translation of the shifted coordinates back to the original coordinates. Let the shifted origin $(\tilde{x}, \tilde{y}) = (0, 0)$ be the point $(x, y) = (0, b)$ in original coordinates. The general relation between coordinate systems is given as follows:

$$x = \tilde{x} \qquad y = \tilde{y} - b \tag{4.5}$$

Substituting the equalities (4.5) into the equation (4.4) just as described in Section 4.1 yields the equation for the line in original coordinates:

$$y - b = mx$$

$$y = mx + b \tag{4.6}$$

Equation (4.6) is the generalized equation for a line. It is a polynomial of one degree. Comparing (4.6) with equation (4.3), $m = a_1$ and $b = a_0$; the slope is given by a_1 and a_0 represents what is commonly known as the y intercept. The y intercept is the value at which the line intersects the y axis.

It is frequently necessary to write the equation of a line between two given points. The problem is posed as follows. Given two points (x_0, y_0) and (x_1, y_1), find the equation of the line passing through them. Solving the problem requires determining the value of the slope, a_1, and constant, a_0. The solution for the slope is given as follows:

$$\Delta y = y_1 - y_0 \qquad \Delta x = x_1 - x_0 \qquad a_1 = \frac{\Delta y}{\Delta x}$$

Once the slope is known, the constant a_0 is determined by placing in values for x, y, and a_1 into the equation of the line and solving for a_0:

$$a_0 = y - a_1 x$$

Either the values (x_0, y_0) may be used or the values (x_1, y_1) may be used to solve for a_0. The answer is the same either way; a_0 represents a vertical shift in coordinate systems and this is the same for every point along the line. An example illustrates this.

Example 4.3   EQUATION OF A LINE

Find the equation of the line passing through the two points $(1, 3)$ and $(5, 6)$.

Solution

$$\Delta y = 6 - 3 = 3 \qquad \Delta x = 5 - 1 = 4$$

$$a_1 = \frac{\Delta y}{\Delta x} = \frac{3}{4} \qquad a_0 = y - a_1 x \tag{4.7}$$

Placing the values $x = 1$ and $y = 3$ into equation (4.7) results in the following value for a_0:

$$a_0 = 3 - \tfrac{3}{4} \times 1 = 2\tfrac{1}{4}$$

Placing the values $x = 5$ and $y = 6$ into equation (4.7) results in the same value for a_0:

$$a_0 = 6 - \tfrac{3}{4} \times 5 = 2\tfrac{1}{4}$$

4.2.2 Parabolas and the Quadratic Equation

Given a line and a point, a parabola is formed by the set of all points that are equidistant from both the line and the point. Consider the point in Cartesian coordinates given by $(0, p)$ and the line given by $\tilde{y} = -p$. A point $(\tilde{x}_A, \tilde{y}_A)$ is on the parabola if the distance to the point $(0, p)$ and the line $\tilde{y} = -p$ is the same (Figure 4.8).

To get an equation for the parabola, proceed as follows. First, find the distance from $(\tilde{x}_A, \tilde{y}_A)$ to the point $(0, p)$. This is found using the Pythagorean theorem:

$$\text{Distance to point } (0, p) = \sqrt{\tilde{x}_A^2 + (\tilde{y}_A - p)^2}$$

Next find the distance from $(\tilde{x}_A, \tilde{y}_A)$ to the line $\tilde{y} = -p$. Using Figure 4.8 it is seen that the distance is the following:

$$\text{Distance to line} = \tilde{y}_A + p$$

Finally, set both distances equal and simplify:

$$\tilde{y}_A + p = \sqrt{\tilde{x}_A^2 + (\tilde{y}_A - p)^2}$$
$$(\tilde{y}_A + p)^2 = \tilde{x}_A^2 + (\tilde{y}_A - p)^2$$

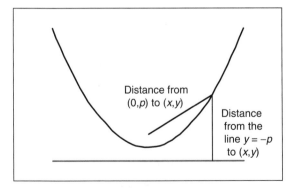

For every point on a parabola, the distances are equal.

Figure 4.8   The parabola.

$$\tilde{y}_A^2 + 2p\tilde{y}_A + p^2 = \tilde{x}_A^2 + \tilde{y}_A^2 - 2p\tilde{y}_A + p^2$$

$$4p\tilde{y}_A = \tilde{x}_A^2$$

$$\tilde{y}_A = \frac{\tilde{x}_A^2}{4p}$$

All points on the parabola satisfy the above relation and so the equation for the parabola is

$$\tilde{y} = \frac{\tilde{x}^2}{4p} \tag{4.8}$$

A family of parabolas is plotted for different values of p (Figure 4.9).

The vertex of the parabola is the parabola's lowest point when p is positive or highest point when p is negative. Note that in the above derivation the vertex lies at the origin $(\tilde{x}, \tilde{y}) = (0, 0)$. Suppose that in the coordinate system of interest the vertex lies, not at the origin, but at a different point, $(x, y) = (a, b)$. Then the general relation between the translated coordinate system and the coordinate system of interest is given by the following:

$$\tilde{x} = x - a \qquad \tilde{y} = y - b \tag{4.9}$$

Substituting the equalities (4.9) into the equation (4.8) just as described in Section 4.1 yields the equation for the parabola in the coordinates of interest:

$$y - b = \frac{(x - a)^2}{4p}$$

$$y - b = \frac{1}{4p}x^2 - \frac{ap}{2}x + \frac{a^2}{4p}$$

$$y = \frac{1}{4p}x^2 - \frac{ap}{2}x + \frac{a^2}{4p} + b \tag{4.10}$$

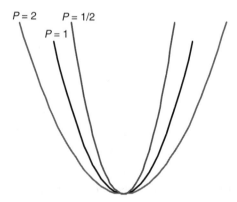

Figure 4.9   Parabolas for different values of p.

This equation is the same as the equation for a second-degree polynomial with the following associations:

$$a_2 = \frac{1}{4p} \qquad a_1 = -\frac{ap}{2} \qquad a_0 = \frac{a^2}{4p} + b \qquad (4.11)$$

We have demonstrated that, given the values p, a, and b that describe a parabola, one obtains a second-order polynomial for its equation. Similarly, one could work backward. If the second-order equation is given so that a_2, a_1, and a_0 are known, the values p, a, and b can be obtained for a parabola using equations (4.11):

$$p = 4a_2 \qquad a = -\frac{2a_1}{p} = -\frac{a_1}{2a_2} \qquad b = a_0 - \frac{a^2}{4p} = a_0 - \frac{a_1^2}{64a_2^2}$$

The conclusion is that a second-order polynomial is the equation of a parabola.

Next the roots of the second-order polynomial are determined; the roots are the corresponding values of x when y is set to zero:

$$a_2 x^2 + a_1 x + a_0 = 0 \qquad (4.12)$$

The solution for x is the well-known quadratic formula that is derived below. A priori, it is assumed that square roots of numbers can be determined, so the final result can be expressed using a square root.

To find the roots, it is possible to assume $a_2 = 1$. How is this so? When $y = 0$, division of equation (4.10) by a_2 yields the following:

$$x^2 + \frac{a_1}{a_2} x + \frac{a_0}{a_2} = 0$$

where a_1/a_2 and a_0/a_2 are arbitrary constants that we take as b_1 and b_0, respectively:

$$x^2 + b_1 x + b_0 = 0 \qquad (4.13)$$

The problem is now to determine the roots for any given values of b_1 and b_0 and then reexpress the roots in terms of the original parameters a_2, a_1, and a_0.

As a first step, let us determine what quadratic equations are solvable. Consider equations of the following form:

$$(x + \lambda)^2 - \gamma = 0 \qquad (4.14)$$

The solution is easily obtained:

$$(x + \lambda)^2 = \gamma$$
$$x + \lambda = \pm\sqrt{\gamma}$$
$$x = -\lambda \pm \sqrt{\gamma} \tag{4.15}$$

The solution is general if it is possible to intelligently choose λ and γ so that expression (4.14) is the same as expression (4.15). Equating the two expressions results in the following:

$$(x + \lambda)^2 - \gamma = x^2 + 2\lambda x + \lambda^2 - \gamma = x^2 + b_1 x + b_0$$

To set the last two expressions equal requires that all the coefficients are the same. The coefficients of the second-order term are both 1. The coefficients of the first-order term are, respectively, 2λ and b_1. The constant terms are, respectively, $\lambda^2 - \gamma$ and b_0.

Setting the first-order coefficients equal and solving for lambda, we have the following:

$$2\lambda = b_1 \qquad \lambda = \tfrac{1}{2}b_1$$

Lambda is now known. Setting the constant terms equal and solving for gamma, we have the following:

$$\lambda^2 - \gamma = b_0$$
$$\gamma = \lambda^2 - b_0 = \left(\tfrac{1}{2}b_1\right)^2 - b_0$$

Substituting the solutions for gamma and lambda into equation (4.15) gives the roots of x in terms of b_1 and b_0:

$$x = -\lambda \pm \sqrt{\gamma} = -\tfrac{1}{2}b_1 \pm \sqrt{\left(\tfrac{1}{2}b_1\right)^2 - b_0}$$

Finally, the quadratic formula is obtained by expressing the roots in terms of the original coefficients, a_2, a_1, and a_0. Recall $b_1 = a_1/a_2$ and $b_0 = a_0/a_2$. Substituting these values into the solution of the roots and then simplifying yield the standard quadratic equation:

$$x = -\frac{b_1}{2} \pm \sqrt{\left(\frac{b_1}{2}\right)^2 - b_0}$$

$$= -\frac{a_1}{2a_2} \pm \sqrt{\left(\frac{a_1}{2a_2}\right)^2 - \frac{a_0}{a_2}}$$

$$= -\frac{a_1}{2a_2} \pm \sqrt{\frac{a_1^2 - 4a_0 a_2}{4a_2^2}}$$

$$= -\frac{a_1}{2a_2} \pm \frac{\sqrt{a_1^2 - 4a_0 a_2}}{\sqrt{4a_2^2}}$$

$$= -\frac{a_1}{2a_2} \pm \frac{\sqrt{a_1^2 - 4a_0 a_2}}{2a_2}$$

Remarks

- There is a significant amount of algebraic manipulation required to derive the quadratic formula. The solution demonstrates the value of mathematical symbols; it is difficult to imagine performing all the manipulation using prose.

- It is possible to relate the algebraic solution of this section with the geometric solution of al-Khwarizmi's as presented in this chapter's introduction. Indeed, the term $x + \lambda$ is the side of the larger square in Figure 4.1 and the area of the larger square is given by γ.

- In general, the value γ may be taken as a positive or negative value. Note that there are two real-valued roots when $\gamma > 0$ and there is one real-valued root when $\gamma = 0$. In the case where γ is negative, solutions are complex valued.

4.3 CIRCLES

While circles are examined in the previous chapter, two issues were left open: the approximation of π and properties of the tangent line to a circle. We address these issues in this section.

4.3.1 Equations for a Circle

A circle is the set of all points equidistant from a given point called the center. Assume that the center is at the origin, $(\tilde{x}, \tilde{y}) = (0, 0)$, and designate the common distance by r. Using the Pythagorean theorem, the distance from all points to its center is given

by the equation

$$\sqrt{\tilde{x}^2 + \tilde{y}^2} = r$$

It is common to square both sides:

$$\tilde{x}^2 + \tilde{y}^2 = r^2$$

This results in the equation of a circle centered at the origin. The general equation for a circle centered elsewhere is found by translation. If in a coordinate system of interest the center is at the point $(x, y) = (a, b)$, the following hold:

$$\tilde{x} = x - a \qquad \tilde{y} = y - b$$

Substituting these relations into the equation for the circle yields the general equation

$$(x - a)^2 + (y - b)^2 = r^2$$

4.3.2   Archimedes and the Value of π

The problem of determining the value of π presents a good opportunity to practice algebra. Additionally, it is a precursor to methods of calculus that are explored in Chapter 7. The approximation for the length of the circumference of a circle with radius 1 follows the approach illustrated in Section 3.1.4. In that section, polygons formed by the bases of inscribed triangles are used to approximate the length of the circumference. In his work *On the Measurement of a Circle* Archimedes examines two sets of polygons, one inscribed within the circle and one circumscribed around the circle (Figure 4.10). The difference in the lengths of the perimeters is a tolerance band for the circumference.

Archimedes' solution is iterative. For the first iterate, he chooses a six-sided polygon. Each successive iterate is obtained by doubling the number of sides of the polygon. Archimedes derives a general method for determining the length of a successive iterate's perimeter from the previous iterate's polygon. The method is explained below.

Figure 4.10   Circumscribed and inscribed polygons.

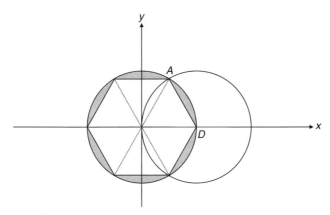

Figure 4.11   Inscribing a hexagon.

For the first iterate, inscribe the circle with a hexagon. This is accomplished using equilateral triangles. The base of each triangle is 1, and accordingly the length of the perimeter is 6. From this, it is possible to deduce that π, which is the length of half the circumference, is greater than 3.

For further iterates, it is necessary to determine the coordinates of the point A in Figure 4.11; this is the intersection between two circles: the initial circle and the circle centered at D with radius 1. The point D has coordinates $(1, 0)$ (see Figure 4.11).

The original circle has the following equation:

$$x^2 + y^2 = 1 \tag{4.16}$$

The circle centered at point D has the equation

$$(x - 1)^2 + y^2 = 1 \tag{4.17}$$

The point A is given by the values of x and y that satisfy both equations. We denote the coordinates by (x_A, y_A). By symmetry it is seen that $x_A = \frac{1}{2}$; x_A must be halfway between the two centers. Placing the value for x_A into either of the two equations (4.16) or (4.17) and simplifying yield the value for y. Using equation (4.16),

$$\left(\tfrac{1}{2}\right)^2 + y^2 = 1$$
$$y^2 = \tfrac{3}{4}$$
$$y = \pm\sqrt{\tfrac{3}{4}} = \pm\tfrac{1}{2}\sqrt{3}$$

There are two values for y, indicating that the circles intersect in two points. We are interested in the point in the first quadrant, $y = \frac{1}{2}\sqrt{3}$.

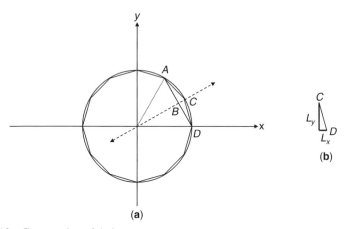

(a)

(b)

Figure 4.12   Construction of dodecagon.

For the second iterate, it is necessary to determine the length of a single side of the 12-sided polygon and then multiply that length by 12 (all sides being equal). The point C of Figure 4.12a represents the intersection between the dashed line and the circle. It is the end point for one side of the 12-sided polygon and it is necessary to find the coordinates. This is accomplished using the three steps below. These steps use the points B, A, and D in Figure 4.12a. The point B is the midpoint of the line segment from A to D.

The steps taken to determine the point C are as follows:

Step 1. Determine the point B. It is the midpoint between the points D and A.

Step 2. Determine the line passing through B and the origin and find the equation (the dashed line of Figure 4.12a).

Step 3. Determine point C as the intersection of the line from step 2 and the original circle.

We next execute these steps:

Step 1. Determine the point B. It is the midpoint between the points D and A.
    Denote the point B by (x_B, y_B). Since it is the midpoint between points D and A, we have the following:

$$x_B = \tfrac{1}{2}(x_A + x_D) = \tfrac{1}{2}\left(\tfrac{1}{2} + 1\right) = \tfrac{3}{4}$$

$$y_B = \tfrac{1}{2}(y_A + y_D) = \tfrac{1}{2}\left(\tfrac{1}{2}\sqrt{3} + 0\right) = \tfrac{1}{4}\sqrt{3}$$

Step 2. Determine the line passing through B and the origin and find the equation (see Figure 4.12a).

Since the line passes through the origin, the equation of the line is given by $y = mx$, where m is the slope.

$$m = \frac{y_B}{x_B} = \frac{\sqrt{3}}{3}$$

Step 3. Determine point C as the intersection of the dashed line and the original circle.

From step 2, the equation for the blue line is the following:

$$y = \tfrac{1}{3}\sqrt{3}x$$

The equation for the circle is the following:

$$x^2 + y^2 = 1$$

The intersection must solve both equations. Substituting the value for y from the equation of the line into the equation for the circle and simplifying give the following value for x:

$$x^2 + \left(\tfrac{1}{3}\sqrt{3}x\right)^2 = 1$$
$$x^2 + \tfrac{1}{3}x^2 = 1$$
$$\tfrac{4}{3}x^2 = 1$$
$$x = \pm\tfrac{1}{2}\sqrt{3}$$

There are two solutions, and we are interested in the positive solution, $x_C = \tfrac{1}{2}\sqrt{3}$. To find the associated value for y, substitute the value of x_C into either the equation for the line or the equation for the circle. It is easier to use the line:

$$y_C = \frac{\sqrt{3}}{3}x_C = \frac{\sqrt{3}}{3}\frac{\sqrt{3}}{2} = \frac{3}{6} = \frac{1}{2}$$

The line segment between the points C and D determines a side of the 12-sided polygon. To get the length of this segment, apply the Pythagorean theorem to the triangle, as shown in Figure 4.12b. The length of the leg along the x axis is given as follows:

$$L_x = x_D - x_C = 1 - \tfrac{1}{2}\sqrt{3}$$

The length of the side parallel to the y axis is as follows:

$$L_y = y_C - y_D = \tfrac{1}{2} - 0 = \tfrac{1}{2}$$

Using the Pythagorean theorem, the length of the segment (the hypotenuse of the triangle) is determined as follows:

$$\text{Length of segment} = \sqrt{L_x^2 + L_y^2}$$

Taking out one's calculator and plugging in the values gives the approximation for the length of the segment as 0.517638.

The perimeter of the 12-sided polygon is 12 times the length of the single segment, approximately 6.211657, and half the perimeter is approximately 3.1058. As π denotes the length of half the circle's circumference and half the perimeter of the 12-sided polygon is less than half the circumference, this gives a lower bound for π, $\pi > 3.1058$.

An upper bound can be determined by calculating the length of the circumscribed 12-sided polygon. While Archimedes performed this calculation, we do not but merely note that the difference between the lower and upper bounds is a tolerance band for π.

To get a better approximation, use a 24-sided polygon. Repeat steps 1, 2, and 3 above with appropriate identification of the points A, B, and C. Assume A is given by the coordinates $(\frac{1}{2}\sqrt{3}, \frac{1}{2})$, the previous iterate's C, and B and C are defined as before but are calculated using the new value of A. One can continue indefinitely. Without the aid of a calculator, Archimedes found the tolerance band $\frac{223}{71} < \pi < \frac{22}{7}$.

Remarks

- Note that the method uses the Pythagorean theorem to determine the distance between two points on a Cartesian plane with the result that the distance between two points A and B is distance $= \sqrt{(x_A - x_B)^2 + (y_A - y_B)^2}$ (see Figure 4.12).

- Using a symmetry argument, one could deduce that the coordinates of points A and C in Figure 4.11 are reversed.

- Following Archimedes, obtaining an upper bound can be accomplished through a resizing argument. Find the resizing factor between the inscribed and circumscribed polygons. Because the perimeter is one dimensional, the resizing factor becomes the ratio between the lengths of the inner and outer perimeters. Applying the ratio to the lower bound of the tolerance band gives an upper bound for the tolerance band. Pi lies between these two bounds.

- Archimedes was not the only mathematician to discover this method. A Chinese mathematician, Zu Chongzhi (429–500), independently found the same method and performed the calculation out to 11 iterates, approximating the circle with a polygon of 12,228 sides. Zu Chongzhi's son, Zu Gengzhi, independently discovered the formula for the volume of a sphere using a different method than Archimedes.

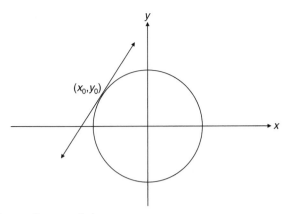

Figure 4.13   Tangent line to a circle.

4.3.3 Tangent Lines to a Circle

In Section 3.2.9 the surface area of the sphere is determined. This is accomplished by transforming the surface into a rectangle. The transformation used the tangent line of a circle. In this section, we determine the equation for the tangent line as illustrated in Figure 4.13. The circle is centered at the origin and has radius 1.

A line is tangent to the circle if it intersects the circle at only one point. Given that the point (x_0, y_0) is on the circle, the point satisfies the equation

$$x_0^2 + y_0^2 = 1 \tag{4.18}$$

The point also satisfies the equation for the tangent line with as-of-yet-undetermined values for a_0 and a_1:

$$y_0 = a_1 x_0 + a_0 \tag{4.19}$$

It is necessary to solve for the values of a_0 and a_1 using the known values of x_0 and y_0. This requires two equations to solve for two unknowns.

Placing the value of y_0 from the equation for the line, equation (4.18), into the value of y_0 for the equation of the circle, equation (4.17), and simplifying yields the following:

$$x_0^2 + (a_1 x + a_0)^2 = 1$$
$$(1 + a_1^2)x_0^2 + 2a_1 a_0 x_0 + a_0^2 - 1 = 0$$

Solving for x_0 using the quadratic equation and simplifying results in the following values for x_0:

$$x_0 = \frac{1}{1 + a_1^2}\left(-a_1 a_0 \pm \sqrt{(a_1 a_0)^2 - (1 + a_1^2)(a_0^2 - 1)}\right)$$

The behavior of this expression is dependent upon the value within the square root, $(a_1 a_0)^2 - (1 + a_1^2)(a_0^2 - 1)$. There are three cases as presented by the table below:

$(a_1 a_0)^2 - (1 + a_1^2)(a_0^2 - 1) > 0$	Two real values for x_0
$(a_1 a_0)^2 - (1 + a_1^2)(a_0^2 - 1) = 0$	One real value for x_0
$(a_1 a_0)^2 - (1 + a_1^2)(a_0^2 - 1) < 0$	No real values for x_0

The case that we are interested in is the case when there is only one real value for x_0; the tangent line intersects only one point on the circle. Accordingly, it is necessary to set the value within the square root to zero:

$$(a_1 a_0)^2 - (1 + a_1^2)(a_0^2 - 1) = 0$$

Simplifying the above expression yields the following:

$$a_1^2 - a_0^2 + 1 = 0 \tag{4.20}$$

This establishes one equation for a_1 and a_0. Because there are two unknowns, another equation is necessary. But the other equation has already been given, the equation for the line, equation (4.18):

$$y_0 = a_1 x_0 + a_0$$

Since x_0 and y_0 are given, it is possible to solve for one of the coefficients; select a_0:

$$a_0 = a_1 x_0 - y_0 \tag{4.21}$$

Substituting for a_0 in equation (4.20) and simplifying yields the following:

$$a_1^2 - (y_0 - a_1 x_0)^2 + 1 = 0$$
$$(1 - x_0^2)a_1^2 + 2x_0 y_0 a_1 + 1 - y_0^2 = 0$$
$$y_0^2 a_1^2 + 2x_0 y_0 a_1 + x_0^2 = 0$$

The final expression uses the fact that since x_0 and y_0 are on the unit circle, $1 - x_0^2 = y_0^2$ and $1 - y_0^2 = x_0^2$. Note that the equation is quadratic in a_1, so the quadratic equation is used to solve for a_1:

$$a_1 = \frac{1}{y_0^2}\left(-x_0 y_0 \pm \sqrt{(x_0 y_0)^2 - x_0^2 y_0^2}\right) = -\frac{x_0}{y_0}$$

The slope of the tangent line is $-x_0/y_0$.

To find a_0, substitute the value of a_1 into equation (4.21) and simplify:

$$a_0 = y_0 - a_1 x_0$$

$$= y_0 + \frac{x_0^2}{y_0}$$

$$= \frac{y_0^2 + x_0^2}{y_0}$$

$$= \frac{1}{y_0}$$

The general equation for the tangent line is the following:

$$y = a_1 x + a_0 = -\frac{x_0}{y_0}x + \frac{1}{y_0}$$

Remark. Finding the slope of the tangent line is the principal objective of differential calculus. In this section, it is hard work. Using differential calculus, this calculation is very simple.

4.4 THE FOUR-DIMENSIONAL SPHERE

It is time for another excursion. In Chapter 3, the surface area and the volume of a sphere are determined. The method equates the volume of the sphere with that of a pyramid using Cavalieri's theorem. This establishes a relation between the volume and the surface area of the sphere, $V = \frac{1}{3}rS$. Finally, the surface area is established, and using the relation between volume and surface area, the volume can be determined. In this section, we generalize the process to higher dimensions and in particular find the volumes (surface volume and four-dimensional volume) for the four-dimensional sphere.

A question concerning presentation may come to mind. Why not present this material in Chapter 3 rather than presenting it in this chapter on algebra? Geometric constructs in higher dimensions are difficult to convey in pure geometric form; Euclid's compass and a straight edge just do not cut it. Coordinate systems in higher dimensions allow for algebraic representations. Then it is possible to perform geometric operations algebraically.

The material in this section is an excursion and as such not essential to the understanding of the rest of the book. The problem itself, finding measurements for a sphere in higher dimensions, arises as a challenge and appears to have little practical application. One of the pleasures of mathematics is finding relations that arise in apparently unrelated problems. As it turns out, the results from this section give results for integral calculus with polynomials, a subject investigated in Chapter 6. Just as exploration of new territories may turn up something interesting, so might a mathematical exploration. This section may also assist the reader in developing some facility for thinking in higher dimensions, a skill most useful as much of mathematics is done in higher dimensions.

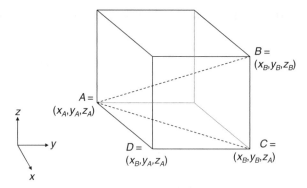

Figure 4.14   Pythagorean theorem in three dimensions.

Prior to beginning, some notation is necessary. In four dimensions, the coordinate axes are given by (x, y, z, w). In higher dimensions, coordinates are given by $(x_1, x_2, x_3, \ldots, x_n)$. A point in space is denoted by a capital letter, that is, A. The coordinates of the point are denoted by (x_A, y_A, z_A, w_A) or, in dimensions greater than 4, $(A_1, A_2, A_3, \ldots, A_n)$.

4.4.1   Pythagorean Theorem in Higher Dimensions

The Pythagorean theorem is used to find the distance between two points. Section 4.2 presents an example in two dimensions where the equation for a circle identifies points of common distance through the Pythagorean theorem.

Figure 4.14 illustrates the geometry in three dimensions. The illustration depicts a rectangular cube with $A = (x_A, y_A, z_A)$ and $B = (x_B, y_B, z_B)$ at opposite vertexes. It is the distance between points A and B that is of interest. The point C lies on the bottom face of the rectangular cube opposite of A; its coordinates are $C = (x_B, y_B, z_A)$. The points A, C, and B form a right triangle, so it follows from the Pythagorean theorem that

$$\text{Distance}^2_{AB} = \text{distance}^2_{AC} + \text{distance}^2_{CB} \tag{4.22}$$

Likewise, the points A, C, and D form a right triangle. The coordinates of the point D are (x_B, y_A, z_A). From this right triangle, it is possible to determine the distance from A to C:

$$\text{Distance}^2_{AC} = \text{distance}^2_{AD} + \text{distance}^2_{DC} = (x_A - x_B)^2 + (y_A - y_B)^2$$

The remaining term in equation (4.22), distance^2_{CB}, is the difference in height between the upper and lower surfaces:

$$\text{Distance}^2_{CB} = (z_A - z_B)^2$$

Putting the results for distance$^2_{AC}$ and distance$^2_{CB}$ into equation (4.22) and taking the square root result in the following:

$$Distance^2_{AB} = (x_A - x_B)^2 + (y_A - y_B)^2 + (z_A - z_B)^2$$

$$Distance_{AB} = \sqrt{(x_A - x_B)^2 + (y_A - y_B)^2 + (z_A - z_B)^2}$$

Note the way that the formula is created. The first term is the difference along the x direction, the second term is difference along the y direction, and the third term is the difference along the z direction.

We next generalize the result to four dimensions as follows. First, it is necessary to consider a four-dimensional rectangular cube. Can we construct a three-dimensional rectangular cube in a way that can be generalized? The three-dimensional rectangular cube is constructed by placing on each coordinate axis two two-dimensional rectangles orthogonal to the axis; these form the faces of the cube (see Figure 4.14). There are three axes and two faces for each axis, making a total of six faces.

A four-dimensional rectangular cube is constructed by placing on each coordinate axis two three-dimensional rectangular cubes; these form the three-dimensional faces of the four-dimensional rectangular cube. There are four axes, making a total of eight faces. This rectangle can be translated so that it is centered anywhere within the four-dimensional space. Given two points A and B with coordinates $A = (x_A, y_A, z_A, w_A)$ and $B = (x_B, y_B, z_B, w_B)$, respectively, it is possible to place a four-dimensional rectangular cube in such a fashion that the points A and B are opposite vertexes of the cube.

Consider the point C given by the coordinates $C = (x_B, y_B, z_A, w_A)$. This is the vertex of the four-dimensional rectangular cube with (x, y, z) values the same as point B and a w value the same as that of point A. Points A, B, and C form a right triangle in four- dimensional space. The triangle itself lies in a plane within the four-dimensional space and the Pythagorean theorem can be applied to this triangle to yield the distance between A and B. Analogous to equation (4.22) the distance is

$$Distance^2_{AB} = distance^2_{AC} + distance^2_{CB}$$

The points A and C lie in a three-dimensional face of the four-dimensional rectangle. Within this face, the value of w is constant; it plays no role in the distance. Therefore, the three-dimensional formula for distance can be applied to the distance from A to C:

$$Distance^2_{AC} = (x_A - x_C)^2 + (y_A - y_C)^2 + (z_A - z_C)^2$$

The points C and B lie along a line segment in which the x, y, and z values are all constant. They play no role in the distance between C and B:

$$Distance^2_{CB} = (w_A - w_B)^2$$

Placing the distance formulas for AC and CB into that of AB and taking the square root result in the distance from A to B:

$$\text{Distance}_{AB}^2 = (x_A - x_B)^2 + (y_A - y_B)^2 + (z_A - z_B)^2 + (w_A - w_B)^2$$

$$\text{Distance}_{AB} = \sqrt{(x_A - x_B)^2 + (y_A - y_B)^2 + (z_A - z_B)^2 + (w_A - w_B)^2}$$

More generally, given two points in n-dimensional space, A and B, with coordinates $A = (A_1, A_2, A_3, \ldots, A_n)$ and $B = (B_1, B_2, B_3, \ldots, B_n)$, the distance between the points is given by the following formula:

$$\text{Distance}_{AB} = \sqrt{(A_1 - B_1)^2 + (A_2 - B_2)^2 + (A_3 - B_3)^2 + \cdots + (A_n - B_n)^2}$$

4.4.2 Measurements in Higher Dimensions and n-Dimensional Cubes

The terms *length*, *area*, and *volume* denote measurements of one-, two-, and three-dimensional objects. It is not feasible to create a new word for measurements of each higher dimensional object, so we are stuck with the term volume. When necessary, to promote clarity, n-dimensional volume is used in reference to the measurement of an n-dimensional object. The symbol V_j is used to denote the volume of a j-dimensional object; that is, V_4 refers to the volume of a four-dimensional object.

In the previous section, a four-dimensional cube is constructed. The concept is generalized to higher dimensions. In n dimensions, a unit cube is constructed by placing two $(n-1)$-dimensional unit cubes through each axis in such a manner that the distance between opposite faces is 1. For an n-dimensional cube, there are n axes giving $2n$ faces. The volume of an n-dimensional object is the maximum number of n-dimensional unit cubes or fractal portions thereof that the object can contain. It follows, for example, that the volume of an n-dimensional rectangle is the product of the lengths of its sides.

4.4.3 Cavalieri's Theorem

Cavalieri's theorem is valid in higher dimensions. To state the theorem, it is necessary to describe a hyperplane. This is a generalization of a plane to higher dimensions, and recall that planes are critical to Cavalieri's theorem in three dimensions. Assume we are in n dimensions. The equation $x_n = k$, with k a constant, describes an $(n-1)$-dimensional surface in which all other variables $x_1, x_2, x_3, \ldots, x_{n-1}$ are not fixed. Such an object is a hyperplane.

Let two n-dimensional objects be situated in n-dimensional space such that their x_n extents are identical; identifying a bottom and a top, the bottom of each object lies at $x_n = 0$ and the top of each object lies at $x_n = a$ for some value of a. If for any arbitrary hyperplane, $x_n = k$, the $(n-1)$-dimensional volumes of each object within that hyperplane are equal, then the n-dimensional volumes of the objects are also equal.

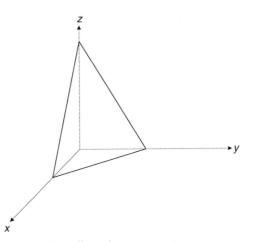

Figure 4.15   A hyperplane in three dimensions.

The extended Cavalieri theorem as stated in Section 3.2.7 also applies. This statement requires two objects A and B and a constant multiple m. As above, let A and B be situated between the hyperplanes $x_n = 0$ and $x_n = a$. If for every hyperplane the $(n-1)$-dimensional volume of object A within each hyperplane is the same constant multiple m of the corresponding $(n-1)$-dimensional volume of object B, then the n-dimensional volume of object A is m times the n-dimensional volume of object B.

Remark. The hyperplane given above is parallel to the base space $x_n = 0$. In general, a hyperplane may have any orientation within the n-dimensional space. The general equation for a hyperplane is the following:

$$a_1 x_1 + a_2 x_2 + a_3 x_3 + \cdots + a_n x_n = 0$$

A hyperplane in three dimensions is shown in Figure 4.15. The equation for the figure is $2x + 3y + z = 1$.

This is a common plane.

4.4.4 Pyramids

The method for determining the volume of n-dimensional pyramids follows the same process as that for three-dimensional pyramids. Each face of an n-dimensional cube becomes the base of an n-dimensional pyramid; as there are $2n$ faces, there will be $2n$ pyramids, two on each axis. We concentrate on one of the two pyramids with its base perpendicular to the x_n axis; all the others are constructed in the same manner. The three-dimensional pyramid of Figure 3.24 is replicated in Figure 4.16 as a visual aid.

The bottom of the pyramid is an $(n-1)$-dimensional unit cube. The x_n coordinate at the base is $-\frac{1}{2}$. The x_n coordinate at the top is zero. We refer to the height of the

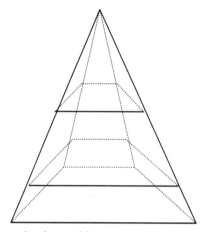

Figure 4.16   A three-dimensional pyramid.

pyramid as the distance from the base to the top and designate it with the letter h. At each point between 0 and $-\frac{1}{2}$ along the x_n coordinate, create an $(n-1)$-dimensional cube with sides of length $-2x_n$. This gives a stack of unit cubes along the x_n axis. The sides of the unit cubes increase in length from 0 to 1 as x_n goes from 0 to $-\frac{1}{2}$. The stack is an n-dimensional pyramid. For example, a three-dimensional pyramid is a stack of squares with sides that increase from top to bottom, a four-dimensional pyramid is a stack of three-dimensional cubes, a five-dimensional pyramid is a stack of four-dimensional cubes, and so on.

Now that we have constructed an n-dimensional pyramid, let us determine its volume. The unit n-dimensional cube of volume 1 has been assembled from $2n$ equally sized pyramids. As the volumes of these pyramids are equal, the volume of each is $1/2n$. The height of each pyramids is $\frac{1}{2}$ and the $(n-1)$-dimensional volume of each base is 1 [the base are all unit $(n-1)$-dimensional cubes]. Putting everything together gives the following formula:

$$V_n(\text{pyramid}) = \frac{1}{n}\left[h V_{n-1}(\text{base})\right] \tag{4.23}$$

While formula (4.23) has been worked out for a pyramid with specific geometry, it can be generalized using the extended Cavalieri theorem as in Section 3.2.8.

4.4.5   The n-Dimensional Sphere as an n-Dimensional Pyramid

In n dimensions, the set of all points that are equidistant from a designated center form an n-dimensional sphere. The common distance from all points to the center is known as the radius and designated by r. The equation for the surface of a sphere centered at the origin is

$$x_1^2 + x_2^2 + x_3^2 + \cdots + x_n^2 = r^2$$

This follows from the Pythagorean theorem in higher dimensions. We call such objects shells. The sphere itself is the collection of all shells centered at the origin that are contained within the surface of the sphere.

Following Section 3.2.9, it is possible to unfold the n-dimensional sphere and obtain an n-dimensional pyramid. Select a radial line from the sphere's center to its surface. At every point of this radial line, unfold the shell that passes through it. The original center of the sphere becomes the pyramid's top and the sphere's outer surface becomes the pyramid's base.

Since the n-dimensional sphere is an n-dimensional pyramid, equation (4.23) holds with the outer surface as the base and the radius as its height:

$$V_n(\text{sphere}) = \frac{1}{n} \left[r V_{n-1}(\text{surface}) \right] \tag{4.24}$$

4.4.6 The Three-Dimensional Volume of the Four-Dimensional Sphere's Surface

This section follows the method of Section 3.10, where the surface area of a conventional sphere is determined and the volume follows by application of equation (4.24). In Section 2.10, the surface of the sphere is described as a stack of circles of latitude. Each of the circles is cut, stretched, and mapped onto the vertical axis of the sphere. Generalizing this method, the surface of the four-dimensional sphere is considered a stack of two-dimensional shells. Each shell is stretched and then mapped onto the vertical axis, taken to be the w axis. As a fist step, it is necessary to determine the area of each shell. Then it is necessary to determine a stretching factor as the shell is removed from the surface of the four-dimensional sphere and mapped down to the w axis. Figure 2.25, replicated in Figure 4.17, presents a visual illustration.

Throughout this section, r denotes the four-dimensional sphere's radius. Consider a shell at the latitude $w = w_0$. The shell is expressed algebraically as follows:

$$x^2 + y^2 + z^2 + w_0^2 = r^2$$
$$x^2 + y^2 + z^2 = r^2 - w_0^2$$

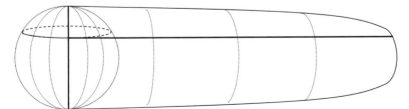

Figure 4.17   Unraveling the surface of the sphere.

This is the equation for the surface of a three-dimensional sphere with radius $r(w_0) = r^2 - w_0^2$. From the results of Section 3.10, the surface area is area $= 4\pi [r(w_0)]^2$. The shell is transformed to an equal-area rectangle. Squeezing all the rectangles off of the surface of the four-dimensional sphere and onto the w axis requires an offsetting stretching of each rectangle. Calculation of the stretching factor is performed in the identical fashion of Section 3.2.10, where the surface area of the three-dimensional sphere is calculated. The calculation yields a stretching factor $s(w)$ with $s(w) = r/r(w)$.

For each point on the w axis, a shell from the four-dimensional sphere has been transformed into an equal area rectangle, stretched, and attached to its associated point along the w axis. The two-dimensional area of each stretched rectangle at the point w is the following:

$$\text{Area }(w) = 4\pi \, (r(w))^2 \, \frac{r}{r(w)}$$

$$= 4\pi r r(w)$$

$$= 4\pi r \sqrt{r^2 - w^2}$$

Align the rectangles so that the w axis meets at their front right corner and choose the dimensions of the rectangles $4\pi r$ units deep and to be $\sqrt{r^2 - w^2}$ units in length. The result is half a cylinder. The entire cylinder is illustrated in Figure 4.18. The axis of the cylinder is perpendicular to the w axis and has length $4\pi r$. The base of the cylinder is a disk with radius r and area πr^2. The volume of the cylinder is twice the volume of the four-dimensional sphere's surface:

$$V_3(\text{surface}) = \tfrac{1}{2} \left(\pi r^2\right) \times (4\pi r) = 2\pi^2 r^3$$

Using the relation between the three-dimensional surface volume and four-dimensional volume, equation (4.24), the volume of the four-dimensional sphere is

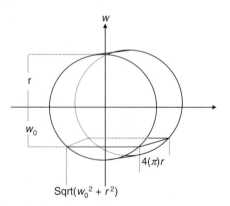

Figure 4.18   The surface of a four-dimensional sphere unravels to a cylinder.

the following:

$$V_4(\text{sphere}) = \tfrac{1}{2}\pi^2 r^4$$

4.5 FINITE SERIES AND INDUCTION

There are individuals who make tremendous scientific contributions in isolation. On the island of Sicily, Archimedes had no sounding boards for his ideas. Isaac Newton developed calculus while alone at his hometown village. Einstein developed special relativity while working as a clerk in a Swiss government patent office. To this list one can add the most outstanding physicist of the middle ages, al-Haytham (965–1040).

Al-Haytham's most productive years occurred while he was under house arrest for displeasing al-Hakim, the Caliph of the Fatim Empire centered in Egypt. Wishing to leave his position as a minister in the Abbasid Caliphate located in Baghdad, al-Haytham devised a plan to reengineer the Nile River and he personally pitched it to al-Hakim. Al-Hakim was so impressed that he placed al-Haytham in charge of the plan's execution as chief engineer. Selling the plan and accepting the commission were the two gravest errors of al-Haytham's life. But these two errors would create the circumstances in which al-Haytham made the most remarkable contributions to a wide range of fields—astronomy, mathematics, mechanics, and optics.

Not long after accepting the commission, al-Haytham recognized the hopelessness of the project. The reality on the ground in Egypt was far different from al-Haytham's vision in Baghdad. Al-Haytham reported his findings to al-Hakim. Souring relations led to the house arrest of al-Haytham. In a desperate attempt to escape a sentence of execution, al-Haytham pretended to be insane. While the ruse worked, al-Haytham remained under house arrest for 10 years until the death of al-Hakim.

During the time of al-Haytham's incarceration, he was prolific. The work for which al-Haytham is most remembered is optics. Al-Haytham pursued two avenues of investigation: the physics of light and the physiology of human sight. Kepler studied al-Haytham's work and expanded upon it to explain the workings of the telescope. And recall, it was Kepler's scientific explanation of the telescope that furnished support to Galileo against critics who claimed that the telescope created chimerical images. Aside from his work in optics, there is a clear line of sight between al-Haytham and the development of calculus. Following the work of al-Karaji, al-Haytham devised a general method for summing a finite series consisting of powers of integers. This section presents the method. In Chapter 6, these sums are used to develop calculus for polynomials. As for the Nile River, it was eventually reengineered along the lines of al-Haytham's vision with the completion of the Aswan Dam in 1970.

Al-Haytham deserves mention for one other accomplishment that is discussed in Chapter 3. Al-Haytham set the first stitch in the thread of mathematics that begins with an attempt to prove the redundancy of the parallel postulate by the method described in Section 3.11 and ends with the creation of non-Euclidean geometry by Lambert.

4.5.1   A Simple Sum

In this section, the sum $1 + 2 + 3 + \cdots + n$ is determined for any number n. By itself this has little consequence. However, it is a starting point for results of use in integral calculus.

Some notation is helpful. Denote the sum of an n-termed expression by S_n and denote each term to be summed by x_j, where the letter j designates the jth term in the sum. Common mathematical notation is to express S_n using a sigma sign, $\sum$, as follows:

$$S_n = \sum_{j=1}^{n} x_j = x_1 + x_2 + x_3 + \cdots + x_n$$

Particular cases of this notation are given below:

$$S_n = \sum_{j=1}^{n} j = 1 + 2 + 3 + \cdots + n$$

$$S_n = \sum_{j=1}^{n} j^2 = 1^2 + 2^2 + 3^2 + \cdots + n^2$$

$$S_n = \sum_{j=1}^{n} 2^j = 2^1 + 2^2 + 2^3 + \cdots + 2^n$$

There is an easy way to determine the sum given at the beginning of the section, $S_n = \sum_{j=1}^{n} j$. The method follows:

$$
\begin{aligned}
S_n &= 1 + 2 + 3 + \cdots + n \\
S_n &= n + (n-1) + (n-2) + \cdots + 1 \\
2S_n &= (n+1) + (n+1) + (n+1) + \cdots + (n+1)
\end{aligned}
$$

The third equation is obtained by summing the first two vertically. Note that there are n identical terms in the third equation, $2S_n = n(n+1)$. Therefore, $S_n = \frac{1}{2}n(n+1)$.

4.5.2   Induction

Suppose that a general formula for S_n is given. Al-Karaji developed the method of induction for proving the correctness of the formula. The method is a two-step algorithm as follows:

Step 1. Demonstrate the formula holds for S_1, the sum containing one term.

Step 2. Demonstrate that if the formula applies to S_{n-1}, it also holds for S_n, where n is a positive integer.

The first step establishes a starting point where the formula is verified. The second step establishes the proof of the formula for each incremental term. Since the formula has been verified for S_1, the second step assures it holds for S_2; since the formula holds for S_2, the second step assures it holds for S_3; since the formula holds for S_3, the second step assures it holds for S_4; and so on.

To illustrate its use, the method of induction is applied to the formula from the preceding section. As in the preceding section, let $S_n = \sum_{j=1}^{n} j$. We demonstrate $S_n = \frac{1}{2}n(n+1)$ using induction:

Step 1. Demonstrate the formula holds for S_1, the sum containing one term $S_1 = 1$. Using the formula, $S_1 = \frac{1}{2}1(1+1) = 1$, the formula is correct for $n = 1$.

Step 2. Demonstrate that if the formula applies to S_{n-1}, it also holds for S_n, where n is a positive integer.

To accomplish this step, it is necessary to assume that the formula holds for S_{n-1} and then establish a result for S_n. If the established result matches the formula, step 2 has been successfully demonstrated and the proof is complete:

$$S_n = 1 + 2 + 3 + \cdots + (n-1) + n$$
$$= S_{n-1} + n$$
$$= \frac{1}{2}(n-1)n + n$$
$$= n\left[\frac{1}{2}(n-1) + 1\right]$$
$$= n\left[\frac{1}{2}n + 1\right]$$
$$= \frac{1}{2}n(n+1)$$

The resulting formula is identical to the original formula, proving its correctness.

4.5.3   Using Induction as a Solution Method

The problem of Section 4.5.1 is solved through a trick that is difficult to generalize. What happens if we take it up a notch and seek a solution for the more difficult sum $S_n = \sum_{j=1}^{n} j^2$? The problem is a purely algebraic; geometry has little to offer. Can we find an algebraic method for this problem? While induction is thought of as a method of proof, it can also guide one to a solution. In this section, induction is used to determine the solution. A benefit of using induction to guide the solution is that proof of the solution's correctness comes along with the solution.

As a starting point, we note that the solution in Section 4.5.1 is a polynomial function of n; it is a second-order polynomial. Let us try a polynomial for the solution of $S_n = \sum_{j=1}^{n} j^2$. Denote the polynomial by $p(n)$. Placing $p(n)$ into step 2 of the inductive method and applying the inductive assumption yield the following:

$$S_n = 1^2 + 2^2 + 3^2 + \cdots + (n-1)^2 + n^2$$
$$= S_{n-1} + n^2$$
$$= p(n-1) + n^2$$
$$= p(n)$$

It is necessary to determine what order polynomial can satisfy the last equality, $p(n) = p(n-1) + n^2$. Clearly, a first-order polynomial cannot; the polynomial has a square term. A second-order polynomial would also not work; the second-order term of $p(n-1) + n^2$ would always be different from the second-order term of $p(n)$. The lowest order polynomial that has a chance is a cubic polynomial. Take $p(n)$ as follows:

$$p(n) = a_3 n^3 + a_2 n^2 + a_1 n + a_0$$

Placing this into the equality $p(n) = p(n-1) + n^2$ and simplifying yield the following:

$$a_3(n-1)^3 + a_2(n-1)^2 + a_1(n-1) + a_0 + n^2 = a_3 n^3 + a_2 n^2 + a_1 n + a_0$$

$$a_3(n^3 - 3n^2 + 3n - 1) + a_2(n^2 - 2n + 1) + a_1(n-1) + a_0 + n^2 =$$
$$a_3 n^3 + a_2 n^2 + a_1 n + a_0$$

$$a_3(-3n^2 + 3n - 1) + a_2(-2n + 1) + a_1(-1) + n^2 = 0$$
$$(1 - 3a_3)n^2 + (3a_3 - 2a_2)n - a_3 + a_2 - a_1 = 0$$

The last line states that all the coefficients of the powers of n on the left-hand side must be set to zero: $1 - 3a_3 = 0$, $3a_3 - 2a_2 = 0$, and $-a_3 + a_2 - a_1 = 0$. This yields the following solution:

$$1 - 3a_3 = 0$$
$$a_3 = \tfrac{1}{3}$$
$$3a_3 - 2a_2 = 0$$
$$a_2 = \tfrac{3}{2}a_3 = \tfrac{1}{2}$$
$$-a_3 + a_2 - a_1 = 0$$
$$a_1 = a_2 - a_3 = \tfrac{1}{6}$$

Using these values results in the following equation for $p(n)$:

$$p(n) = \tfrac{1}{3}n^3 + \tfrac{1}{2}n^2 + \tfrac{1}{6}n + a_0$$

We have found a polynomial that satisfies the inductive step, step 2 of the inductive proof. All that remains is to verify that it is possible to satisfy the initial step. Note that there is still a constant, a_0, that can be set to any value required by step 1.

Setting n to 1 and verifying step 1 yield the following:

$$S_1 = 1^2$$
$$= p(1)$$
$$= \tfrac{1}{3} + \tfrac{1}{2} + \tfrac{1}{6} + a_0$$
$$= 1 + a_0$$

The solution follows by setting $a_0 = 0$:

$$S_n = \tfrac{1}{3}n^3 + \tfrac{1}{2}n^2 + \tfrac{1}{6}n$$

Remarks

- Al-Haytham used the method to determine $S_n = \sum_{j=1}^{n} j^k$ for any positive integer value of k. The solution is a polynomial of order $k + 1$ and the coefficient a_{k+1} is always $1/(k + 1)$.

- To perform the inductive step, it is necessary to expand $(n - 1)^m$ for integer values of m up to $k + 1$. There is a clever schema that allows one to tabulate the expansion rather quickly. The schema is known as Pascal's triangle, named for Blaise Pascal (1623–1662), who developed it. Pascal was in good company; both al-Haytham and Jia Xian (circa 1010–1070) had discovered the triangle centuries before Pascal.

- Induction can be used to determine the volume of the n-dimensional sphere.

4.6   LINEAR ALGEBRA IN TWO DIMENSIONS

It is time to take stock of our progress and map out the remaining course. Our ultimate objective is to derive Kepler's elliptical planetary motion using both Newton's calculus as expressed by Leibniz and Newton's laws of motion as written by Lambert. What mathematics have we developed to accomplish this and what more needs to be considered? Euclidean geometry, the space in which these calculations take place, has been introduced. Geometric measurements within Euclidean geometry have also been performed. Cartesian coordinates, an approach to integrating geometry with algebra, have been considered along with several examples utilizing the coordinate system. We have also become acquainted with finite sums that arise in integral calculus.

What still needs to be accomplished? Of course, calculus is necessary. It might be possible to jump right into calculus at this point, but there is benefit to be gained in addressing other shortcomings prior to moving onto calculus. One shortcoming that comes to mind is that the ellipse has not yet been introduced. Another shortcoming is that we have not yet related the geometric concepts and the Cartesian approach with motion. Finally, trigonometry that arises in the solution to the ellipse has not yet been introduced.

It is with these latter goals in mind that we set a course for linear algebra. In the West, linear algebra was developed after calculus; strictly speaking, it is possible to address all our needs without it. However, I believe the clearest route to our objective lies through linear algebra. With the tools of linear algebra at hand, the ellipse is as easy as the circle. Concepts of motion are most naturally expressed using vectors, the fundamental elements of linear algebra. Finally, basic identities of trigonometry become apparent with the use of linear algebra.

The field of linear algebra is very broad. It is the starting point for the study of abstract algebra, but this is not the course we will follow. We restrict ourselves to the essentials that are required for the purposes identified above. The primary emphasis is on clarity of exposition as opposed to abstraction and generalization. With this goal in mind, for the most part the material is presented in a two-dimensional setting.

4.6.1   Vectors

Vectors are often used to represent various physical phenomena; two common uses are as identifiers of relative positions in Euclidean space and as representations of an object's velocity. It is common to depict a vector as a line segment with a specified direction using an arrowhead. In the case of identifying relative positions, the length is the distance between the positions and the arrow points from the initial point toward the direction of the terminal point. In the case of representing velocity, the arrow points in the direction of travel and the length of the line segment is the speed. The length of a vector is more commonly referred to as the vector's magnitude. In this book, the notation for a vector is a letter with an arrow above it, that is, $\vec{V}$.

Vectors may be added, subtracted, and multiplied by a constant to yield another vector. Figure 4.19 presents a diagram of algebraic operations on vectors. Addition of the vectors, denoted by $\vec{A} + \vec{B} = \vec{C}$, yields a new vector, as illustrated in Figure 4.19a. Multiplication of a vector by a positive quantity is a resizing of the length by the quantity. In Figure 4.19b vector $\vec{A}$ is multiplied by 2; the result is denoted by $2\vec{A} = \vec{C}$. Multiplying a vector by a negative quantity is a resizing and reversal of direction. In Figure 4.19c, vector $\vec{A}$ is multiplied by -1. Subtracting one vector from another requires reversing the direction of the subtracted vector and adding the result to the remaining vector; that is, the subtracted vector is multiplied by -1 and then added as illustrated in Figure 4.19d. The notation to indicate subtraction is $\vec{A} - \vec{B} = \vec{C}$. This is operationally the same as $\vec{A} + (-\vec{B})$.

Given a coordinate system for a Cartesian plane, one represents a vector in two dimensions by providing two numbers in a column atop one another, as in the vectors $\vec{V}$ and $\vec{W}$ below:

$$\vec{V} = \begin{pmatrix} 1 \\ 2 \end{pmatrix} \qquad \vec{W} = \begin{pmatrix} -3 \\ 5 \end{pmatrix}$$

The first number is the vector's extent along the x coordinate and the second number is the vector's extent along the y coordinate. If the vector represents a point in the plane relative to the origin, the entries are the vector's coordinates. The operations

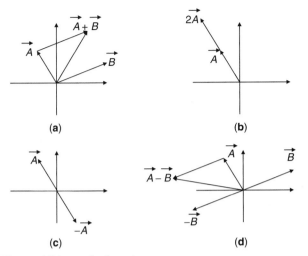

Figure 4.19   Vector addition and subtraction.

of addition, subtraction, and multiplication by a scalar are all performed on each component of the vectors, as seen in the following examples. Notationally, vectors are often presented in rows, with a superscript T that indicates the rows are actually transposed columns. For example, the vector $\vec{V}$ above is written as $\vec{V} = (1\ 2)^{\mathrm{T}}$.

Example 4.4   ADDITION

$$\vec{V} + \vec{W} = \begin{pmatrix} 1 \\ 2 \end{pmatrix} + \begin{pmatrix} -3 \\ 5 \end{pmatrix} = \begin{pmatrix} -2 \\ 7 \end{pmatrix}$$

Example 4.5   SUBTRACTION

$$\vec{V} - \vec{W} = \begin{pmatrix} 1 \\ 2 \end{pmatrix} - \begin{pmatrix} -3 \\ 5 \end{pmatrix} = \begin{pmatrix} 4 \\ -3 \end{pmatrix}$$

Example 4.6   SCALAR MULTIPLICATION

$$5\vec{V} = 5\begin{pmatrix} 1 \\ 2 \end{pmatrix} = \begin{pmatrix} 5 \\ 10 \end{pmatrix}$$

The magnitude of a vector is found using the Pythagorean theorem and is designated by enclosing the symbol in double bars; that is, $\|\vec{V}\|$ is the magnitude of the vector $\vec{V}$. The magnitudes of the above vectors, $\vec{V}$ and $\vec{W}$, are given below:

$$\|\vec{V}\| = \sqrt{1^2 + 2^2} = \sqrt{5}$$
$$\|\vec{W}\| = \sqrt{(-3)^2 + 5^2} = \sqrt{34}$$

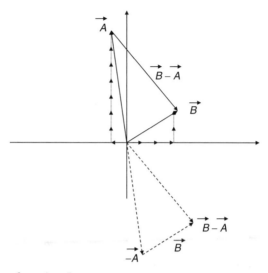

Figure 4.20   Vector from A to B.

As vectors are presented as abstract objects, one gains familiarity by using them in concrete examples. Some examples are given below.

Example 4.7   AS THE CROW FLIES

A crow located at the grid point $(-1, 7)$ wishes to fly to the grid point $(3, 2)$. Determine the vector that gives the direction and distance that the crow must fly assuming that the crow maintains a constant altitude.

The solution is illustrated in Figure 4.20, where $\vec{C} = \vec{B} - \vec{A}$ is the required vector. The initial position of the crow is given by the vector $\vec{A} = \begin{pmatrix} -1 \\ 7 \end{pmatrix}$ and the destination is given by the vector $\vec{B} = \begin{pmatrix} 3 \\ 2 \end{pmatrix}$.

$$\vec{C} = \begin{pmatrix} 3 \\ 2 \end{pmatrix} - \begin{pmatrix} -1 \\ 7 \end{pmatrix} = \begin{pmatrix} 4 \\ -5 \end{pmatrix}$$

The distance that the crow must fly is the magnitude of $\vec{C}$, $\| \vec{C} \| = \sqrt{4^2 + 5^2} = \sqrt{41}$.

Example 4.8   RUNNING IN CIRCLES

A dog runs clockwise around a circle with radius 100 m at uniform speed. It takes 1 min to complete a circuit. Determine the vector that gives the speed and direction of the dog when the dog is at the position $(x_0, y_0) = (-50, \sqrt{7500})$.

The solution is illustrated in Figure 4.21, where $\vec{V} = (v_1 \ v_2)^{\mathrm{T}}$ is the required vector. The direction of motion at any point on the circle is tangent to the circle. Therefore, the ratio v_2/v_1 is the same as the slope of the tangent line. By the

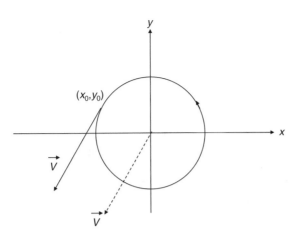

Figure 4.21   Velocity vector for circular motion.

results of Section 4.3.3, the slope of the tangent line is $-y/x = -\sqrt{7500}/50 = -\sqrt{3}$. Additionally, we know that the magnitude of $\vec{V}$ is the speed of the dog. Since the dog travels a distance of one circuit in 1 min, the dog's speed is 200π m/min. Two relations for the quantities v_1 and v_2 have been determined:

$$\frac{v_2}{v_1} = -\sqrt{3} \qquad v_1^2 + v_2^2 = 4000\pi^2$$

The first equation relates the slope of the tangent line to the vector entries and the second equation relates the square of the velocity to the vector entries. Solving the two equations yields the values for v_1 and v_2:

$$v_2 = -\sqrt{3}v_1$$
$$v_1^2 + 3v_1^2 = 4000\pi^2$$
$$v_1^2 = 1000\pi^2$$
$$v_1 = \sqrt{1000}\pi$$
$$v_2 = -\sqrt{3000}\pi$$

Remarks

- Note that the equation $v_1^2 = 1000\pi^2$ has a positive and a negative solution for v_1. The positive solution is the relevant one as the dog runs clockwise around the circle and the x component of the velocity is positive at the point $(x, y) = (-50, \sqrt{7500})$.
- The units of measurement associated with $\vec{V}$ are meters per minute. The units apply to both components of $\vec{V}$.
- Both examples illustrate that a vector does not necessarily originate at the origin but may be translated so that its beginning point does lie at the origin.

4.6.2 The Span of Vectors

Given a set of vectors $(\vec{V}_1, \vec{V}_2, \vec{V}_3, \ldots, \vec{V}_n)$ the following expression is a linear combination of the vectors:

$$a_1 \vec{V}_1 + a_2 \vec{V}_2 + a_3 \vec{V}_3 + \cdots + a_n \vec{V}_n$$

where the coefficients a_j are any real numbers.

The span of a set of vectors is the set of all vectors that are linear combinations of the set. For example, if $\vec{B}_1 = (1 \ 0)^{\mathrm{T}}$ and $\vec{B}_2 = (0 \ 1)^{\mathrm{T}}$ is a set of vectors, then the span of the vectors is the entire Cartesian plane. This is seen as follows. Given any vector in the Cartesian plane, $\vec{V} = (v_1 \ v_2)^{\mathrm{T}}$, for some real values v_1 and v_2, $\vec{V}$ can be expressed as a linear combination of the vectors $\vec{B}_1$ and $\vec{B}_2$:

$$\vec{V} = \begin{pmatrix} v_1 \\ v_2 \end{pmatrix} = v_1 \begin{pmatrix} 1 \\ 0 \end{pmatrix} + v_2 \begin{pmatrix} 0 \\ 1 \end{pmatrix} = v_1 \vec{B}_1 + v_2 \vec{B}_2$$

In one dimension, a single vector spans the entire line. In two dimensions, two vectors are required to span the plane, although not any two vectors suffice. Three vectors are required to span all of the three-dimensional Euclidean space, although once again not any three vectors suffice. In general, n vectors are required to span all of the Euclidean n-dimensional space, although not any n vectors suffice.

Let us examine the two-dimensional case. Given two initial vectors $\vec{A}$ and $\vec{B}$ as depicted in Figure 4.22a, any arbitrary vector $\vec{C}$ can be expressed as a linear combination of the initial vectors. However, the vector $\vec{C}$ cannot be expressed as a linear combination of two vectors $\vec{A}$ and $\vec{B}$ that are collinear as in Figure 4.22b. Only vectors on the common line of $\vec{A}$ and $\vec{B}$ can be expressed in terms of $\vec{A}$ and $\vec{B}$.

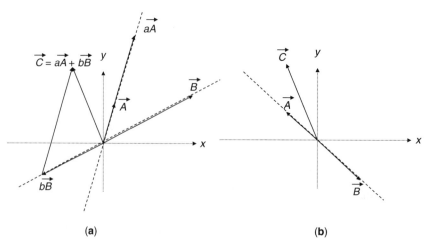

(a) (b)

Figure 4.22 Span of vectors.

The results above are generalized in an intuitive manner. Vectors that are collinear span a one-dimensional space; vectors that are coplanar span a two-dimensional space; and if m is the lowest dimensional Euclidean space that a set of vectors shares, the vectors span the m-dimensional space.

4.6.3 Linear Transformations of the Plane Onto Itself

A vector transformation is a generalization of a function of a variable. Just as a function takes one value to another value, a vector transformation takes one vector to another vector. We present some examples.

Example 4.9

$$\mathbf{T}\tilde{V} = \begin{pmatrix} v_1^2 + 3v_2 \\ v_1 + v_2^3 \end{pmatrix}$$

$$\mathbf{T}\begin{pmatrix} 1 \\ 2 \end{pmatrix} = \begin{pmatrix} 1^2 + (3 \times 2) \\ 1 + 2^3 \end{pmatrix} = \begin{pmatrix} 7 \\ 9 \end{pmatrix}$$

Example 4.10

$$\mathbf{T}\tilde{V} = \begin{pmatrix} \sqrt{|v_1|} \\ v_1 v_2 - v_2 \end{pmatrix}$$

$$\mathbf{T}\begin{pmatrix} -4 \\ 7 \end{pmatrix} = \begin{pmatrix} \sqrt{|-4|} \\ (-4 \times 7) - 7 \end{pmatrix} = \begin{pmatrix} 2 \\ -35 \end{pmatrix}$$

Example 4.11

$$\mathbf{T}\tilde{V} = \begin{pmatrix} 3v_1 - 7v_2 \\ v_1 + 2v_2 \end{pmatrix}$$

$$\mathbf{T}\begin{pmatrix} 2 \\ -3 \end{pmatrix} = \begin{pmatrix} (3 \times 2) - (7 \times (-3)) \\ 2 + (2 \times (-3)) \end{pmatrix} = \begin{pmatrix} 27 \\ -4 \end{pmatrix}$$

In general, a transformation of the plane onto itself is given by the following:

$$\mathbf{T}\tilde{V} = \begin{pmatrix} f_1(v_1, v_2) \\ f_2(v_1, v_2) \end{pmatrix}$$

When both functions $f_1(v_1, v_2)$ and $f_2(v_1, v_2)$ have only linear terms, as in the final example above, the transformation is a linear transformation. In two dimensions, the general form of a linear transformation is given as follows:

$$\mathbf{T}\tilde{V} = \begin{pmatrix} a_{11}v_1 + a_{12}v_2 \\ a_{21}v_1 + a_{22}v_2 \end{pmatrix} \tag{4.25}$$

for real values of $a_{11}, a_{12}, a_{21},$ and a_{22}.

While one may venture the guess that a linear transformation is so named because the transformation contains only linear terms, there is another more fundamental explanation. A linear transformation satisfies the following linearity property:

$$\mathbf{T}(a\tilde{V} + b\tilde{W}) = a\mathbf{T}\tilde{V} + b\mathbf{T}\tilde{W} \tag{4.26}$$

One can directly verify that transformations given by equation (4.25) satisfy the linearity property of equation (4.26):

$$\mathbf{T}(a\tilde{V} + b\tilde{W}) = \mathbf{T}\begin{pmatrix} av_1 + bw_1 \\ av_2 + bw_2 \end{pmatrix}$$

$$\mathbf{T}(a\tilde{V} + b\tilde{W}) = \mathbf{T}\begin{pmatrix} av_1 + bw_1 \\ av_2 + bw_2 \end{pmatrix}$$

$$= \begin{pmatrix} a_{11}(av_1 + bw_1) + a_{12}(av_2 + bw_2) \\ a_{21}(av_1 + bw_1) + a_{22}(av_2 + bw_2) \end{pmatrix}$$

$$= \begin{pmatrix} a_{11}av_1 + a_{12}av_2 \\ a_{21}av_1 + a_{22}av_2 \end{pmatrix} + \begin{pmatrix} a_{11}bw_1 + a_{12}bw_2 \\ a_{21}bw_1 + a_{22}bw_2 \end{pmatrix}$$

$$= a\begin{pmatrix} a_{11}v_1 + a_{12}v_2 \\ a_{21}av + a_{22}v_2 \end{pmatrix} + b\begin{pmatrix} a_{11}w_1 + a_{12}w_2 \\ a_{21}bw + a_{22}w_2 \end{pmatrix}$$

$$= a\mathbf{T}\tilde{V} + b\mathbf{T}\tilde{W}$$

A consequence of the linearity property, equation (4.26), in two dimensions is that, by knowing the result of the transformation on two vectors that span the plane, it is possible to determine the transformation on any vector in the plane. This is most easily conveyed by an example.

Example 4.12   LINEAR TRANSFORMATION

Suppose that $\mathbf{T}\begin{pmatrix} 1 \\ 0 \end{pmatrix} = \begin{pmatrix} 1 \\ 2 \end{pmatrix}$ and $\mathbf{T}\begin{pmatrix} 0 \\ 1 \end{pmatrix} = \begin{pmatrix} -3 \\ 1 \end{pmatrix}$. Find $\mathbf{T}\begin{pmatrix} 4 \\ -2 \end{pmatrix}$.

Solution   Let $\vec{V} = \begin{pmatrix} 1 \\ 0 \end{pmatrix}$ and $\vec{W} = \begin{pmatrix} 0 \\ 1 \end{pmatrix}$. Then $\begin{pmatrix} 4 \\ -2 \end{pmatrix} = 4\vec{V} - 2\vec{W}$.

Using the linearity property, equation (4.26),

$$\mathbf{T}(4\vec{V} - 2\vec{W}) = 4\mathbf{T}\vec{V} - 2\mathbf{T}\vec{W}$$

$$= 4\begin{pmatrix} 1 \\ 2 \end{pmatrix} - 2\begin{pmatrix} -3 \\ 1 \end{pmatrix}$$

$$= \begin{pmatrix} 10 \\ 6 \end{pmatrix}$$

An elegant notation is used to represent linear transformations. In two dimensions, transformations are given by a matrix of the coefficients:

$$M = \begin{pmatrix} a_{11} & a_{12} \\ a_{21} & a_{22} \end{pmatrix}$$

One applies the matrix to the vector $\vec{V}$ to get the resulting transformed vector, $\vec{W}$, as follows:

$$M\tilde{V} = \tilde{W} = \begin{pmatrix} w_1 \\ w_2 \end{pmatrix}$$

$$w_i = \sum_{j=1}^{2} a_{ij} v_j$$

Expanding the summation for each component of $\vec{W}$ results in the following:

$$w_1 = a_{11}v_1 + a_{12}v_2$$

$$w_2 = a_{21}v_1 + a_{22}v_2$$

The expression for $M\vec{V}$ may also be written in the following notation:

$$M\vec{V} = \begin{pmatrix} a_{11} & a_{12} \\ a_{21} & a_{22} \end{pmatrix} \begin{pmatrix} v_1 \\ v_2 \end{pmatrix}$$

$$= \begin{pmatrix} a_{11}v_1 + a_{12}v_2 \\ a_{21}v_1 + a_{22}v_2 \end{pmatrix}$$

We adopt the matrix notation in the remainder of this text.
Let us interpret the coefficients within the matrix using the above example.

Example 4.13   COEFFICIENTS USING THE TRANSFORMATION OF EXAMPLE 4.12

Using matrix notation for the transformation of Example 4.12 where $M\begin{pmatrix} 1 \\ 0 \end{pmatrix} = \begin{pmatrix} 1 \\ 2 \end{pmatrix}$ and $M\begin{pmatrix} 0 \\ 1 \end{pmatrix} = \begin{pmatrix} -3 \\ 1 \end{pmatrix}$ results in the following:

$$M\begin{pmatrix} 1 \\ 0 \end{pmatrix} = \begin{pmatrix} a_{11} & a_{12} \\ a_{21} & a_{22} \end{pmatrix}\begin{pmatrix} 1 \\ 0 \end{pmatrix}$$

$$= \begin{pmatrix} (a_{11} \times 1) + (a_{12} \times 0) \\ (a_{21} \times 1) + (a_{22} \times 0) \end{pmatrix}$$

$$= \begin{pmatrix} a_{11} \\ a_{21} \end{pmatrix} = \begin{pmatrix} 1 \\ 2 \end{pmatrix}$$

$$M\begin{pmatrix} 0 \\ 1 \end{pmatrix} = \begin{pmatrix} a_{11} & a_{12} \\ a_{21} & a_{22} \end{pmatrix}\begin{pmatrix} 0 \\ 1 \end{pmatrix}$$

$$= \begin{pmatrix} (a_{11} \times 0) + (a_{12} \times 1) \\ (a_{21} \times 0) + (a_{22} \times 1) \end{pmatrix}$$

$$= \begin{pmatrix} a_{12} \\ a_{22} \end{pmatrix} = \begin{pmatrix} -3 \\ 1 \end{pmatrix}$$

$$M = \begin{pmatrix} 1 & -3 \\ 2 & 1 \end{pmatrix}$$

Example 4.13 illustrates that the matrix coefficients are obtained by the transformation's behavior upon the vectors $(1\ 0)^T$ and $(0\ 1)^T$; the first column of the matrix is the result of applying the transformation to the vector $(1\ 0)^T$ and the second column of the matrix is the result of applying the transformation to the vector $(0\ 1)^T$.

In the next example, the matrix notation is illustrated on a vector.

Example 4.14   TRANSFORMATION USING MATRIX NOTATION

Applying the matrix of Example 4.13,

$$M = \begin{pmatrix} 1 & -3 \\ 2 & 1 \end{pmatrix}$$

to the vector of Example 4.4,

$$\vec{V} = \begin{pmatrix} 4 \\ -2 \end{pmatrix}$$

gives the following:

$$M\vec{V} = \begin{pmatrix} a_{11} & a_{12} \\ a_{21} & a_{22} \end{pmatrix} \begin{pmatrix} v_1 \\ v_2 \end{pmatrix}$$

$$= \begin{pmatrix} a_{11}v_1 + a_{12}v_2 \\ a_{21}v_1 + a_{22}v_2 \end{pmatrix}$$

$$= \begin{pmatrix} (1 \times 4) + ((-3) \times (-2)) \\ (2 \times 4) + (1 \times (-2)) \end{pmatrix}$$

$$= \begin{pmatrix} 10 \\ 6 \end{pmatrix}$$

The result is the same as the answer in Example 4.12.

Remarks

- The results of this section can be generalized to linear transformations from n-dimensional Euclidean space back to itself. The generalizations are given below.

- A linear transformation from n-dimensional Euclidean space back to itself satisfies the linearity property, equation (4.26).

- By the linearity property, the transformation on a set of n vectors that span all of n-dimensional Euclidean space determines the transformation over all of n-dimensional Euclidean space.

- There is a matrix associated with the transformation. The matrix has n rows and n columns. The jth column of the matrix is given by determining the transformation on the vector having a 1 in the jth position and zeroes in all other positions.

4.6.4 The Inverse of a Linear Transformation

We continue to consider linear transformations from the plane back to the plane. The following diagram depicts a linear transformation acting on a vector; the original vector, $\vec{V}$, is mapped to a new vector, $\vec{W}$. In this section, the following questions are addressed (Figure 4.23):

Question 1. If $\vec{W}$ is any arbitrary vector in the plane, is there a vector $\vec{V}$ that gets mapped to $\vec{W}$?

Question 2. If the answer to question 1 is affirmative, how can $\vec{V}$ be established for any arbitrary $\vec{W}$?

The specific form of the matrix M is used to answer question 1. As established in Section 4.6.3, the matrix M is composed of two columns, and each column is a vector. The first column (vector) is the result of the linear operator acting on the vector

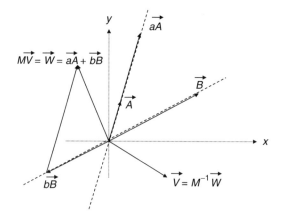

$\vec{A}$ and $\vec{B}$ are column vectors for the matrix M

Figure 4.23   Solution to linear equations.

$(1\ 0)^{\mathrm{T}}$ and the second column (vector) is the result of the linear operator acting on the vector $(0\ 1)^{\mathrm{T}}$. Suppose that these two columns (vectors), $\vec{A}$ and $\vec{B}$, are collinear; their span lies on a line as depicted in Figure 4.22b and all vectors are mapped onto that line. If a vector $\vec{W}$ is not on the line spanned by the columns (vectors), it is not possible to find a vector $\vec{V}$ that is mapped to $\vec{W}$. (See Figure 4.22b where the vector $\vec{C}$ takes the role of $\vec{W}$.)

Alternatively, suppose that the two columns (vectors) $\vec{A}$ and $\vec{B}$ are not collinear. As depicted in Figure 4.23, the span of the two columns (vectors) is the entire plane. Associated with every vector $\vec{W}$ is a vector $\vec{V}$ that is mapped to $\vec{W}$ by the linear transformation. In fact, the vector $\vec{V}$ can be specified. As depicted, $\vec{W} = a\vec{A} + b\vec{B}$, where $\vec{A}$ is the first column of the matrix M and $\vec{B}$ is the second column of the matrix M. Then the vector $\vec{V} = a \begin{pmatrix} 1 \\ 0 \end{pmatrix} + b \begin{pmatrix} 0 \\ 1 \end{pmatrix}$ is mapped to $\vec{W}$. Indeed, we have

$$M\vec{V} = M \left(a \begin{pmatrix} 1 \\ 0 \end{pmatrix} + b \begin{pmatrix} 0 \\ 1 \end{pmatrix} \right)$$

$$= aM \begin{pmatrix} 1 \\ 0 \end{pmatrix} + bM \begin{pmatrix} 0 \\ 1 \end{pmatrix}$$

$$= a\vec{A} + b\vec{B}$$

$$= \vec{W}$$

With the construction of $\vec{V}$, question 1 is answered. As long as the columns (vectors) of M are not collinear, every vector $\vec{W}$ has a vector $\vec{V}$ that is mapped to $\vec{W}$.

We next turn our attention to question 2. There are some observations in the way that $\vec{V}$ was constructed that are helpful. First, note that $\vec{W}$ is a unique linear combination of the vectors $\vec{A}$ and $\vec{B}$; that is, there is a unique pair of coefficients a and b such that $\vec{W} = a\vec{A} + b\vec{B}$. As such, there is only one vector, $\vec{V}$, that is mapped to $\vec{W}$.

Let us use this observation to reverse our perspective. Instead of thinking of $\vec{W}$ as a transformation of $\vec{V}$, think of $\vec{V}$ as a transformation of $\vec{W}$; any given $\vec{W}$ is mapped to a vector $\vec{V}$. We will denote this transformation by $\mathbf{T}^{-1}$. Using this notation, $\vec{W} = \mathbf{T}\vec{V}$ and $\vec{V} = \mathbf{T}^{-1}\vec{W}$, where $\mathbf{T}^{-1}$ is known as the inverse of $\mathbf{T}$.

An observation of note is that $\mathbf{T}^{-1}$ is a linear transformation. Let us test this out. Let $\vec{W}$ and $\vec{X}$ be two vectors, $\vec{W} = a_w\vec{A} + b_w\vec{B}$ and $\vec{X} = a_x\vec{A} + b_x\vec{B}$, where $\vec{A}$ and $\vec{B}$ are the column vectors of the matrix associated with $\mathbf{T}$. Using the construction of the inverse vector, $\mathbf{T}^{-1}\vec{W} = a_w \begin{pmatrix} 1 \\ 0 \end{pmatrix} + b_w \begin{pmatrix} 0 \\ 1 \end{pmatrix}$ and $\mathbf{T}^{-1}\vec{X} = a_x \begin{pmatrix} 1 \\ 0 \end{pmatrix} + b_x \begin{pmatrix} 0 \\ 1 \end{pmatrix}$. We have to show that the linearity property, equation (4.26), is satisfied:

$$\mathbf{T}^{-1}(\alpha\vec{W} + \beta\vec{X}) = \mathbf{T}^{-1}\left(\alpha(a_w\vec{A} + b_w\vec{B}) + \beta(a_x\vec{A} + b_x\vec{B})\right)$$

$$= \mathbf{T}^{-1}\left((\alpha a_w + \beta a_x)\vec{A} + (\alpha b_w + \beta b_x)\vec{B}\right)$$

$$= (\alpha a_w + \beta a_x)\begin{pmatrix} 1 \\ 0 \end{pmatrix} + (\alpha b_w + \beta b_x)\begin{pmatrix} 0 \\ 1 \end{pmatrix}$$

$$= \alpha\left[a_w\begin{pmatrix} 1 \\ 0 \end{pmatrix} + b_w\begin{pmatrix} 0 \\ 1 \end{pmatrix}\right] + \beta\left[a_x\begin{pmatrix} 1 \\ 0 \end{pmatrix} + b_x\begin{pmatrix} 0 \\ 1 \end{pmatrix}\right]$$

$$= \alpha\mathbf{T}^{-1}\vec{W} + \beta\mathbf{T}^{-1}\vec{X}$$

Note that the third equality follows from the construction of the inverse vector.

Since $\mathbf{T}^{-1}$ is a linear transformation, it has an associated matrix that we denote by M^{-1}. One arrives at a complete answer to question 2 by determining the components of M^{-1}. The first column of M^{-1} is the result of the mapping $\mathbf{T}^{-1}$ on $(1\ 0)^{\mathrm{T}}$ and the second column of M^{-1} is the result of the mapping $\mathbf{T}^{-1}$ on $(0\ 1)^{\mathrm{T}}$. Using the process for constructing the inverse vector, we have the following. Let

$$\begin{pmatrix} 1 \\ 0 \end{pmatrix} = a_1\vec{A} + b_1\vec{B}$$

$$= a_1\begin{pmatrix} a_{11} \\ a_{21} \end{pmatrix} + b_1\begin{pmatrix} a_{12} \\ a_{22} \end{pmatrix} \tag{4.27}$$

$$\begin{pmatrix} 0 \\ 1 \end{pmatrix} = a_2\vec{A} + b_2\vec{B}$$

$$= a_2\begin{pmatrix} a_{11} \\ a_{21} \end{pmatrix} + b_2\begin{pmatrix} a_{12} \\ a_{22} \end{pmatrix} \tag{4.28}$$

where

- $\vec{A}$ is the first column of the matrix M
- $\vec{B}$ is the second column of the matrix M

• $M = \begin{pmatrix} a_{11} & a_{12} \\ a_{21} & a_{22} \end{pmatrix}$

$$\mathbf{T}^{-1}\begin{pmatrix} 1 \\ 0 \end{pmatrix} = a_1 \mathbf{T}^{-1}\vec{A} + b_1 \mathbf{T}^{-1}\vec{B}$$

$$= a_1 \begin{pmatrix} 1 \\ 0 \end{pmatrix} + b_1 \begin{pmatrix} 0 \\ 1 \end{pmatrix}$$

$$= \begin{pmatrix} a_1 \\ b_1 \end{pmatrix}$$

Similarly,

$$\mathbf{T}^{-1}\begin{pmatrix} 0 \\ 1 \end{pmatrix} = a_2 \mathbf{T}^{-1}\vec{A} + b_2 \mathbf{T}^{-1}\vec{B}$$

$$= a_2 \begin{pmatrix} 1 \\ 0 \end{pmatrix} + b_2 \begin{pmatrix} 0 \\ 1 \end{pmatrix}$$

$$= \begin{pmatrix} a_2 \\ b_2 \end{pmatrix}$$

The matrix M^{-1} is then given by the following equality:

$$M^{-1} = \begin{pmatrix} a_1 & a_2 \\ b_1 & b_2 \end{pmatrix}$$

The solution to M^{-1} is found by determining the quantities a_1, b_1, a_2, and b_2 in terms of the original matrix M. Equating both components of the vector in equation (4.27) results in the following two equations:

$$a_{11}a_1 + a_{12}b_1 = 1$$
$$a_{21}a_1 + a_{22}b_1 = 0$$

We solve these equations simultaneously. From the second equation

$$b_1 = -\frac{a_{21}a_1}{a_{22}}$$

Substituting into the first equation and simplifying yield the result for a_1:

$$a_{11}a_1 - a_{12}\frac{a_{21}a_1}{a_{22}} = 1$$

$$\frac{a_{11}a_{22} - a_{12}a_{21}}{a_{22}}a_1 = 1$$

$$a_1 = \frac{a_{22}}{a_{11}a_{22} - a_{12}a_{21}}$$

Substitution of a_1 into the expression for b_1 yields the result for b_1:

$$b_1 = \frac{-a_{21}}{a_{11}a_{22} - a_{12}a_{21}}$$

A similar process applied to equations (4.28) yields the results for a_2 and b_2:

$$a_2 = \frac{-a_{12}}{a_{11}a_{22} - a_{12}a_{21}} \qquad b_2 = \frac{a_{11}}{a_{11}a_{22} - a_{12}a_{21}}$$

Placing the values for a_1, b_1, a_2, and b_2 into the expression for M^{-1} and simplifying result in the expression for M^{-1}:

$$M^{-1} = \frac{1}{a_{11}a_{22} - a_{12}a_{21}} \begin{pmatrix} a_{22} & -a_{12} \\ -a_{21} & a_{11} \end{pmatrix} \qquad (4.29)$$

Note that each term inside the matrix is multiplied by the factor in front of the matrix.

Example 4.15   NONINVERTIBLE MATRIX

Let

$$M = \begin{pmatrix} 1 & -2 \\ -3 & 6 \end{pmatrix}$$

Then

$$\vec{A} = \begin{pmatrix} 1 \\ -3 \end{pmatrix} \qquad \vec{B} = \begin{pmatrix} -2 \\ 6 \end{pmatrix}$$

Since $\vec{B} = -2\vec{A}$, $\vec{A}$ and $\vec{B}$ are collinear, as shown in Figure 4.22b. It was demonstrated that inverses only occur when the column vectors are not collinear. Accordingly, the matrix of this example has no inverse. Note that the term $a_{11}a_{22} - a_{12}a_{21}$ that occurs in the denominator of the factor for the inverse matrix evaluates to zero for this example:

$$a_{11}a_{22} - a_{12}a_{21} = (1 \times 6) - [(-2) \times (-3)] = 0$$

The expression for M^{-1} indicates that M is noninvertible as division by zero is undefined.

Example 4.16   INVERTIBLE MATRIX

Let

$$M = \begin{pmatrix} 1 & -3 \\ 2 & 5 \end{pmatrix}$$

Then

$$\vec{A} = \begin{pmatrix} 1 \\ 2 \end{pmatrix} \qquad \vec{B} = \begin{pmatrix} -3 \\ 5 \end{pmatrix}$$

Assume $\vec{A}$ and $\vec{B}$ are not collinear as in Figure 4.22b, so the inverse exists. Using equation (4.29), $a_{11}a_{22} - a_{12}a_{21} = (5 \times 1) - (-3 \times 2) = 11$ and

$$M^{-1} = \frac{1}{11} \begin{pmatrix} 5 & 3 \\ -2 & 1 \end{pmatrix}$$

Select a vector $\vec{V} = \begin{pmatrix} 3 \\ 4 \end{pmatrix}$. Applying the matrix M to $\vec{V}$ yields the following:

$$M\vec{V} = \begin{pmatrix} 1 & -3 \\ 2 & 5 \end{pmatrix} \begin{pmatrix} 3 \\ 4 \end{pmatrix}$$

$$= \begin{pmatrix} (1 \times 3) - (3 \times 4) \\ (2 \times 3) + (5 \times 4) \end{pmatrix}$$

$$= \begin{pmatrix} -9 \\ 26 \end{pmatrix}$$

If everything is correct, applying M^{-1} to

$$M\vec{V} = \begin{pmatrix} -9 \\ 26 \end{pmatrix}$$

should result in the original vector $\vec{V} = \begin{pmatrix} 3 \\ 4 \end{pmatrix}$. The calculation is verified below:

$$M^{-1} \begin{pmatrix} 9 \\ 26 \end{pmatrix} = \frac{1}{11} \begin{pmatrix} 5 & 3 \\ -2 & 1 \end{pmatrix} \begin{pmatrix} -9 \\ 26 \end{pmatrix}$$

$$= \frac{1}{11} \begin{pmatrix} (5 \times (-9)) + (3 \times 26) \\ (-2 \times (-9)) + (1 \times 26) \end{pmatrix}$$

$$= \frac{1}{11} \begin{pmatrix} 33 \\ 44 \end{pmatrix}$$

$$= \begin{pmatrix} 3 \\ 4 \end{pmatrix}$$

$$= \vec{V}$$

Example 4.16 presents a specific example of a general property. If M^{-1} does exist and $\vec{V}$ gets mapped to $\vec{W}$, the following relations hold:

$$M^{-1}M\vec{V} = M^{-1}\vec{W} = \vec{V}$$
$$MM^{-1}\vec{W} = M\vec{V} = \vec{W} \tag{4.30}$$

These relations can be used to solve systems of equations.

Example 4.17   SOLVING A SYSTEM OF EQUATIONS

Let the matrix M be as in the previous example. Solve for $\vec{V}$ in the equation

$$M\vec{V} = \begin{pmatrix} 55 \\ -66 \end{pmatrix}$$

Using relations (4.30),

$$M\vec{V} = \begin{pmatrix} 55 \\ -66 \end{pmatrix}$$

$$M^{-1}M\vec{V} = M^{-1}\begin{pmatrix} 55 \\ -66 \end{pmatrix}$$

$$\vec{V} = M^{-1}\begin{pmatrix} 55 \\ -66 \end{pmatrix}$$

$$= \frac{1}{11}\begin{pmatrix} 5 & 3 \\ -2 & 1 \end{pmatrix}\begin{pmatrix} 55 \\ -66 \end{pmatrix}$$

$$= \begin{pmatrix} 5 & 3 \\ -2 & 1 \end{pmatrix}\begin{pmatrix} 5 \\ -6 \end{pmatrix}$$

$$= \begin{pmatrix} (5 \times 5) + (3 \times (-6)) \\ (-2 \times 5) + (1 \times (-6)) \end{pmatrix}$$

$$= \begin{pmatrix} 7 \\ -16 \end{pmatrix}$$

Remarks

- In general, the condition that a linear transform from n dimensions to n dimensions has an inverse is that the column vectors of the associated matrix span n-dimensional Euclidean space.

- Inverting a transformed object occurs in many settings. One setting is encoding and decoding; messages are often coded to preserve confidentiality. The code transforms characters from a character set to other characters, perhaps from the same character set. The result is an incomprehensible message. Applying the inverse code to the encoded message restores the original message.

- The relations expressed by equation (4.30) are often used to define the inverse of a matrix. That is, the inverse must satisfy both relations. In mathematics, it is also of interest to find the inverse of common functions. The functions and their inverses must satisfy relations that are analogous to equations (4.30); $f^{-1}(f(x)) = x$ and $f(f^{-1}(y)) = y$.

4.6.5 The Determinant

In Section 4.6.4, the formula for the inverse of a matrix, equation (4.29), contains the term $a_{11}a_{22} - a_{12}a_{21}$. This term is known as the determinant. The determinant has the following geometric interpretation. If an object in the plane has area A, then the image of the object under the linear transformation represented by the matrix M has area $| a_{11}a_{22} - a_{12}a_{21} | \times A$. This interpretation is used in the following sections. It is demonstrated in Chapter 5.

4.7 THE ELLIPSE

The human obsession with beauty borders on disease. Perhaps the best indication of our desire for beauty is the marketplace where we spend our money. Merchants must have beautiful packaging for their products that themselves must be beautiful. Does anyone remember the AMC Pacer? Probably not—the ugly car was not in production very long. There are two types of beauty, the type that must be pursued and the type that just exists. Music, art, and fashion lie within the first category. The creator is in specific pursuit of beauty. Athletics and mathematics are in the second category. The athlete does not pursue beauty, the athlete heads a corner kick to score a goal and, if the timing and the placement are right, the result just happens to be beautiful. Likewise, the mathematician creates mathematics and beauty comes with the territory. Perhaps it is watching this beauty unfold that motivates mathematicians.

Apollonius' (c.200BC) work *Conics* contains brilliant and beautiful mathematics. The work, which earned Apollonius the title as the greatest geometer, comes in eight volumes; the first seven have survived. There is not a trace of utility in any of the surviving seven volumes. Instead, seductive and elegant geometric constructions underlie beautiful theorems. As an example, the fifth volume presents theorems that give the maximum and minimum distances from any planar point, the plane within which the conic section lies, to the conic section. The work exquisitely applies the notion of orthogonality as the basic indicator of a maximum or a minimum distance. The work also details the intricate calculations required to uncover the distances. But Apollonius presents not even the slightest hint that there is any functionality behind these theorems.

In fact, there are practical applications of Apollonius' work. Apollonius develops foundational knowledge in the field of optimization. Additionally, al-Haytham and later Kepler would use properties of the parabola and ellipse that are stated to describe reflections of light sources on both parabolic and elliptic surfaces. It is not conceivable that history's leading authority on conic sections was unaware of these broader, more practical implications, yet he is mum on purporting any use at all. The work is beautiful and that alone justifies both the writer's and the reader's efforts.

In his book, *The Optical Part of Astronomy*, Kepler includes a chapter on the conic sections. (The main objective of the text is to expand upon al-Haytham's work in optics.) Apollonius had not quite exhausted all the properties of the conic sections and Kepler demonstrates his mastery of the curves by discovering yet one more property. Additionally, Kepler presents interesting interpretations of Apollonius' work. In particular, his view of the relation between the foci of the sections and the sections themselves could be viewed as a pioneering effort in bifurcation theory; the curves adopt new shapes as the foci pass through thresholds.

Kepler's (1993) work *New Astronomy* is a historian's dream. Unlike other scientific work, it is very personal, part diary and part science. This affords a unique entry point into the mind of a genius. What can one expect? Is not Captain Spock the quintessential model of a genius? Should not Kepler, this brilliant intellect, have a clear path toward his objective? Should not his drive be based upon pure logic, unclouded by emotional content? If these are the preconceptions of the genius mind, Kepler dispels them all. Kepler is not a guided missile; he is a fireworks display. He has no clear path toward his objective, but ideas percolate through his mind sending him in all different directions. And with each new possibility, there is childlike excitement in pursuing it to its end. Even when Kepler has intellectual certainty that one of his ideas will end in failure, he doggedly pursues it to the end and seems to take an enormous amount of pleasure in the magnificence of the failure as well as its public disclosure. But never does he dwell upon the failure at length for as soon as one idea has exploded, another surfaces, and well there he goes again. This pattern went on for 6 years, interrupted by legal proceedings, negotiations, and his other seminal work *The Optical Part of Astronomy*.

Kepler had a great sense of humor, and despite the many tragedies that he confronted, his life could also inspire comedy. Imagine a movie in which a genius is working on a project and due to a legal snag is forced to temporarily abandon it and take up something else. Everyone in the audience knows that the seemingly unrelated projects are united by a single element, say the ellipse. The genius having learned everything about the ellipse in one of the projects just does not pick up on it for his other project. He pursues one failure after another, but he pursues it with such joy and in the process pokes so much fun at himself that despite this tortuous display of cluelessness the audience loves the genius. Indeed, his cluelessness is one of his endearing features. But he is no clown, for accompanying his cluelessness is penetrating insight, unequaled dedication, and remarkable honesty. The honesty is what really strikes the audience. The genius could at any time pull the wool over the eyes of his contemporaries and present his many failures as a success. But instead he chooses to expose his own personal failures, learn from them, and move on. It is this quality that earns the genius immortality; his honest confrontation of his failures leads him to the correct conclusion that is enduring. The movie ends with the genius' epiphany that what he needed all along was readily available to him, Apollonius' beautiful ellipse. What a great movie. Should it ever be made, it would be the movie of Kepler's greatest discovery.

Kepler was the first to propose a force between the sun and its planets and he posited that the force lessens with distance. A consequence of this view is that all points equidistant from the sun experience the same force. The intuitive orbit that results from this view is the circle of Copernicus; the planet would experience an equal force at all points of a circular orbit and this equal force would keep the planet an equal distance from the sun. There is a peculiarity with the ellipse; it is elongated along a designated direction. This is counterintuitive as a possibility for a planet's orbit. The sun's force is circularly symmetric; it does not have a preferred direction that would account for the elongation of a planet's orbit. But the circular symmetry of the solar force expresses itself in a less transparent fashion through conservation of angular momentum and Kepler's planetary pathway around the ellipse satisfies the conservation of angular momentum. Kepler was ahead of his time. There was no concept of angular momentum, let alone the understanding of symmetry conditions that preserve the angular momentum. Indeed, the general relation between symmetry and constants of motion would come nearly 300 years after Kepler is discovered by David Hilbert (1862–1943) and Emmy Noether (1882–1935).

This section closes our chapter on algebra with a brief introduction to the ellipse. The equations of the ellipse are derived using a linear transformation of the circle. Afterward, Kepler's property of the foci is presented. Some find the results beautiful, some find them useful and beautiful, but only the unaware can find neither beauty nor use.

4.7.1   The Ellipse as a Linear Transformation of a Circle

As a starting point, we define the ellipse as a linear transformation of the circle with radius 1. Specifically, let the matrix of the transformation be given by

$$M = \begin{pmatrix} a & 0 \\ 0 & b \end{pmatrix} \tag{4.31}$$

in which a and b are positive numbers.

If a is less than 1, the transformation compresses the x axis; alternatively, if a is greater than 1, the x axis is stretched. Similarly, the transformation compresses or stretches the y axis depending upon whether b is less than or greater than 1. Letting $a = \frac{4}{3}$ and $b = \frac{1}{2}$ and applying M to the circle with radius 1 yield the ellipse as illustrated in Figure 4.24. As illustrated by the figure, the length of the major axis is twice the larger of a and b, while the length of the minor axis is twice the lesser of a and b.

Using the geometric interpretation of the determinant as presented in Section 4.6.4, the area of the ellipse is found by multiplying the determinant with the area of the corresponding circle. Noting that the determinant of M is ab results in the following:

$$\text{Area} = \pi ab$$

4.7.2   The Equation of an Ellipse

An equation for the ellipse is obtained via the equation for the circle. The points on the ellipse are mapped back to the circle and placed into the equation of the circle. Figure 4.24 illustrates the process that is expressed algebraically below.

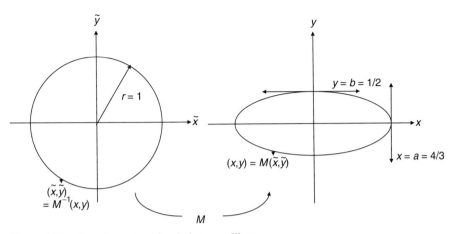

Figure 4.24   Transformation of a circle to an ellipse.

Let $(\tilde{x}, \tilde{y})$ be a point on the circle that is mapped by the matrix M of equation (4.31) onto a point (x, y) of an ellipse:

$$M \begin{pmatrix} \tilde{x} \\ \tilde{y} \end{pmatrix} = \begin{pmatrix} x \\ y \end{pmatrix}$$

Note that M is invertible as its determinant is nonzero. Applying equation (4.29) results in the inverse:

$$M^{-1} = \frac{1}{ab} \begin{pmatrix} b & 0 \\ 0 & a \end{pmatrix} = \begin{pmatrix} \frac{1}{a} & 0 \\ 0 & \frac{1}{b} \end{pmatrix}$$

Applying M^{-1} to the point (x, y) on the ellipse gives the associated point on the circle, $(\tilde{x}, \tilde{y})$:

$$M^{-1} \begin{pmatrix} x \\ y \end{pmatrix} = \begin{pmatrix} \frac{x}{a} \\ \frac{y}{b} \end{pmatrix} = \begin{pmatrix} \tilde{x} \\ \tilde{y} \end{pmatrix} \tag{4.32}$$

Since the point $(\tilde{x}, \tilde{y})$ is on the unit circle, it satisfies the equation for the circle:

$$\tilde{x}^2 + \tilde{y}^2 = 1$$

Substituting for $(\tilde{x}, \tilde{y})$ using the equality $(\tilde{x}, \tilde{y}) = (x/a, y/b)$ yields the equation for the ellipse:

$$\left(\frac{x}{a}\right)^2 + \left(\frac{y}{b}\right)^2 = 1 \tag{4.33}$$

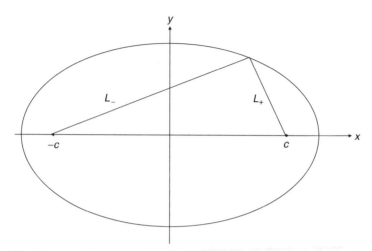

Figure 4.25   The sum of the lengths from the foci to points on the ellipse is constant

4.7.3   An Excursion into the Foci of an Ellipse

Kepler discovered that the sum of the distances from each focus to any point on the ellipse is equal to the length of the major axis of the ellipse. This property is illustrated in Figure 4.25. In the figure, the lengths of the line segment to the positive and negative foci are correspondingly denoted by L_+ and L_-. The property states that $L_+ + L_- = 2a$. A derivation of the property is presented. The derivation is in a sense unsatisfactory; it is an algebraic exercise that does not offer much insight into why the property holds. However, after going through the exercise, one appreciates Kepler all the more. How did he intuit his results without our modern notation, without Cartesian coordinates, without equations for the ellipse, and without our standard algebraic operations? As with other excursions, this section is unnecessary for the understanding of the remaining material in the book.

For an ellipse given by equation (4.33), with $a > b > 0$, the foci are points that lie along the x axis in the positions $(c, 0)$ and $(-c, 0)$ with

$$c^2 = a^2 - b^2 \tag{4.34}$$

We now verify the property $L_+ + L_- = 2a$:

$$L_+ + L_- = \sqrt{(x - c)^2 + y^2} + \sqrt{(x + c)^2 + y^2} \tag{4.35}$$

Using equation (4.33) to solve for y^2, substituting the result into equation (4.35), and simplifying result in the following:

$$y^2 = b^2 \left[1 - \left(\frac{x}{a} \right)^2 \right]$$

$$L_+ + L_- = \sqrt{(x-c)^2 + b^2 \left[1 - \left(\frac{x}{a}\right)^2\right]} + \sqrt{(x+c)^2 + b^2 \left[1 - \left(\frac{x}{a}\right)^2\right]}$$

$$= \sqrt{x^2 - 2xc + c^2 + b^2 - \left(\frac{bx}{a}\right)^2} + \sqrt{x^2 + 2xc + c^2 + b^2 - \left(\frac{bx}{a}\right)^2}$$

Recall that $c^2 = a^2 - b^2$ and substitute for c^2 to obtain the following:

$$L_+ + L_- = \sqrt{x^2 - 2xc + a^2 - \left(\frac{bx}{a}\right)^2} + \sqrt{x^2 + 2xc + a^2 - \left(\frac{bx}{a}\right)^2}$$

$$= \sqrt{\left[1 - \left(\frac{b}{a}\right)^2\right] x^2 - 2xc + a^2} + \sqrt{\left[1 - \left(\frac{b}{a}\right)^2\right] x^2 + 2xc + a^2}$$

Note that from equation (4.34), $1 - (b/a)^2 = (c/a)^2$. Substituting into the above expression and simplifying yield the result:

$$L_+ + L_- = \sqrt{\left(\frac{c}{a}\right)^2 x^2 - 2xc + a^2} + \sqrt{\left(\frac{c}{a}\right)^2 x^2 + 2xc + a^2}$$

$$= \sqrt{\left(a - \frac{c}{a}x\right)^2} + \sqrt{\left(a + \frac{c}{a}x\right)^2}$$

$$= \left(a - \frac{c}{a}x\right) + \left(a + \frac{c}{a}x\right)$$

$$= 2a$$

Kepler's result allows one to draw an ellipse using a string. Let the string be of length $2a$ and tape the ends of the string onto the focal points. With a pencil, pull the string taut in all directions while marking off the points where the string reaches its maximum extent.

CHAPTER 5

THE UNIVERSAL TOOL: TRIGONOMETRY

A microphone planted by the U.S. Navy senses noise from an underwater object entering the North Sea through the Skagerrak Strait. Through a series of communication relays, the signal is transmitted to a computer that resides in the continental United States. The computer performs a Fourier transform on the noise; the Fourier transform decomposes the original noise into a spectrum of constituent trigonometric functions. The computer then analyzes the resulting spectrum and compares it with a stored library of known signals. The computer identifies the noise as a whale and a naval officer concurs after visually inspecting the spectrum. The story of trigonometry begins with a very elementary problem, triangulation. It then traverses ground into a completely unrelated applied problem, the heat equation; spurs research in theoretical mathematics; and then finds application in ever more areas of engineering—such as signal processing, which allows the U.S. Navy to identify the source of a noise that is sensed by an undersea microphone. It is not a coincidence that many of the same men who are central to our story of the ellipse are also central to the development of trigonometry and its applications.

The field of astronomy has motivated mathematical development across cultures and time. Many of the early astronomers devoted considerable effort toward a mathematical description of the pathways of heavenly bodies that were divided into four types: stars, planets, the moon, and the sun. Underlying their efforts were the assumptions of Aristotle's universe: The earth is dominant at the universe's center and is heavier than all the heavenly bodies. Aristarchus made no such assumptions but sought answers. Aristarchus devised a method to determine the relative sizes of the earth, moon, and sun. Using his method, Aristarchus could either confirm Aristotle or bring Aristotle's model into question.

Aristarchus understood the dynamics of a lunar eclipse and used this understanding to answer his question. The basic premise is illustrated in Figure 5.1. From this figure, it is seen that the Aristotelian configuration in which the earth is larger than the sun causes the earth to cast a larger shadow than the alternative case. Aristarchus determined the geometry of the earth's shadow using measurements he had taken during a lunar eclipse; the moon falls within the shadow during such events, providing an

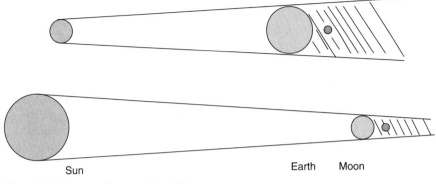

Figure 5.1   Lunar eclipse, small and big sun.

opportunity to take measurements. Using the measurements along with the assistance of a known trigonometric inequality, Aristarchus determined that the sun's diameter is between 19/3 and 43/6 times the diameter of the earth. Due to errors in measurement, Aristarchus' quantitative result is inaccurate; the sun's diameter is roughly 110 times that of the earth. Nevertheless, his geometric argument is impeccable and it was accepted by later Greek astronomers. But the result, other than in Aristarchus' mind, did not cause a reassessment of geocentrism. Aristachus' analysis could not be taken seriously until Copernicus revived heliocentrism nearly 1300 years later.

Ptolemy's achievements in astronomy, mathematics, and geography were worthy of the acclaim that they received. Unlike Aristarchus, Ptolemy did not have to wait 1300 years for his ideas to take hold. In the field of astronomy, Ptolemy followed the Aristotelian program. Nobody devoted more effort toward providing a geocentric description of the heavenly bodies than Ptolemy. Far from a failure, the description proved quite accurate. Truth be told, it was more accurate in describing the motion of the planets from the perspective of an earthbound observer than the proposal of Copernicus a millennium and three centuries later. A related body of Ptolemy's work is a comprehensive table of chord lengths associated with angles from half a degree through 180°. Today, we view the tables as providing trigonometric relations from 0° to 90° in quarter-degree increments. Among other uses for the tables, they certainly were of assistance for determining the positions of equants and eccentrics as well as the dimensions of epicycles.

Both Aristarchus' and Ptolemy's application indicate the importance of trigonometry to astronomy as well as its primary application, triangulation, in both Euclidean and spherical coordinates. In the West, there were a few contributions to the field of trigonometry between Ptolemy and Newton–Leibniz. Both Newton and Leibniz independently determined power series expansions for trigonometric functions. [The Indian mathematician and astronomer Madhava Sanganagrana (1350–1425) made this discovery two and a half centuries before Newton and Leibniz.] The expansions allow one to approximate trigonometric functions as polynomials with the approximation improving as one considers polynomials of higher order.

Newton and Leibniz' discovery would lead to an investigation of power series, polynomials of indefinite degree. A magnificent body of mathematical theory was developed. The theory culminated two centuries later with the triumphant outcome of Weierstrass (1815–1897), who in 1885 demonstrated that a large class of functions can be expressed as power series.

But trigonometric functions, originally established for triangulation, held other secrets that would not be fully uncovered until the beginning of the twentieth century. In 1822, Fourier published a solution to a problem of heat transfer across a rod. It is assumed that the rod has an initial arbitrary temperature distribution, not uniform, and that heat cannot escape from the rod. The problem is to find the temperature distribution across the rod at any time. Using calculus, it was possible to derive an equation that determine the evolution of the temperature at each point on the rod. Fourier showed that trigonometric functions satisfy the equation and that solutions are additive; multiples of solutions could be added together and the result is also a solution to the heat equation. To complete his analysis, Fourier had to demonstrate that he could find a composition of trigonometric solutions that satisfy the initial temperature distribution. Toward this end, he produced a method for describing the initial temperature distribution with a series expansion of trigonometric functions that he claimed was in fact equal to the initial temperature distribution. Nobody could blame Fourier for not being able to prove his claim. Fourier's series initiated a flurry of research into this issue that involved nearly all of the most capable mathematicians of the nineteenth century. No single individual can lay credit to fully solving the problem. The research culminated in a validation of Fourier's approach in the earlier part of the twentieth century. The Fourier transform, which allows the U.S. Navy to determine if a noise relayed by an undersea microphone is from a whale or a submarine, is an application that follows from this highly theoretical research.

This chapter follows the path of Ptolemy from a modern perspective. A table of trigonometric functions is developed. Along the way, definitions are introduced and identities that assist with finding table entries are also demonstrated. Afterward, an application is considered, and Aristarchus' calculation is revisited. In addition, we attend to unfinished business; the determinant of a two-dimensional transformation is demonstrated to have the property claimed in Section 4.6.5. In addition to the above material, one excursion is taken. The Greeks were able to perform amazing constructs with a compass and a straight edge. One achievement is the inscription of a pentagon into a sphere; this construction assisted Ptolemy in the development of his table of chords. We will demonstrate the construction. Finally, an understanding of trigonometry allows us to relate polar coordinates, a method of coordinating the plane using radial and angular measurements, to standard Cartesian coordinates. This chapter demonstrates the relation that is critical for uncovering the ellipse.

5.1 TRIGONOMETRIC FUNCTIONS

5.1.1 Basic Definitions

This section presents the basic definitions of the trigonometric functions. Let a unit circle be given over a Cartesian plane with the usual (x, y) coordinates. Let an angle

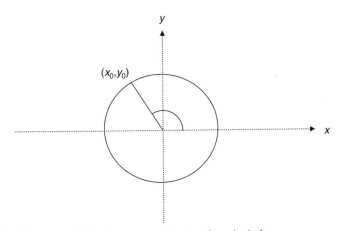

Figure 5.2   Trigonometric functions as coordinates of a unit circle.

θ_0 be given by a point (x_0, y_0) on the circle; θ_0 is the angle formed along the arc of the circle from the x axis to the point (x_0, y_0) (Figure 5.2). The trigonometric functions are functions of the angle. The table below presents their definition for an arbitrary angle θ with corresponding coordinates (x, y) on the unit circle. Note that the functions are given in terms of the (x, y) coordinates.

Function Name	Notation	Definition
Sine	$\sin(\theta)$	y
Cosine	$\cos(\theta)$	x
Tangent	$\tan(\theta)$	$\dfrac{y}{x}$
Cosecant	$\csc(\theta)$	$\dfrac{1}{y}$
Secant	$\sec(\theta)$	$\dfrac{1}{x}$
Cotangent	$\cot(\theta)$	$\dfrac{x}{y}$

The angle θ is sometimes expressed in radians and sometimes expressed in degrees. It is worth one's effort to be able to use both units of measurement.

5.1.2 Triangles

For angles between $0°$ and $90°$ (from 0 to $\pi/2$ radians), the trigonometric functions correspond to ratios of right triangles as illustrated in Figure 5.3.

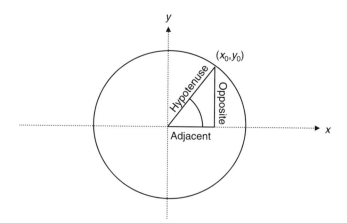

Figure 5.3   Trigonometric functions as triangular ratios.

Notation	Definition
$\sin(\theta)$	$\dfrac{\text{opposite}}{\text{hypotenuse}}$
$\cos(\theta)$	$\dfrac{\text{adjacent}}{\text{hypotenuse}}$
$\tan(\theta)$	$\dfrac{\text{opposite}}{\text{adjacent}}$
$\csc(\theta)$	$\dfrac{\text{hypotenuse}}{\text{opposite}}$
$\sec(\theta)$	$\dfrac{\text{hypotenuse}}{\text{adjacent}}$
$\cot(\theta)$	$\dfrac{\text{adjacent}}{\text{opposite}}$

The equivalence between the definitions in the preceding two tables is seen by noting the triangle formed between a point on the unit circle, the origin, and the point along the x axis given by the x coordinate of the original point. The ratios that define the trigonometric functions are the same for all similar triangles.

5.1.3 Examples

Using the definitions, it is possible to determine the trigonometric functions for some values of θ. Examples are given below. These examples are the first entries into a trigonometric table that is further developed in subsequent sections.

Example 5.1

Determine the trigonometric for the value $\theta = \pi/4$ rad ($45°$).

Solution   When the angle θ is $\pi/4$ rad, $x = y$ along the unit circle (see Figure 5.4). With the assistance of the Pythagorean theorem, the values for x and y and the

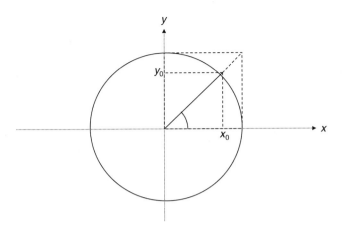

Figure 5.4   Trigonometric functions for 45° angle.

trigonometric functions are obtained:

$$x^2 + y^2 = 1$$
$$x^2 + x^2 = 1$$
$$2x^2 = 1$$
$$x^2 = \tfrac{1}{2}$$
$$x = \sqrt{\tfrac{1}{2}}$$
$$y = \sqrt{\tfrac{1}{2}}$$

θ (deg)	θ (rad)	$\sin(\theta)$	$\cos(\theta)$	$\tan(\theta)$	$\csc(\theta)$	$\sec(\theta)$	$\cot(\theta)$
45	$\pi/4$	$\sqrt{\tfrac{1}{2}}$	$\sqrt{\tfrac{1}{2}}$	1	$\sqrt{2}$	$\sqrt{2}$	1

Example 5.2

Determine the trigonometric functions for the value $\theta = \pi/3$ rad (60°).

As noted in Section 4.3.2, a hexagon (six-sided polygon) with sides of equal length may be inscribed into the unit circle. Figure 5.5a shows the case when a vertex is on the x axis. The point (x_A, y_A) corresponds with the angle $\pi/3$ (60°). The triangle with (x_A, y_A) at the apex is an equilateral triangle. Bisecting the base of the triangle results in the value $x_A = \tfrac{1}{2}$. To find the value of y_A, use the Pythagorean theorem:

$$x_A^2 + y_A^2 = 1$$
$$y_A^2 = 1 - x_A^2$$
$$y_A = \sqrt{1 - x_A^2}$$

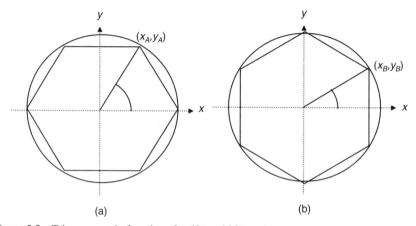

Figure 5.5   Triagonometric functions for 60° and 30° angles.

$$= \sqrt{1 - \left(\tfrac{1}{2}\right)^2}$$

$$= \sqrt{\tfrac{3}{4}}$$

$$= \tfrac{1}{2}\sqrt{3}$$

θ (deg)	θ (rad)	$\sin(\theta)$	$\cos(\theta)$	$\tan(\theta)$	$\csc(\theta)$	$\sec(\theta)$	$\cot(\theta)$
60	$\tfrac{1}{3}\pi$	$\tfrac{1}{2}\sqrt{3}$	$\tfrac{1}{2}$	$\sqrt{3}$	$2/\sqrt{3}$	2	$1/\sqrt{3}$

Example 5.3

Determine the trigonometric functions for the value $\theta = \pi/6$ rad (30°)

Inscribe the hexagon in the unit circle so that the x axis bisects a side of the polygon (Figure 5.5b). The point corresponds to the angle $\pi/6$ (30°). From the geometry, it is apparent that $x_B = y_A$ and $y_B = x_A$. Indeed, the argument of Example 5.2 applies with x_B replacing y_A and y_B replacing x_A.

θ (deg)	θ (rad)	$\sin(\theta)$	$\cos(\theta)$	$\tan(\theta)$	$\csc(\theta)$	$\sec(\theta)$	$\cot(\theta)$
30	$\tfrac{1}{6}\pi$	$\tfrac{1}{2}$	$\tfrac{1}{2}\sqrt{3}$	$1/\sqrt{3}$	2	$2/\sqrt{3}$	$\sqrt{3}$

Example 5.4   OTHER QUADRANTS

Knowledge of the trigonometric functions for an angle that lies in the first quadrant allows one to determine the trigonometric functions of associated angles in the other quadrants. The associated angles, along with their coordinates corresponding to the unit circle, are illustrated in Figure 5.6.

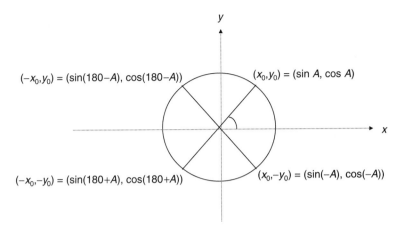

Figure 5.6   Trigonometric functions generated from first quadrant.

As an example, the associated angles with $\pi/6$ rad ($30°$) and their corresponding coordinates are presented in the table below.

θ	$\pi/6$	$5\pi/6$	$7\pi/6$	$11\pi/6$
(x, y)	$\left(\frac{1}{2}\sqrt{3}, \frac{1}{2}\right)$	$\left(-\frac{1}{2}, \frac{1}{2}\sqrt{3}\right)$	$\left(-\frac{1}{2}\sqrt{3}, -\frac{1}{2}\right)$	$\left(\frac{1}{2}, -\frac{1}{2}\sqrt{3}\right)$

In a similar fashion, the trigonometric functions for the angles associated with $\pi/4$ and $\pi/3$ can be determined. The known trigonometric functions are presented in the following table.

θ (deg)	θ (rad)	$\sin(\theta)$	$\cos(\theta)$	$\tan(\theta)$	$\csc(\theta)$	$\sec(\theta)$	$\cot(\theta)$
0	0	0	1	0	Undefined	1	Undefined
30	$\frac{\pi}{6}$	$\frac{1}{2}$	$\frac{\sqrt{3}}{2}$	$\frac{1}{\sqrt{3}}$	2	$\frac{2}{\sqrt{3}}$	$\sqrt{3}$
45	$\frac{\pi}{4}$	$\sqrt{\frac{1}{2}}$	$\sqrt{\frac{1}{2}}$	1	$\sqrt{2}$	$\sqrt{2}$	1
60	$\frac{\pi}{3}$	$\frac{\sqrt{3}}{2}$	$\frac{1}{2}$	$\sqrt{3}$	$\frac{2}{\sqrt{3}}$	2	$\frac{1}{\sqrt{3}}$
90	$\frac{\pi}{2}$	1	0	Undefined	1	Undefined	0
120	$\frac{2\pi}{3}$	$\frac{\sqrt{3}}{2}$	$-\frac{1}{2}$	$-\sqrt{3}$	$\frac{2}{\sqrt{3}}$	-2	$-\frac{1}{\sqrt{3}}$
135	$\frac{3\pi}{4}$	$\sqrt{\frac{1}{2}}$	$-\sqrt{\frac{1}{2}}$	-1	$\sqrt{2}$	$-\sqrt{2}$	-1

(*continued*)

Table (*Continued*)

θ (deg)	θ (rad)	$\sin(\theta)$	$\cos(\theta)$	$\tan(\theta)$	$\csc(\theta)$	$\sec(\theta)$	$\cot(\theta)$
150	$\dfrac{5\pi}{6}$	$\dfrac{1}{2}$	$-\dfrac{\sqrt{3}}{2}$	$-\dfrac{1}{\sqrt{3}}$	2	$-\dfrac{2}{\sqrt{3}}$	$-\sqrt{3}$
180	π	0	-1	0	Undefined	-1	Undefined
210	$\dfrac{7\pi}{6}$	$-\dfrac{1}{2}$	$-\dfrac{\sqrt{3}}{2}$	$\dfrac{1}{\sqrt{3}}$	-2	$-\dfrac{2}{\sqrt{3}}$	$\sqrt{3}$
225	$\dfrac{5\pi}{4}$	$-\sqrt{\dfrac{1}{2}}$	$-\sqrt{\dfrac{1}{2}}$	1	$-\sqrt{2}$	$-\sqrt{2}$	1
240	$\dfrac{4\pi}{3}$	$-\dfrac{\sqrt{3}}{2}$	$-\dfrac{1}{2}$	$\sqrt{3}$	$-\dfrac{2}{\sqrt{3}}$	-2	$\dfrac{1}{\sqrt{3}}$
270	$\dfrac{3\pi}{2}$	0	-1	0	Undefined	-1	Undefined
300	$\dfrac{5\pi}{3}$	$-\dfrac{\sqrt{3}}{2}$	$\dfrac{1}{2}$	$-\sqrt{3}$	$-\dfrac{2}{\sqrt{3}}$	2	$-\dfrac{1}{\sqrt{3}}$
315	$\dfrac{7\pi}{4}$	$-\sqrt{\dfrac{1}{2}}$	$\sqrt{\dfrac{1}{2}}$	-1	$-\sqrt{2}$	$\sqrt{2}$	-1
330	$\dfrac{11\pi}{6}$	$-\dfrac{1}{2}$	$\dfrac{\sqrt{3}}{2}$	$-\dfrac{1}{\sqrt{3}}$	-2	$\dfrac{2}{\sqrt{3}}$	$-\sqrt{3}$
360	2π	0	1	0	Undefined	1	Undefined

Example 5.5 THE DIRECTION OF A VECTOR

Determine the angle associated with the vector $\vec{v} = \begin{pmatrix} 1 \\ -\sqrt{3} \end{pmatrix}$.

Solution The vector lies in the fourth quadrant (Figure 5.7). The tangent of the angle by the ratio $v_2/v_1 = -\sqrt{3}$. From the table of the previous example, the angle having a tangent of $-\sqrt{3}$ is $5\pi/3$ (300°).

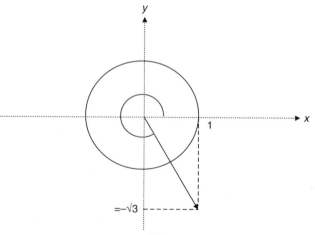

Figure 5.7 The angle satisfying $\tan A = -\sqrt{3}$.

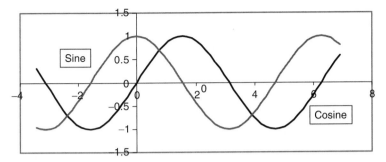

Figure 5.8   Graphs of sine and cosine functions.

5.2 GRAPHS OF THE SINE, COSINE, AND TANGENT FUNCTIONS

Plotting the points of the table from Example 5.4 on a Cartesian plane and interpolating between points result in a rough sketch of the graphs of the sine, cosine, and tangent functions. The results are given in Figure 5.8. Of note is that the sine and cosine curves repeat a pattern every 2π radians. The pattern is the result of executing one cycle about the circle in 2π radians. The pattern is repeated indefinitely in both the positive and negative directions as the circle is indefinitely traversed counterclockwise or clockwise.

5.3 ROTATIONS

Rotations are linear transformations as illustrated in Figure 5.9. The vector $\vec{A} = a\vec{V} + b\vec{W}$ is rotated through an angle θ by a transformation $\mathbf{R}_\theta$. The result is the same as the sum of rotations on the components of $\vec{A}$, $\mathbf{R}_\theta \vec{A} = a\mathbf{R}_\theta \vec{V} + b\mathbf{R}_\theta \vec{W}$. Accordingly, the rotation satisfies the linearity property given in equation (4.26).

The rotation can be represented by a matrix M_θ. Recall the first column of M_θ is obtained by applying the rotation to the vector $(1\ 0)^T$, while the second column of the matrix M_θ is obtained by applying the rotation to the vector $(0\ 1)^T$. Figure 5.10 illustrates the rotation on these two basis vectors with the following result:

$$\mathbf{R}_\theta \begin{pmatrix} 1 \\ 0 \end{pmatrix} = \begin{pmatrix} \cos(\theta) \\ \sin(\theta) \end{pmatrix} \qquad \mathbf{R}_\theta \begin{pmatrix} 0 \\ 1 \end{pmatrix} = \begin{pmatrix} -\sin(\theta) \\ \cos(\theta) \end{pmatrix}$$

The matrix M_θ is then

$$M_\theta = \begin{pmatrix} \cos(\theta) & -\sin(\theta) \\ \sin(\theta) & \cos(\theta) \end{pmatrix}$$

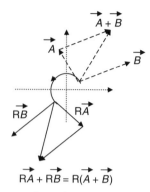

Figure 5.9 Rotations as linear transformations.

Example 5.6 ROTATING A VECTOR

Rotate the vector $\vec{V} = \binom{2\sqrt{3}}{2}$ through an angle of $\frac{5}{6}\pi$ radians (150°).

Solution Using the trigonometric table from Example 5.3, the entries in the rotation matrix M_θ may be specified:

$$M_\theta = \begin{pmatrix} \cos\left(\frac{5}{6}\pi\right) & -\sin\left(\frac{5}{6}\pi\right) \\ \sin\left(\frac{5}{6}\pi\right) & \cos\left(\frac{5}{6}\pi\right) \end{pmatrix}$$

$$= \begin{pmatrix} -\frac{1}{2}\sqrt{3} & -\frac{1}{2} \\ \frac{1}{2} & -\frac{1}{2}\sqrt{3} \end{pmatrix}$$

Applying M_θ to the vector $\vec{V}$ results in the solution

$$M_\theta\vec{V} = \begin{pmatrix} -\frac{1}{2}\sqrt{3} & -\frac{1}{2} \\ \frac{1}{2} & -\frac{1}{2}\sqrt{3} \end{pmatrix} \begin{pmatrix} 2\sqrt{3} \\ 2 \end{pmatrix}$$

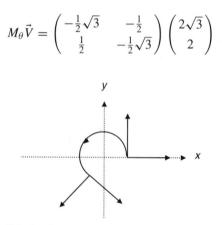

Figure 5.10 Rotation of the basis vectors.

$$= \begin{pmatrix} \left(-\frac{1}{2}\sqrt{3}\right)(2\sqrt{3}) + \left(\frac{1}{2}\right)2 \\ \left(\frac{1}{2}\right)(2\sqrt{3}) - \left(\frac{1}{2}\sqrt{3}\right)2 \end{pmatrix}$$

$$= \begin{pmatrix} -2 \\ 0 \end{pmatrix}$$

The solution is easily verified. The initial vector has length 2 and points in the direction of a $30°$ angle from the x axis. Rotating the vector through $150°$ maintains its length, but its angle is now $30° + 150° = 180°$. Accordingly, the vector of length 2 points along the x axis in the negative direction.

5.4   IDENTITIES

This section presents several trigonometric identities. The identities are later used to extend the trigonometric table.

5.4.1   Pythagorean Identity

Applying the Pythagorean theorem to any values of x and y on the unit circle results in the following equality:

$$x^2 + y^2 = 1$$

Replacing x and y by their trigonometric values yields the following identity:

$$\cos^2(\theta) + \sin^2(\theta) = 1$$

A notational convention has been used. More commonly $[\cos(\theta)]^2$ is written as $\cos^2(\theta)$. Similarly, $[\sin(\theta)]^2$ is more commonly written as $\sin^2(\theta)$.

5.4.2   Negative of an Angle

From the definitions of the trigonometric functions and Figure 5.6, it is readily seen that the following equalities hold:

$$\cos(-\theta) = \cos(\theta) \quad \sin(-\theta) = -\sin(\theta) \quad \tan(-\theta) = -\tan(\theta)$$

$$\sec(-\theta) = \sec(\theta) \quad \csc(-\theta) = -\csc(\theta) \quad \cot(-\theta) = -\cot(\theta)$$

Note the use of these equalities in Example 5.4 where the trigonometric functions were extended to other quadrants.

5.4.3 Tan(θ) in Terms of Sin(θ) and Cos(θ)

Also, from the definitions of the trigonometric functions, the following holds:

$$\tan(\theta) = \frac{y}{x} = \frac{\sin(\theta)}{\cos(\theta)}$$

5.4.4 Sines and Cosines of Sums of Angles

Let two angles α and β be given. As illustrated in Figure 5.11, let (x_α, y_α) be the point on the unit circle aligned with angle α and $(x_{\alpha+\beta}, y_{\alpha+\beta})$ be the point on the unit circle aligned with the angle $\alpha + \beta$. Note that the point $(x_{\alpha+\beta}, y_{\alpha+\beta})$ is a rotation of the point (x_α, y_α) by the angle β.

Using the rotation matrix M_β, the point $(x_{\alpha+\beta}, y_{\alpha+\beta})$ can be expressed in terms of the point (x_α, y_α):

$$\begin{pmatrix} x_{\alpha+\beta} \\ y_{\alpha+\beta} \end{pmatrix} = M_\beta \begin{pmatrix} x_\alpha \\ y_\alpha \end{pmatrix}$$

$$= \begin{pmatrix} \cos(\beta) & -\sin(\beta) \\ \sin(\beta) & \cos(\beta) \end{pmatrix} \begin{pmatrix} x_\alpha \\ y_\alpha \end{pmatrix}$$

$$= \begin{pmatrix} \cos(\beta)x_\alpha - \sin(\beta)y_\alpha \\ \sin(\beta)x_\alpha + \cos(\beta)y_\alpha \end{pmatrix}$$

$$= \begin{pmatrix} \cos(\alpha)\cos(\beta) - \sin(\alpha)\sin(\beta) \\ \cos(\alpha)\sin(\beta) + \sin(\alpha)\cos(\beta) \end{pmatrix}$$

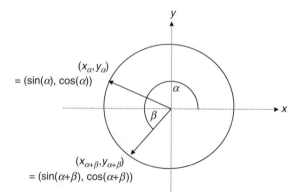

Figure 5.11   Trigonometric functions of the angle $\alpha + \beta$.

Substituting for $x_{\alpha+\beta}$ and $y_{\alpha+\beta}$ results in the following two summation identities:

$$\cos(\alpha + \beta) = \cos(\alpha)\cos(\beta) - \sin(\alpha)\sin(\beta)$$

$$\sin(\alpha + \beta) = \cos(\alpha)\sin(\beta) + \sin(\alpha)\cos(\beta)$$

5.4.5 Difference Formulas

Using the negative-angle formulas of Section 5.4.2 in the summation identities of Section 5.4.4 results in the difference formulas

$$\cos(\alpha - \beta) = \cos(\alpha)\cos(-\beta) - \sin(\alpha)\sin(-\beta) = \cos(\alpha)\cos(\beta) + \sin(\alpha)\sin(\beta)$$

$$\sin(\alpha - \beta) = \cos(\alpha)\sin(-\beta) + \sin(\alpha)\cos(-\beta) = -\cos(\alpha)\sin(\beta) + \sin(\alpha)\cos(\beta)$$

5.4.6 Double-Angle Formulas

Setting β equal to α in the summation identities of Section 5.4.4 results in the following double-angle formulas:

$$\cos(2\alpha) = \cos^2(\alpha) - \sin^2(\alpha)$$

$$\sin(2\alpha) = 2\sin(\alpha)\cos(\beta)$$

Note that applying the Pythagorean theorem to the cosine formula yields an alternative form for the cosine double-angle formula:

$$\cos(2\alpha) = \cos^2(\alpha) - \sin^2(\alpha)$$
$$= \cos^2(\alpha) - (1 - \cos^2(\alpha))$$
$$= 2\cos^2(\alpha) - 1$$

5.4.7 Half-Angle Formulas

Setting $2\alpha = \theta$ in the final cosine double-angle formula of Section 5.4.6 and solving for $\cos(\theta/2)$ result in the cosine half-angle formulas:

$$2\cos^2(\alpha) - 1 = \cos(2\alpha)$$
$$2\cos^2\left(\tfrac{1}{2}\theta\right) - 1 = \cos(\theta)$$
$$2\cos^2\left(\tfrac{1}{2}\theta\right) = 1 + \cos(\theta)$$
$$\cos^2\left(\tfrac{1}{2}\theta\right) = \left(\tfrac{1}{2}\right)[1 + \cos(\theta)]$$
$$\cos\left(\tfrac{1}{2}\theta\right) = \pm\sqrt{\left(\tfrac{1}{2}\right)[1 + \cos(\theta)]}$$

The sign associated with the square root is determined by the quadrant of the angle $\theta/2$.

The corresponding half-angle formula for the sine function is obtained with the use of the half-angle cosine identity and the Pythagorean identity.

$$\cos^2\left(\tfrac{1}{2}\theta\right) = \left(\tfrac{1}{2}\right)[1 + \cos(\theta)]$$

$$\sin^2\left(\tfrac{1}{2}\theta\right) + \cos^2\left(\tfrac{1}{2}\theta\right) = \sin^2\left(\tfrac{1}{2}\theta\right) + \tfrac{1}{2}[1 + \cos(\theta)]$$

$$1 = \sin^2\left(\tfrac{1}{2}\theta\right) + \tfrac{1}{2}[1 + \cos(\theta)]$$

$$\sin^2\left(\tfrac{1}{2}\theta\right) = 1 - \tfrac{1}{2}[1 + \cos(\theta)]$$

$$\sin^2\left(\tfrac{1}{2}\theta\right) = \tfrac{1}{2}[1 - \cos(\theta)]$$

$$\sin\left(\tfrac{1}{2}\theta\right) = \pm\sqrt{\tfrac{1}{2}[1 - \cos(\theta)]}$$

Once again, the sign associated with the root is determined by the quadrant of the angle $\theta/2$.

5.5 LUCKY 72

The angle $72°$ has a nice property that allows one to determine its trigonometric functions. Let (x_{36}, y_{36}), (x_{72}, y_{72}), and (x_{144}, y_{144}) all be points on the unit circle associated with the angles of $36°$, $72°$, and $144°$, respectively. Figure 5.12 illustrates relations between these points. These relations, along with identities established in Section 5.4, allow for the solutions of x_{72} and y_{72}.

The values for (x_{144}, y_{144}) are attained in terms of the values (x_{72}, y_{72}) in two ways. A formula for (x_{72}, y_{72}) arises by equating the two resulting expressions. For the first expression, begin by using the half-angle cosine formula from Section 5.4.7

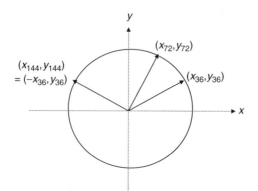

Figure 5.12   Relations between angles of $72°$, $36°$, and $144°$.

and noting the relation between x_{36} and x_{144}:

$$x_{144} = -x_{36} = -\sqrt{\tfrac{1}{2}(1 + x_{72})}$$

Observing that $y_{144} = y_{36}$ and applying the half-angle formulas result in the following vector equality, which is our first expression:

$$\begin{pmatrix} x_{144} \\ y_{144} \end{pmatrix} = \begin{pmatrix} -x_{36} \\ y_{36} \end{pmatrix} = \begin{pmatrix} -\sqrt{\tfrac{1}{2}(1 + x_{72})} \\ \sqrt{\tfrac{1}{2}(1 - x_{72})} \end{pmatrix}$$

Obtain the second expression by applying the double-angle formula from Section 5.4.6 to the angle 72°:

$$\begin{pmatrix} x_{144} \\ y_{144} \end{pmatrix} = \begin{pmatrix} 2x_{72}^2 - 1 \\ 2x_{72}y_{72} \end{pmatrix}$$

$$= \begin{pmatrix} 2x_{72}^2 - 1 \\ 2x_{72}\sqrt{1 - x_{72}^2} \end{pmatrix}$$

The Pythagorean formula was used to solve for y_{72} in the final expression. Equating the above two expressions for $\begin{pmatrix} x_{144} \\ y_{144} \end{pmatrix}$ results in two equations for x_{72}, one for each component of the vector:

$$\begin{pmatrix} -\sqrt{\tfrac{1}{2}(1 + x_{72})} \\ \sqrt{\tfrac{1}{2}(1 - x_{72})} \end{pmatrix} = \begin{pmatrix} 2x_{72}^2 - 1 \\ 2x_{72}\sqrt{1 - x_{72}^2} \end{pmatrix} \tag{5.1}$$

From this point on, it is an algebraic exercise to determine solutions to equation (5.1). The solution route given below takes us through two cubic polynomials. Subtraction of one cubic from the other yields a quadratic polynomial for which a solution is available.

We arrive at one of the cubic polynomials by squaring both sides of the second component of equation (5.1) and simplifying:

$$\sqrt{\tfrac{1}{2}(1 - x_{72})} = 2x_{72}\sqrt{1 - x_{72}^2}$$

$$\tfrac{1}{2}(1 - x_{72}) = 4x_{72}^2\left(1 - x_{72}^2\right)$$

$$\tfrac{1}{2}(1 - x_{72}) = 4x_{72}^2(1 - x_{72})(1 + x_{72}) \tag{5.2}$$

$$\tfrac{1}{2} = 4x_{72}^2(1 + x_{72})$$

$$4x_{72}^3 + 4x_{72}^2 - \tfrac{1}{2} = 0$$

We arrive at the other cubic polynomial by multiplying both components of equation (5.1) with one another and simplifying:

$$-\sqrt{\tfrac{1}{2}(1+x_{72})}\sqrt{\tfrac{1}{2}(1-x_{72})} = \left(2x_{72}^2 - 1\right)\left(2x_{72}\sqrt{1-x_{72}^2}\right)$$

$$-\sqrt{\tfrac{1}{4}(1+x_{72})(1-x_{72})} = \left(2x_{72}^2 - 1\right)\left(2x_{72}\sqrt{1-x_{72}^2}\right)$$

$$-\tfrac{1}{2}\sqrt{1-x_{72}^2} = \left(2x_{72}^2 - 1\right)\left(2x_{72}\sqrt{1-x_{72}^2}\right) \qquad (5.3)$$

$$-\tfrac{1}{2} = \left(2x_{72}^2 - 1\right)2x_{72}$$

$$4x_{72}^3 - 2x_{72} + \tfrac{1}{2} = 0$$

The value x_{72} satisfies both of the above cubic equations, equations (5.2) and (5.3); it must also satisfy their difference. Subtracting the two equations and solving the resulting quadratic polynomial yield the following solution for $x_{72} = \cos(72°)$:

$$\left(4x_{72}^3 + 4x_{72}^2 - \tfrac{1}{2}\right) - \left(4x_{72}^3 - 2x_{72} + \tfrac{1}{2}\right) = 0$$

$$4x_{72}^2 + 2x_{72} - 1 = 0$$

$$x_{72}^2 + \tfrac{1}{2}x_{72} - \tfrac{1}{4} = 0$$

$$x_{72} = \tfrac{1}{2}\left(-\tfrac{1}{2} + \sqrt{\tfrac{1}{4}+1}\right)$$

$$= \tfrac{1}{2}\left(-\tfrac{1}{2} + \sqrt{\tfrac{5}{4}}\right)$$

$$= \tfrac{1}{4}\left(-1 + \sqrt{5}\right)$$

Note that the value $x_{72} = \cos(72°) = \tfrac{1}{4}\left(-1 + \sqrt{5}\right)$ is positive since the angle 72° lies in the first quadrant. Accordingly, the positive sign is assigned to the square root.

Using the Pythagorean theorem and simplifying, one finds the sine of 72°:

$$\sin(72°) = \sqrt{1 - \cos^2(72°)}$$

$$= \sqrt{1 - \tfrac{1}{16}\left(-1 + \sqrt{5}\right)^2}$$

$$= \sqrt{\tfrac{1}{8}\left(5 + \sqrt{5}\right)}$$

Remarks

- The negative square root with $x = -\frac{1}{4}\left(1 + \sqrt{5}\right)$ gives the cosine for the angle $216°$. A drawing analogous to Figure 5.12 can be drawn using the corresponding point to $216°$; the half-angle is $108°$ and the double angle is $432°$ shares a common point on the circle with $72°$. As these three points share the same double- and half-angle relations depicted in Figure 5.12, they share the same equations (5.2) and (5.3). Hence, the equations yield a solution for the angle $216°$ using different signs for the square root.

- The value x_{72} could have been solved from a single cubic equation. Doing so would have required finding the roots of the cubic equation, which is a tedious process. As such, we introduced the second cubic equation.

5.6 PTOLEMY AND ARISTARCHUS

5.6.1 Construction of Ptolemy's Table

Ptolemy's table of chords gives the lengths of chords with end points at increments of half a degree. It is Ptolemy's work that cements the standard of angle measurement through degrees, with the circle a full $360°$. In this regard, Ptolemy is influenced by the Mesopotamians, who had assembled an impressive body of mathematical knowledge centuries before the Greeks.

The Mesopotamians used a base-60 number system complete with sexagesimal representations of numbers. (In a sexagesimal system, the symbol 15.2 has the base 10, meaning $65\frac{2}{60}$.) One can only conjecture why it was that the Mesopotamians used a base-60 system, but the coincidences with the calendar and geometric properties of the circle afford a good guess. The Mesopotamians had a good estimate for the length of the year in days. Rounding the estimate to 360 yields a nice number for the number of degrees in a circle. The number 360 is divisible by 2, 3, 4, 5, 6, 8, 9, 10, 12, and more, so proportioning the circle into halves, thirds, fourths, and so on, is easily expressed in whole degrees. For example, one-twelfth of the circle is $30°$.

A property of the circle that we have used to inscribe a hexagon is that the radius of the hexagon is the same as the lengths of its sides. This is an elegant feature unique to the hexagon that the Mesopotamians were aware of. Using the Mesopotamian's system of measuring degrees in angles, the arc between the sides of the hexagon spans $60°$, and it is one-sixth of the circle. The Mesopotamians certainly noted this and it could well be the basis for their decision to use a base-60 system.

Because of the Mesopotamian base-60 system, each degree is further divided into 60 units called minutes and each minute is divided into 60 units called seconds. Aside from measurements of angles, this system became the basis for time measurement. Hence, we have 60 seconds in a minute and 60 minutes in an hour.

The system begs an answer to the question, were the Mesopotamians the original Copernicans? While Aristarchus goes on record as a heliocentrist, the choices of the

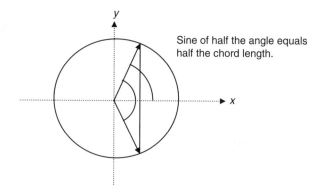

Figure 5.13   Ptolemy's chord length and the sine function.

Mesopotamians indicate that perhaps someone among them held a heliocentric view of the universe. With their system of measuring angles, which is reflected in their number system, the earth would travel roughly through 1° of its circular journey around the sun everyday. That the Mesopotamians chose an angular measurement of the circle, 360°, which is divisible by 12, has other implications. There are roughly 12 lunar cycles in a year corresponding to a 12-month calendar. Each month has roughly 30 days, (360/12), so the earth travels through roughly 30° of its orbit each month.

Let us leave conjecture to others and discuss what has been established. It is known that the Mesopotamian system for measuring angles was common in Ptolemy's time. Ptolemy used this system, and due to the popularity of Ptolemy's works, the system became an international standard. Ptolemy's table presents chord lengths corresponding to angles from half a degree to 180° in half-degree increments. The relation between the chord lengths and the sine of the angle is illustrated in Figure 5.13; $\sin(\theta/2)$ equals half the chord length. From this relation, it is seen that Ptolemy's table is equivalent to a sine table from 0° to 90° in quarter-degree increments. It is this latter table that we develop.

The starting point for our table is the trigonometric table from Section 5.1. This includes the trigonometric functions for the angles in all quadrants. Henceforth, we determine values only in the first quadrant. The table can be extended into the other quadrants using the same techniques as illustrated in Example 5.4.

The results of the previous section allow us to enter the angle 72° into the table. Using the half-angle formulas from Section 5.4.7, one can also include the angles 36°, 18°, 9°, and 15° (half of 30°, which is already in the table). In addition, by applying the difference formulas from Section 5.4.5 to the angles 72° and 60°, the angle 12° may be included in the table. A further application of the half-angle formula allows one to enter in succession the angles 6°, 3°, $1\frac{1}{2}^{\circ}$, and $\frac{3}{4}^{\circ}$.

By repeated use of the sum formula from Section 5.4.4, one can fill out the table for all remaining angles that are multiples of $\frac{3}{4}^{\circ}$ (that is, $30\frac{3}{4} = 30 + \frac{3}{4}$,

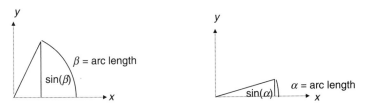

Figure 5.14   The inequality $\sin(\beta)/\beta < \sin(\alpha)/\alpha$.

$31\frac{1}{2} = 30\frac{3}{4} + \frac{3}{4}$). After accomplishing this, the equivalent of one-third of Ptolemy's table is filled; Ptolemy's table presents the sine of angles for multiples of $\frac{1}{4}^\circ$.

To fill out the table in the manner of Ptolemy, another application of the half-angle formula to the angle $\frac{3}{4}^\circ$ determines the sine of $\frac{3}{8}^\circ$. Then apply an interpolation method based upon the following inequality, which was established prior to Aristarchus. The inequality is expressed in terms of radians, so the angles are given in their radian measurements:

$$\frac{\sin(\beta)}{\beta} < \frac{\sin(\alpha)}{\alpha} < 1 \quad \text{whenever } 0 < \alpha < \beta < \frac{\pi}{2} \tag{5.4}$$

Because the angles are expressed in radians, α and β refer to arc lengths. Figure 5.14 illustrates the inequality. Geometrically, it states that the ratio of a point's y coordinate to the arc length between the x axis and the point decreases as the angle increases. So as one traverses the circle from angle α to angle β, the ratio decreases and the inequality holds. Visually, this seems obvious and for now we accept the inequality. In Chapter 6, we will apply calculus to prove this statement.

Applying the ratio $180/\pi$ to each side of the inequality converts the radian measurement to degree measurement. And so the inequality remains valid if degree rather than radian measurements are used.

The inequality is applied twice. First, let $\alpha = \frac{1}{2}$ and $\beta = \frac{3}{4}$:

$$\frac{\sin(\beta)}{\beta} < \frac{\sin(\alpha)}{\alpha}$$

$$\frac{\alpha \sin(\beta)}{\beta} < \sin(\alpha)$$

$$\frac{\frac{1}{2}\sin\left(\frac{3}{4}^\circ\right)}{\frac{3}{4}} < \sin\left(\frac{1}{2}^\circ\right)$$

$$\frac{2}{3}\sin\left(\frac{3}{4}^\circ\right) < \sin\left(\frac{1}{2}^\circ\right)$$

Next, let $\alpha = \frac{3}{8}$ and $\beta = \frac{1}{2}$:

$$\frac{\sin(\beta)}{\beta} < \frac{\sin(\alpha)}{\alpha}$$

$$\sin(\beta) < \frac{\beta \sin(\alpha)}{\alpha}$$

$$\sin\left(\tfrac{1}{2}^\circ\right) < \frac{\tfrac{1}{2}\sin\left(\tfrac{3}{8}^\circ\right)}{\tfrac{3}{8}}$$

$$\sin\left(\tfrac{1}{2}^\circ\right) < \tfrac{4}{3}\sin\left(\tfrac{3}{8}^\circ\right)$$

Putting the two inequalities together results in the following bounds for $\sin(\tfrac{1}{2}^\circ)$:

$$\tfrac{2}{3}\sin\left(\tfrac{3}{4}^\circ\right) < \sin\left(\tfrac{1}{2}^\circ\right) < \tfrac{4}{3}\sin\left(\tfrac{3}{8}^\circ\right)$$

We will cheat and use a calculator to obtain the left-hand and right-hand bounds. For Ptolemy, it was a long, arduous calculation involving many iterates of complicated arithmetic operations, in particular many square roots, that led him to the bounds for $\sin(\tfrac{1}{2}^\circ)$:

$$\tfrac{2}{3}\sin\left(\tfrac{3}{4}^\circ\right) = 0.00872639 < \sin\left(\tfrac{1}{2}^\circ\right) < \tfrac{4}{3}\sin\left(\tfrac{3}{8}^\circ\right) = 0.00872658$$

From the above bounds, the value of $\sin(\tfrac{1}{2}^\circ)$ out to six decimal places is $\sin(\tfrac{1}{2}^\circ) = 0.008726$:

Once the trigonometric functions for the angle $\tfrac{1}{2}^\circ$ is established, using the summation formulas from Section 5.4.4, all remaining multiples of $\tfrac{1}{2}^\circ$ can be filled. Upon completing this, two-thirds of the table entries have been made. The only values that remain are multiples of $\tfrac{1}{4}^\circ$ that have not yet been filled in. Ptolemy filled in these values by interpolating between their nearest values. For example, the value at $34\tfrac{1}{4}^\circ$ remains unfilled. Take $\sin(34\tfrac{1}{4}^\circ)$ to be the following value:

$$\sin\left(34\tfrac{1}{4}^\circ\right) = \tfrac{1}{2}\left[\sin\left(34^\circ\right) + \sin\left(34\tfrac{1}{2}^\circ\right)\right]$$

Remarks

- Archimedes knew all of the trigonometric identities necessary to establish Ptolemy's table nearly three centuries before Ptolemy, but there is no evidence that Archimedes set upon this task.

- Ptolemy could have used the difference relations to determine the chord lengths for $\frac{1}{2}^{\circ}$ (trigonometric functions for $\frac{1}{4}^{\circ}$). An interpolation provides the degree of accuracy that Ptolemy deemed acceptable.

5.6.2 Remake of Aristarchus

This section presents a modern rendition of Aristarchus' calculation for the relative size of the sun and the earth. A substantial degree of license is taken in this remake. For a presentation that more accurately reflects Aristarchus' original work, consult Heath's excellent book *Aristarchus of Samos: The Ancient Copernicus* (1920). What is striking about the method is the coexistence of that which Aristarchus has full control over with that which he has no control over. He had control over the analytic thought process and it is elegant. However, the process involves two measurements one taken during a lunar eclipse and another taken at half moon. While a half moon occurs every month, a lunar eclipse is far less frequent. Once the process was determined, all Aristarchus could do was sit around, wait for an eclipse, and then use poor instruments to make a rough measurement.

Aristarchus' method uses the similar triangles that are illustrated in Figure 5.15. The figure depicts the configuration of the sun, earth, and moon during an eclipse (not to scale). During the eclipse, the earth casts a shadow that extends to the point P in the figure. This shadow is in the shape of a cone with the circumference of a circle of the earth acting as the base. The illustration depicts three similar triangles, one with the radius of the sun as base, one with the radius of the earth as base, and one with a base that extends from the center of the moon to the shadow's boundary. Denote these lengths by R_S, R_E, and R, respectively. Also, denote the distance from the center of the sun to the point P by D_P, the distance from the earth to the sun by D_S, and the distance from the earth to the moon by D_M.

Ratios of similar sides of the two triangles (R_s to P and R_E to P) result in the following equality:

$$\frac{R_E}{R_S} = \frac{D_P - D_S}{D_P} = 1 - \frac{D_S}{D_P} \tag{5.5}$$

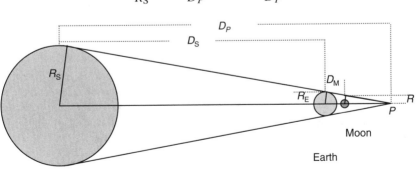

Figure 5.15   Similar triangles used by Aristarchus.

Similarly, ratios using triangles with the moon's entry and exit points yield the following equality:

$$\frac{R}{R_S} = \frac{D_P - D_S - D_M + \varepsilon}{D_P}$$

$$= \frac{R_E}{R_S} - \frac{D_M}{D_P} \tag{5.6}$$

Solving for D_P in equation (5.6), placing the result into equation (5.5), and simplifying yield the following equations:

$$\frac{R}{R_S} = \frac{R_E}{R_S} - \frac{D_M - \varepsilon}{D_P}$$

$$\frac{D_M - \varepsilon}{D_P} = \frac{R_E - R}{R_S}$$

$$D_P = \frac{(D_M - \varepsilon) R_S}{R_E - R}$$

$$\frac{R_E}{R_S} = 1 - \frac{D_S}{D_P}$$

$$\frac{R_E}{R_S} = 1 - \frac{D_S (R_E - R)}{(D_M - \varepsilon) R_S}$$

$$\left(1 + \frac{D_S}{(D_M - \varepsilon)}\right) \frac{R_E}{R_S} = 1 + \frac{D_S R}{(D_M - \varepsilon) R} \tag{5.7}$$

$$\frac{R_E}{R_S} = \frac{1 + D_S R/[(D_M - \varepsilon) R_S]}{1 + D_S/(D_M - \varepsilon)}$$

Aristarchus expresses the quantity R in terms of the moon's radius, R_M, so that $R = \lambda R_M$. It is the quantity λ that Aristarchus must determine during an eclipse. With the substitution $R = \lambda R_M$, the equation for the ratio R_E/R_S becomes the following:

$$\frac{R_E}{R_S} = \frac{1 + D_S \lambda R_M/[(D_M - \varepsilon) R_S]}{1 + D_S/(D_M - \varepsilon)} \tag{5.8}$$

Aristarchus again uses similar triangles to express the ratio R_M/R_S in terms of D_M and D_S. Aristarchus notes that to the earthbound observer the moon and the sun occupy equal areas of the sky. Figure 5.16 illustrates the geometry, and once again using similar triangles, it is seen that $R_M/R_S = D_M/D_S$. Placing this equality into the final expression of equation (5.8) results in an expression for R_E/R_S:

$$\frac{R_E}{R_S} = \frac{1 + \lambda D_M/(D_M - \varepsilon)}{1 + D_S/(D_M - \varepsilon)}$$

$$= \frac{(1 + \lambda) D_M - \varepsilon}{D_M + D_S - \varepsilon}$$

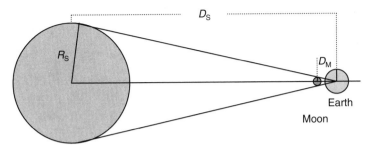

Figure 5.16   Similar triangles.

$$= \frac{D_{\mathrm{M}} + D_{\mathrm{S}} - \varepsilon + \lambda D_{\mathrm{M}} - D_{\mathrm{S}}}{D_{\mathrm{M}} + D_{\mathrm{S}} - \varepsilon} \tag{5.9}$$

$$= 1 + \frac{\lambda D_{\mathrm{M}} - D_{\mathrm{S}}}{D_{\mathrm{M}} + D_{\mathrm{S}} - \varepsilon}$$

$$= 1 + \frac{\lambda - D_{\mathrm{S}}/D_{\mathrm{M}}}{1 + (D_{\mathrm{S}} - \varepsilon)/D_{\mathrm{M}}}$$

Aristarchus has reduced the problem of determining the ratio between the earth and the sun to three measurements, the ratio $D_{\mathrm{S}}/D_{\mathrm{M}}$, the value ε, and the value λ. To determine the ratio $D_{\mathrm{S}}/D_{\mathrm{M}}$, Aristarchus turns to trigonometry. He notes that when an earthbound observer sees a half moon, the earth, sun, and moon form a right triangle with the moon at the apex of the right angle (Figure 5.17). The quantity $D_{\mathrm{S}}/D_{\mathrm{M}}$ is the cosecant of the angle α, $D_{\mathrm{S}}/D_{\mathrm{M}} = 1/\sin(\alpha)$.

Aristarchus inaccurately assigns the angle α the value $\alpha = 3°$. If he were able to consult with Ptolemy or perhaps Archimedes, he could find $\sin(3°)$ quite accurately. But both Archimedes and Ptolemy would follow Aristarchus. Instead, Aristarchus had the inequality [equation (5.4)] available that he used to demonstrate that $\frac{1}{20} < \sin(3°) < \frac{1}{18}$. Finally, using an argument that we do not reproduce, Aristarchus demonstrates that $0 < \varepsilon < D_{\mathrm{M}}/675$. With his bounds on the value of

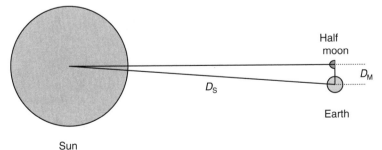

Figure 5.17   Right triangle during half moon.

$\sin(3°)$, Aristarchus could place bounds on the ratio R_E/R_S. We first treat the upper bound. Noting that $D_S/D_M = 1/\sin(\alpha) > 18$, equation (5.9) yields the following upper bound:

$$\frac{R_E}{R_S} = 1 + \frac{\lambda - D_S/D_M}{1 + (D_S - \varepsilon)/D_M}$$

$$< 1 + \frac{\lambda - D_S/D_M}{1 + D_S/D_M}$$

$$< \frac{1 + \lambda}{1 + D_S/D_M}$$

$$< \frac{1 + \lambda}{19}$$

For the lower bound, use $D_S/D_M = 1/\sin(\alpha) < 20$ and $\varepsilon < D_M/675$ in equation (5.9):

$$\frac{R_E}{R_S} = 1 + \frac{\lambda - D_S/D_M}{1 + (D_S - \varepsilon)/D_M}$$

$$> 1 + \frac{\lambda - D_S/D_M}{1 + D_S/D_M - \frac{1}{675}}$$

$$> 1 + \frac{\lambda - 20}{21 - \frac{1}{675}}$$

Placing the two bounds together results in the one-way inequalities,

$$1 + \frac{\lambda - 20}{21 - \frac{1}{675}} < \frac{R_E}{R_S} < \frac{1 + \lambda}{19} \tag{5.10}$$

The only thing left is for Aristarchus to await a lunar eclipse so that he can measure λ using the method illustrated in Figure 5.15. The quantity λ is the ratio between the value R and the radius of the moon. In his work *On the Sizes and Distances of the Sun and Moon*, Aristarchus demonstrates $\frac{88}{45} < \lambda < 2$. Appropriately placing the bounds of λ into the inequalities (5.10) yields the result:

$$1 + \frac{\frac{88}{45} - 20}{21 - \frac{1}{675}} < \frac{R_E}{R_S} < \frac{1 + 2}{19}$$

$$\frac{8973}{63,783} < \frac{R_E}{R_S} < \frac{3}{19}$$

Aristarchus loosens his bounds to find a simpler expression. After demonstrating that $\frac{6}{43} < \frac{8973}{63,783}$, Aristarchus presents his final result:

$$\frac{6}{43} < \frac{R_E}{R_S} < \frac{3}{19}$$

Remarks

- Aristarchus' method and final result give the appearance that he is more interested in the mathematical analysis than in the astronomy. He goes through a significant effort to account for the circular motion of the moon as it passes through the earth's shadow. The mathematical rigor required to accomplish this is not trivial. In doing so, Aristarchus develops clever estimation techniques of interest to mathematics. However, the consequences on the result are immaterial. It would have been far simpler to assume that the motion is a straight line; because of the very small distance traveled, the assumption would have had very little impact on the result. In the analysis above, this means that $\varepsilon = 0$.

- Remarkably, while Aristarchus took great care to include the somewhat insignificant parameter ε, he did not take such care with the most critical parameter, α. While Aristarchus sets the angle at $3°$, the angle α is closer to 10 min $\left(\frac{1}{6}^{\circ}\right)$. As a result, his approximation for $D_S/D_M = 1/\sin(\alpha)$ is ridiculously far off. Rather than $18 < D_S/D_M < 20$ as Aristarchus claims, in reality, the ratio D_S/D_M is closer to 400. It is rather surprising that Aristarchus is not more careful. After all, the observation is available to him on a monthly basis. He could have taken several measurements while awaiting the event of an eclipse. It is also notable that in a review of Aristarchus' work Archimedes sets the angle α at a more plausible $\frac{1}{2}^{\circ}$.

5.7 DRAWING A PENTAGON

One of the more artful accomplishments of Greek geometers was their ability to draw pentagons using only a compass and a straight edge. And the pentagons were drawn with perfection. One wonders how the construction was ever developed prior to the advent of trigonometry. Perhaps after many trials it was discovered by chance or perhaps brilliant intuition guided a talented individual, whose name is unknown, to its construction. In this section, the results of the cosine of $72°$ are used to construct the pentagon.

First use a compass to draw a circle and then add the x and y axes. Consider the point $(1, 0)$ to be the first vertex of the pentagon, V_1, and then $\cos(72°)$ is the x coordinate of both neighboring vertices.

The result of Section 5.5 is $\cos(72°) = \frac{1}{4}\left(-1 + \sqrt{5}\right)$. It is easy to locate the point $\left(-\frac{1}{4}, 0\right)$ along the x coordinate; simply bifurcate the x axis between $(-1, 0)$ and the origin two times (Figure 5.18). To this point, it is necessary to add the value $\frac{1}{4}\sqrt{5}$ along the x coordinate and locate the point $\left(\frac{-1+\sqrt{5}}{4}, 0\right)$. This is easily accomplished using the triangles depicted in Figure 5.19. The triangle on the left (Figure 5.19a) has base 1 and height 2. Using the Pythagorean theorem, the hypotenuse has length $\sqrt{5}$. The inscribed triangle is similar with the lengths of the base, height, and hypotenuse, respectively, $\frac{1}{4}$, $\frac{1}{2}$, and $\frac{1}{4}\sqrt{5}$.

The triangle has vertices $\left(-\frac{1}{4}, 0\right)$, $\left(0, \frac{1}{2}\right)$, and the origin (Figure 5.19b) is congruent with the triangle of Figure 5.19a, so the distance between the points $\left(-\frac{1}{4}, 0\right)$ and $\left(0, \frac{1}{2}\right)$ is $\frac{1}{4}\sqrt{5}$. Using a compass, set the point $\left(-\frac{1}{4}, 0\right)$ as the

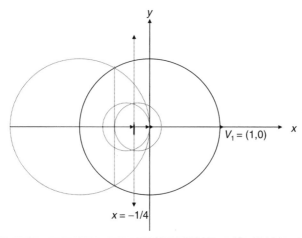

Figure 5.18   Locating $x = -1/4$.

center of the circle and adjust the radius so that the point $\left(0, \frac{1}{2}\right)$ is at the circle. Next arc the compass through the x axis to determine the point $\left[\frac{1}{4}(-1 + \sqrt{5}), 0\right]$, as depicted in Figure 5.19b.

Next determine a line perpendicular to the x axis at the point $\left(\frac{1}{4}(-1 + \sqrt{5}), 0\right)$ and mark off the intersections of this line with the original circle. These intersections are the neighboring vertices to the first vertex at $(1, 0)$, V_2 and V_5 (Figure 5.19b).

Next center the compass at the vertex V_2 in the first quadrant and set the radius as the distance to the original vertex V_1. While maintaining the center and radius, determine where the arc intersects the original circle. This intersection gives the fourth vertex V_3. Perform the same operation on the vertex in the third quadrant to find the final vertex V_4 (Figure 5.20).

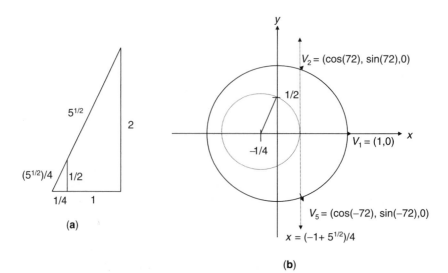

Figure 5.19   Locating V_2 and V_5.

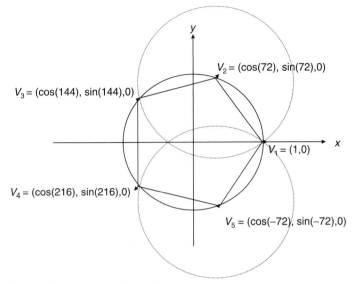

Figure 5.20   Locating the remaining vertices.

Finally, using a straight edge, draw lines between neighboring vertices to obtain a pentagon.

Remarks

- The initial step in the construction is to set up the axes. Then one draws a line perpendicular to the x axis at the point $\left(\frac{1}{4}(-1 + \sqrt{5}), 0\right)$. It also requires the bisection of a line segment. After some experimentation, the reader can accomplish these operations using only a straight edge and a compass.

- While the method given above does display the relation between $\cos(72°)$ and the construction clearly, it is not the only construction. There are more efficient constructions that complete the figure using fewer steps.

- It is not possible to construct every polygon using a compass and a straight edge. The outstanding mathematician Carl Gauss (1777–1855) discovered a construction for a 17-sided polygon. He then developed conditions that if satisfied guarantee that the construction of a particular polynomial is feasible. Pierre Wantzel (1814–1848) then proved that polygons not satisfying Gauss' sufficiency conditions are not constructable using only a compass and a straight edge.

5.8 POLAR COORDINATES

In Chapter 4, the Cartesian coordinate system is introduced. The system is very useful for identifying points in space and indicating functional relations, but there are other coordinate systems used for the same purpose. In this section, polar coordinates are

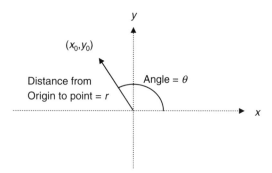

Figure 5.21   Polar coordinates.

introduced on a plane. Polar coordinates identify a point by its distance from the origin and the angle that is made between the point, the origin, and the x axis. Often, because of the geometry of a specific problem, polar coordinates are a more natural candidate for expressing relations. In the case of the governing equations for planetary motion, the equations are most simply expressed in polar coordinates because the forcing function as determined by the force from the sun is identical for points equidistant from the sun. The objective of this section is to transform expressions between standard Cartesian coordinates and polar coordinates.

Polar coordinates are given by the pair (r, θ) in which r represents the distance from the origin and θ represents the angle as illustrated in Figure 5.21. Note that r is always zero or a positive value. The relationship between the polar coordinates of a point and the standard Cartesian coordinates is given by the following equations (see Figure 5.21):

$$r = \sqrt{x^2 + y^2} \qquad r\cos(\theta) = x \qquad r\sin(\theta) = y \qquad (5.11)$$

These expressions can be used to transform equations from Cartesian to polar coordinates as illustrated by the following examples.

Example 5.7

Write the equation for a circle centered at the origin in polar coordinates.

The equation in Cartesian coordinates is given by $x^2 + y^2 = c^2$, where c represents the radius of the circle. Using the above relations, $x^2 + y^2 = r^2$, which when substituted into the equation of the circle yields $r^2 = c^2$, or equivalently $r = c$. The geometry of the problem allows for a simpler expression in polar coordinates.

Example 5.8

Write the equation for an ellipse centered at the origin in polar coordinates.

The equation in Cartesian coordinates is given by $x^2/a^2 + y^2/b^2 = 1$. Simplifying the relation and applying the relations of equation (5.11) give an expression in polar coordinates:

$$\frac{x^2}{a^2} + \frac{y^2}{b^2} = 1$$

$$\frac{x^2}{a^2} - \frac{x^2}{b^2} + \frac{x^2 + y^2}{b^2} = 1$$

$$\frac{b^2 - a^2}{a^2 b^2} x^2 + \frac{r^2}{b^2} = 1$$

$$\frac{b^2 - a^2}{a^2 b^2} r^2 \cos^2(\theta) + \frac{r^2}{b^2} = 1$$

$$\frac{r^2}{b^2} \left(1 - \frac{a^2 - b^2}{a^2} \cos^2(\theta) \right) = 1$$

There are other possibilities. For example, the solution could be expressed in terms of $\sin(\theta)$ by eliminating x instead of y. Or, both x and y can be replaced directly and a corresponding expression involving both $\sin(\theta)$ and $\cos(\theta)$ results.

Example 5.9

Write the equation for an ellipse centered at its left-hand focal point in polar coordinates.

The equation in Cartesian coordinates is given by $\left[(x - \sqrt{\alpha^2 - \beta^2})^2 / \alpha^2 \right] +$ $(y^2 / \beta^2) = 1$, with the focal length given by $\sqrt{\alpha^2 - \beta^2}$. As above, simplifying the expression and applying the relations of equation (5.11) yield an expression in polar coordinates. The expression is chosen to match the result required for uncovering the ellipse in Chapter 7:

$$\frac{\left(x - \sqrt{\alpha^2 - \beta^2} \right)^2}{\alpha^2} + \frac{y^2}{\beta^2} = 1$$

$$\left(x - \sqrt{\alpha^2 - \beta^2} \right)^2 + \frac{\alpha^2}{\beta^2} y^2 = \alpha^2$$

$$\left(x - \sqrt{\alpha^2 - \beta^2} \right)^2 - \left(\alpha^2 - \beta^2 \right) + \frac{\alpha^2}{\beta^2} y^2 = \beta^2$$

$$\left(x^2 - 2\sqrt{\alpha^2 - \beta^2} x \right) + \frac{\alpha^2}{\beta^2} y^2 = \beta^2$$

$$\frac{1}{\beta^2} x^2 - 2\sqrt{\left(\frac{\alpha}{\beta^2} \right)^2 - \frac{1}{\beta^2} x} + \left(\frac{\alpha}{\beta^2} \right)^2 y^2 = 1$$

$$\frac{1}{\beta^2} x^2 - 2\sqrt{\left(\frac{\alpha}{\beta^2} \right)^2 - \frac{1}{\beta^2} x} + \left(\frac{\alpha}{\beta^2} \right)^2 \left(x^2 + y^2 \right) = 1 + \left(\frac{\alpha}{\beta^2} \right)^2 x^2$$

$$\left(\frac{\alpha}{\beta^2}\right)^2\left(x^2+y^2\right)=1+2\sqrt{\left(\frac{\alpha}{\beta^2}\right)^2-\frac{1}{\beta^2}}x+\left(\left(\frac{\alpha}{\beta^2}\right)^2-\frac{1}{\beta^2}\right)x^2$$

$$\left(\frac{\alpha}{\beta^2}\right)^2 r^2=1+2\sqrt{\left(\frac{\alpha}{\beta^2}\right)^2-\frac{1}{\beta^2}}x+\left(\left(\frac{\alpha}{\beta^2}\right)^2-\frac{1}{\beta^2}\right)r^2\cos^2(\theta)$$

$$\left(\frac{\alpha}{\beta^2}\right)^2 r^2=\left(1+\sqrt{\left(\frac{\alpha}{\beta^2}\right)^2-\frac{1}{\beta^2}}r\cos(\theta)\right)^2$$

$$\frac{\alpha}{\beta^2}r=1+\sqrt{\left(\frac{\alpha}{\beta^2}\right)^2-\frac{1}{\beta^2}}r\cos(\theta)$$

$$\left(-\sqrt{\left(\frac{\alpha}{\beta^2}\right)^2-\frac{1}{\beta^2}}\cos(\theta)+\frac{\alpha}{\beta^2}\right)r=1$$

The next point of interest is to express a vector in polar coordinates. The setup is illustrated in Figure 5.22 where two sets of coordinate axes are illustrated, (x, y) and (S_r, t_θ). There a vector $\vec{v}$ attached to the point (x_0, y_0) is given. The vector may be thought of as providing the velocity of an object located at the point (x_0, y_0). We wish to find the components of the vector in polar coordinates, v_r and v_θ, where v_r is the s_r component of the vector and v_θ is the t_θ component of the vector. Note that the s_r axis points in the radial component of the point (x_0, y_0) and the t_θ axis is perpendicular to the s_r axis and points in the direction of an increasing angle. Also, note that the

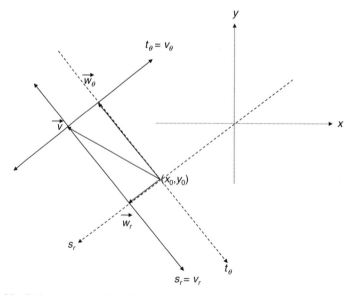

Figure 5.22 Polar representation of a vector.

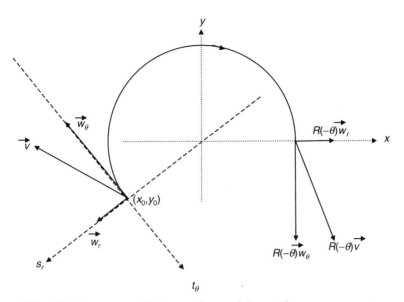

Figure 5.23   Aligning a vector with the coordinate axis by rotation.

(s_r, t_θ) coordinate axes are centered at the point (x_0, y_0). Finally, note that the vector $\vec{w}_r$ is the component of $\vec{v}$ that is aligned with the s_r axis, while the vector $\vec{w}_\theta$ is the component of $\vec{v}$ that is aligned with the t_θ axis so that $\vec{v} = \vec{w}_r + \vec{w}_\theta$.

In Figure 5.23, the vectors $\vec{v}$, $\vec{w}_r$, and $\vec{w}_\theta$ are rotated through the angle $-\theta$. The result is that after rotation the direction of the vector $\vec{w}_r$ is along the x axis and the direction of the vector $\vec{w}_\theta$ is along the y axis. As illustrated, the x component of the rotated vector is identical to the radial component of the original vector. Similarly, the y component of the rotated vector is identical to the angular component of the original vector. This gives the following relation:

$$\begin{pmatrix} v_r \\ v_\theta \end{pmatrix} = R(-\theta)\vec{v} = R(-\theta) \begin{pmatrix} v_x \\ v_y \end{pmatrix}$$

in which $R(-\theta)$ is the rotation matrix through the angle $-\theta$:

$$R(-\theta) = \begin{pmatrix} \cos(-\theta) & -\sin(-\theta) \\ \sin(-\theta) & \cos(-\theta) \end{pmatrix} = \begin{pmatrix} \cos(\theta) & \sin(\theta) \\ -\sin(\theta) & \cos(\theta) \end{pmatrix}$$

The relation is not to be thought of as a rotation of the vector $\vec{v}$. Indeed, the original vector remains the same. The relation is merely a means to express the polar representation of the vector $\vec{v}$ given the Cartesian components.

The inverse operation of rotating through a given angle in one direction is merely rotating through the angle in the opposite direction. Therefore, $R^{-1}(-\theta) = R(\theta)$. From this it is possible to determine a vector's representation in Cartesian coordinates if the vector in polar coordinates is known:

$$\begin{pmatrix} v_r \\ v_\theta \end{pmatrix} = R(-\theta) \begin{pmatrix} v_x \\ v_y \end{pmatrix}$$

$$R(\theta) \begin{pmatrix} v_r \\ v_\theta \end{pmatrix} = R(\theta)R(-\theta) \begin{pmatrix} v_x \\ v_y \end{pmatrix}$$

$$R(\theta) \begin{pmatrix} v_r \\ v_\theta \end{pmatrix} = \begin{pmatrix} v_x \\ v_y \end{pmatrix}$$

5.9 THE DETERMINANT

The claim that the area of an image that is a linear mapping of an original object is equal to the absolute value of the determinant times the area of the original object is made in Section 4.6.5. In this section, we demonstrate the claim.

It has already been noted in Section 4.6.5 that a matrix with determinant zero maps onto a line (or a point if everything gets mapped to the origin). Accordingly, whenever the determinant is zero, all mapped objects have zero area. It is only left to consider the case when the determinant is nonzero.

The approach that this argument takes is to demonstrate the claim for a simple case and then show that if the claim holds for the simple case, it holds in general. For the simple case, the initial object is a square aligned with the x and y axes. To move toward the general case, there is an intermediate stage—a demonstration that the claim holds for a mesh of squares having bases aligned with the x axis. Finally, the general case can be demonstrated. Below, the detail for the simple case is presented, while the intermediate and final stages are sketched.

Figure 5.24 shows the simple case, a square aligned with the x and y axes; the square has sides of length s. Note that the square is mapped by a transformation with matrix M into a parallelogram denoted by P_s. The square with base given by the unit vector is also mapped into a parallelogram, denoted by P_1, and the two parallelograms are similar; just as the original square is a resizing of the unit square with a resizing coefficient of s, its associated parallelogram P_s is a resizing of the mapped unit square P_1 with a resizing factor s. Using a dimensionality argument, the parallelogram P_s has area equal to s^2 times the area of the parallelogram P_1, $\text{Area}(P_s) = s^2 \times \text{Area}(P_1)$.

For the simple case, the claim is demonstrated provided that $\text{Area}(P_1) = \text{determinant}(M)$, where M is the matrix of the linear transformation. Recall the following relations:

$$M = \begin{pmatrix} a_{11} & a_{12} \\ a_{21} & a_{22} \end{pmatrix}$$

$$M \begin{pmatrix} 1 \\ 0 \end{pmatrix} = \begin{pmatrix} a_{11} \\ a_{21} \end{pmatrix} \qquad M \begin{pmatrix} 0 \\ 1 \end{pmatrix} = \begin{pmatrix} a_{12} \\ a_{22} \end{pmatrix}$$

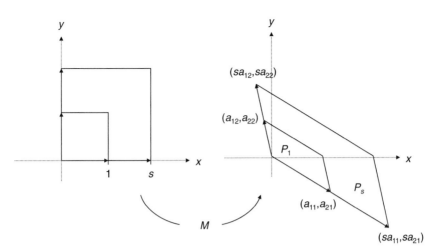

Figure 5.24   Transformation of squares into parallelograms.

Figure 5.25 illustrates that the transformation applied to the unit square with sides given by the vectors $(1\ 0)^T$ and $(0\ 1)^T$ results in a parallelogram with sides given by the vectors $(a_{11}\ a_{21})^T$ and $(a_{12}\ a_{22})^T$. It is necessary to determine the area of the parallelogram.

The general formula for the area of a parallelogram with base b and height h is Area $= bh$. For the height, the definition of the sine function produces the equality $h = c\ |\sin(\theta)\ |$, where c is the length of the diagonal side as illustrated in Figure 5.25. In terms of the sides and the angle between them, the area is given by the formula Area $= bc\ |\sin(\theta)\ |$.

The length b is given by the length of the vector $(a_{11}\ a_{21})^T$, while c is given by the length of the vector $(a_{12}\ a_{22})^T$. Figure 5.24 illustrates that $\theta = \beta - \alpha$. Applying the difference formula from Section 5.4.5 and expressing the sine and cosine functions

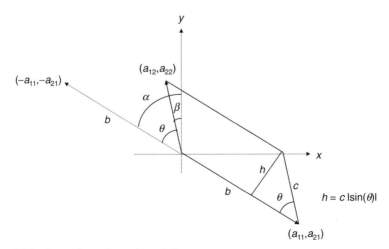

Figure 5.25   Area of transformed parallelogram.

in terms of the entries to the matrix M result in the following equality:

$$|\sin(\theta)| = |\sin(\alpha)\cos(\beta) - \cos(\alpha)\sin(\beta)|$$

$$= |\frac{-a_{11}}{b}\frac{a_{22}}{c} + \frac{a_{21}}{b}\frac{a_{12}}{c}|$$

$$= \frac{1}{bc}|a_{22}a_{11} - a_{12}a_{21}|$$

Placing the value for $|\sin(\theta)|$ into the formula for the area of a parallelogram furnishes the result.

$$\text{Area}(P_1) = bc\,|\sin(\theta)|$$

$$= bc\frac{1}{bc}|a_{22}a_{11} - a_{12}a_{21}|$$

$$= |a_{22}a_{11} - a_{12}a_{21}|$$

$$= |\text{determinant}(M)|$$

This completes the argument that the result holds when the original object is a square aligned with the x and y axes.

The intermediate step is to show that the result holds for a mesh of squares; each square in the mesh has the same area. We do not present a rigorous calculation. Instead, we demonstrate this pictorially through Figure 5.26. All squares in the mesh have equal area and the square mesh is transformed into a mesh of parallelograms all with equal area.

The final step is to take an arbitrary object with well-defined area and demonstrate the claim; the area of the mapped image is equal to the area of the initial object times the absolute value of the determinant of the transformation matrix. The area of the object can be approximated by a mesh with mesh elements sufficiently small as indicated in Figure 5.27. The set of mesh elements entirely within the object produces

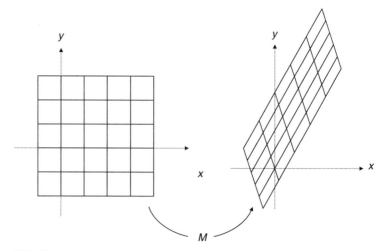

Figure 5.26   Transformation of a mesh of squares to a mesh of parallelograms.

Figure 5.27   Inner and outer mesh areas.

an underestimate of the area, while the set of mesh elements that intersect the object either wholly or partially produces an overestimate. This is expressed by the following inequalities:

$$\text{Inner mesh area} \leq \text{object area} \leq \text{outer mesh area}$$

The area of the object's image is approximated by the images of the mesh elements that are all parallelograms of equal area. Then the following inequality holds:

$$\text{Inner image area} \leq \text{image area} \leq \text{outer image area}$$

But using the intermediate step, the inner (outer) image area is equal to the product of the absolute value of the determinant and the inner (outer) mesh area, Inner image area $=|$ determinant$(M)| \times$(inner mesh area) and similarly for the outer image area. Substitution into the inequalities for the image area results in the following inequalities:

$$|\text{Determinant}(M)| \times (\text{inner mesh area}) \leq \text{Image area}$$
$$\leq |\text{determinant}(M)| \times (\text{outer mesh area}) \quad (5.12)$$

Imagine a sequence of meshes (mesh 1, mesh 2, mesh 3, and so on) in which the number of mesh elements increases and the size of the mesh elements approaches zero as one proceeds along the sequence of meshes. The approximations given by the inequalities get better and better as we move along the mesh sequence. Since the area of the initial object is well defined, the inner and outer mesh areas converge to the actual object's area as the mesh is further refined. So both the left and right sides of the equality converge to the same quantity, $|$ determinant$(M)| \times$(object area), providing the following inequalities:

$$|\text{determinant}(M)| \times (\text{object area}) \leq \text{image area}$$
$$\leq |\text{determinant}(M)| \times (\text{object area})$$

But the left-hand and right-hand sides are the same value, so in fact, $|$ determinant$(M)| \times$(object area) $=$ image area, demonstrating the claim.

We use a similar construct to determine areas in calculus, the topic of Chapter 6.

THE SLAYER: CALCULUS

It is not coincidental that calculus was discovered by both Newton and Leibniz within 10 years of one another. History granted an opportune moment that each of these towering intellects was able to seize. On the cultural side, the church ceased to interfere in the scientific teachings of secular universities; there was a freedom of thought and expression that had not been previously experienced. After Galileo's persecution, Descartes was hesitant but forthcoming. The mathematicians who followed Descartes, including Huygens, Bernoulli, Gregory, Mercator, and Barrow, had no reason to hesitate. Unlike Kepler and Galileo, both Newton and Leibniz were educated in an environment that encouraged scientific inquiry and where religious censorship in the sciences was a foreign and irrelevant concept.

On the mathematical side, the stage had been set for a breakthrough; Archimedes' pioneering work was not the only prop on the stage. Algebraic notation and usage had matured allowing for complex operations with equations. In addition, geometric and algebraic concepts had been united in the Cartesian framework. With these innovations, several problems that lie within the realm of calculus had been solved. Among the more notable achievements are those of Isaac Barrow, who devised a method for determining the tangent to a limited set of curves and also discovered the fundamental theorem of calculus.

The interaction between mathematics and physics was also critical to the discovery of calculus. The cross-fertilization between these disciplines has been a recurring scientific theme. At times, mathematical theory precedes physical application. At other times, physical theory and mathematical development are contemporaneous. Yet at other times physical theory initiates mathematical research. Our story of mathematical development contains all three possibilities.

Among the ancient Greeks, Archimedes stands out as both a mathematician and a physicist. Regarding both disciplines, his works were so far ahead of their time that they could not make the contribution they warranted; the rest of the world was not prepared to understand these works. While Archimedes' contemporaries utilized mathematics to describe observations, Archimedes used and developed mathematics to describe fundamental physical principles. Similarly, Archimedes used his insight into fundamental physical principles to solve mathematical problems.

An example of the latter is Archimedes' method for discovering the formula for the volume of a sphere, his method and the mathematical proof were distinct. Archimedes imagined how a sphere could be placed into balance on a scale using other figures whose volume he could determine. Through this exercise he developed the physical insight to solve the problem. On the other hand, the proof of his formula is purely mathematical; there are no arguments about balancing different bodies on a scale.

Archimedes' investigation of the stability positions for floating paraboloids makes impressive contributions to both physics and mathematics, contributions that were not appreciated until more than a millenium and a half later. Archimedes develops a physical law to determine stable positions in terms of the center of mass, specific gravity, and geometry of an object. This is a problem in the field of hydrostatics that has engineering implications. In particular, Archimedes' analysis is central to the naval engineer's task of designing a boat that will not capsize. Archimedes then turns to the mathematical problem of determining the center of mass of the paraboloid. In accomplishing this feat, Archimedes develops a new mathematical methodology that is a precursor to calculus.

Archimedes (2002) was also a pioneer in the subject of the study of motion. In his treatise *On Spirals*, he performs an analysis of a particle moving along a spiral path. The particle revolves around a central point at a uniform rate while concurrently moving away from the central point at a uniform radial speed. This is the first known example in which the velocity is parsed into two perpendicular components, the radial direction outward from the center and the angular direction along a series of circles with the central point as a common center. Archimedes relates the particle's velocity to other measurements providing a stunning set of theorems. One theorem is a direct precursor to Kepler and calculus. Archimedes determines the area between the trajectory and the center that is swept out in any given time interval.

Like Archimedes, Kepler had both brilliant physical and mathematical insight. As noted in Chapter 2, Kepler set himself upon a mission to salvage Copernicus' heliocentric theory. Kepler reasoned that the sun applies a force to the planets and that the force depends upon distance; each circle about the sun experiences a force of equal strength and the force decreases as the radius of the circle increases. Following this reasoning, a circular orbit seems most natural; the force remains constant, keeping the planet in its circular path. Kepler's physical instincts led him to believe that if the orbit is not circular, some property of circular motion must result from the symmetry of the sun's force. Toward this end, Kepler examined the properties of circular motion and noted that the area swept out by the orbit's path and a line connected to the circle's center is proportional to the time of the measurement. In modern-day language, Kepler noted that the angular momentum is a constant and proposed that the elliptical orbits of the planets must satisfy this property. Kepler's study of angular momentum led him to a general mathematical method for determining the area swept out by an orbit. It is a method that Archimedes would have approved of, for indeed Kepler rediscovered what Archimedes developed in his treatise *on spirals*.

Galileo was another individual who traveled down a path laid by Archimedes. In Galileo's study of the motion of falling objects, Galileo parsed the object's motion along two distinct perpendicular directions, the vertical direction and the horizontal direction. Galileo's insight was that this parsing allows for the independent study of the motion in each direction. The motion in the vertical direction is influenced by gravity, but this is not the case for the motion in the horizontal direction. Since gravity is the only force on the object, Galileo recognized that the object traverses freely along the horizontal direction; its horizontal movement is independent of its vertical motion until the object strikes the ground. Galileo shows that the resulting trajectory of these two independent motions is the parabola. Later, the ability of Newton to replicate Galileo's conclusion using his laws of motion and calculus furnished confirmation of the correctness of Newton's approach.

What was it that allowed Newton to become the highest authority in matters of physics over a 200-year period while introducing a new mathematics that has been a foundation for mathematical research since its inception? Certainly, Newton possessed a rare combination of instinct and intelligence that permitted penetrating insight into his subjects of investigation. While this combination is rare, before Newton as well as after Newton, there have been men who possess such talent. For those who came before Newton, the time was not right, but Kepler, Galileo, and Descartes had endowed Newton with the knowledge to make his breakthrough.

As with Archimedes, Newton's formidable physical intuition guided his mathematics. His route to calculus directly followed his insights into the motion of free bodies. Indeed, Newton developed laws of motion that are expressible in the vocabulary of calculus. Furthermore, calculus can be applied to solve the trajectories of moving objects that obey Newton's laws of motion. One such instance is the parabolic motion of a falling object as described by Galileo. But the most inspiring application was the use of calculus and the laws of motion to describe the trajectory of a planet about the sun. The result of this beautiful theory that married physics with a new branch of mathematics perfectly matches Kepler's ellipse. The result closed the case of the geocentric versus heliocentric dispute that Aristarchus initiated and once more resurfaced with Copernicus. And the result permanently changed the attitudes of men toward science. Never again would science be looked upon as a heretical activity; it would be viewed as a way to serve mankind's causes.

Ten years after Newton's accomplishment, Leibniz paved a different path to calculus. Leibniz' formal education was in law, and he received little university training in either mathematics or physics. However, Leibniz possessed a curiosity that throughout his life caused him to pursue many disciplines. He had a tremendous intellectual capacity along with a gregarious personality that he used to gain access to Europe's most prestigious intellectual circles. His curiosity led him to mathematics and his personality attracted the attention of Christiaan Huygens (1629–1695), one of Europe's finest mathematicians. After impressing Huygens by solving very difficult problems, Huygens agreed to tutor Leibniz.

Through the tutelage of Huygens, Leibniz became familiar with the works of the modern mathematicians as well as those of the ancient Greeks, including Archimedes and Apollonius. Leibniz' strength was in abstraction and communication, and at one point he envisioned the development of a universal language that would encompass all disciplines. It is from this perspective that he distilled the works of his contemporaries and Archimedes and generalized the results into calculus. Leibniz' calculus was technically identical to Newton's, but while Newton focused on physical implications, Leibniz focused on his strengths, abstraction and communication. The result of Leibniz' work was an intuitive notation in symbols that could be manipulated with known algebraic operations. Leibniz' notation as well as his terminology has become the international convention.

It was Leibniz who coined the basic vocabulary associated with calculus, including the word *calculus* as well as the terms *differential calculus* and *integral calculus*. By contrast, Newton's naming convention, the theory of fluxions, is not widely known. Similarly, while Leibniz' notational conventions lend themselves to generalization for a broad set of applications, Newton's notational conventions are far more limited and not as widely used. The historical review of Chapter 2 closes with Leibniz' well-known phrase, "we live in the best of all possible worlds." The phrase is certainly applicable to the invention of calculus. Through Newton and his remarkable physical insights, we have a mathematical theory that reflects the physical world we live in. Newton showed how to use calculus to describe our world. And through Leibniz with his incomparable skills at abstraction and communication, we have symbolic conventions that are intuitive and in accord with the broader set of mathematical notation. Historical perspective has brought us to this realization, but this was far from obvious to the men central to its development.

After Newton's bouts with Hooke concerning his work on optics, Newton locked his scientific gems in a vault; only a small circle viewed them. For 10 years from 1674 through 1683, Newton, one of history's greatest physicists, pursued the field of alchemy with the same dedication that he earlier pursued mathematics; his scientifically explosive mind became dormant. As for Leibniz, he was a very busy man whose wide-ranging interests left him a bit unfocused. Leibniz had an unmatched intellectual curiosity about everything and fluttered between disciplines. In addition to his pursuits in mathematics, including the discovery of calculus in 1674, he had engaged himself in an array of projects that could have easily occupied 10 men around the clock. As an example of how his mind wandered, one such project (a folly reminiscent of al-Haytham's plan to divert the Nile) was an effort to drain local silver mines of water using a series of windmills as pumps. The project caused Leibniz to study windmills and he became thoroughly immersed with the science and history of wind power. Concerning the history of wind power, Leibniz became familiar with different fashions of sail boats, including those of the Chinese. He then became fascinated with Chinese history and contributions to science, becoming a world authority and authoring a book on the subject. With these commitments, Leibniz had little time for putting his calculus results in publishable form. But the year 1684 was a banner year for both Leibniz and Newton. Each published works that they held privately for years.

Through his nasty disposition, Hooke was one cause of Newton's self-cocooning, and by coincidence, Hooke was in some ways instrumental in getting Newton to return to scientific endeavors, although the bulk of the credit goes to Edmond Halley (1656–1742). One day, Hooke and Halley, both distinguished members of the Royal Society of Scientists, along with the architect Christopher Wren (1632–1723) were sitting around in a coffee shop and talking shop. The topic of discussion was the shape of the trajectory for a comet orbiting the sun. Halley believed it would be an ellipse but did not have any argument in support of his view. Hooke claimed that he had solved the problem using an inverse square law as the binding force between the comet and the sun, and the force is inversely proportional to the square of the distance from the sun. Hooke was unable to substantiate his claim.

The matter must have nagged at Halley, who apparently took an unusual interest in comets; Halley's comet is named for this man. Halley discussed the issue with several other members of the Society and through the grapevine heard that Newton may be of some assistance. To satisfy his longing for a solution, Halley took a trip to Cambridge for a visit with Newton.

Halley was the opposite of Hooke. He was a gentleman who could bring out the best in others. Upon meeting Newton, he posed his problem concerning the path taken by a comet. Without hesitation, Newton stated that the orbit would be an ellipse. The certainty of Newton's response must have registered with Halley, and Halley was able to use his charm to elicit an explanation. Newton claimed that he had long ago solved the problem but lost the results. Newton did promise to furnish an explanation and, just like that, Newton abandoned alchemy and once more pursued mathematics and physics with a fierce dedication—the volcano was once again active. Newton returned to his unfinished investigations of planetary motion that he had initiated during the plague years of the 1660s. Newton wrote a treatise that derived Kepler's elliptic orbit using the inverse square law and mailed the treatise to Halley.

Halley was a well-trained scientist and he knew a treasure when he saw it. An argument could be made that this work was the most important scientific work up to that point in history. But through Halley's persistence it would soon be surpassed by the most influential scientific work ever written before or after. From every perspective, Halley's response to his encounter with Newton was admirable. Halley's objective was to make Newton's work available to the scientific community. With Newton's permission, Halley reported the results to the Royal Society in December of 1684 and after some revision they were published. Halley then went further; he persuaded the Royal Society to sponsor a project so that Newton could expand upon his work. Both the Royal Society and Newton agreed, but when funding for the publication was not forthcoming, Halley personally financed the work. Through Halley's intervention, the most influential scientific work ever written, *Principia*, came about.

In *Principia*, Newton (1995) lays out his laws of motion and using the laws performs an analysis of several motions. Included in three volumes and several revisions are investigations of planetary motion, the moon's orbit, tides, and the shape of the earth. Concerning the latter, Newton demonstrates the earth is not quite round but oblate; the earth's rotation causes it to bulge at the equator. So impressive is *Principia* that Newton was deified throughout Europe during the eighteenth century. The French scientist Pierre de Maupertuis, inspired by Newton's theory of an oblate earth, set upon an Arctic expedition to make measurements that would confirm Newton's theory. The philosopher Voltaire, no friend of Maupertuis, took a swipe at his fellow countryman while singing Newton's praises with his remark that Maupertuis went to the ends of the earth to find that which Newton discovered from his desk.

Newton's methods would be central to every investigation in dynamics over the next two centuries. It is central to many scientific and engineering disciplines: classical mechanics, fluid mechanics, structural engineering, aeronautical engineering, thermodynamics, and electrostatics. Electrodynamics and relativity are an outgrowth of Newtonian thought. Even quantum mechanics pays homage to Newton. A key principle of quantum mechanics is its consistency with Newton's classical theory in the limit of many interacting particles.

It took Newton 3 years to finalize his first version of *Principia*. When *Principia* was published in 1687, one element of Newton's analysis was conspicuously missing, calculus. While Newton used calculus to attain his results, he did not use calculus as an explanatory vehicle. This was not an oversight; it was intentional. The central theme of *Principia* was the dynamics of motion and it was all new material. Perhaps Newton believed that the book would have lost its focus had it included an additional body of material on new mathematical methods. Nearly three centuries later, von Neumann proposed his version of relativistic quantum mechanics that included new mathematical results in the area of functional analysis. The scientific community preferred Dirac's work in the same field as it focused on the physics. From a historical perspective, Newton's judgment in omitting a presentation of calculus was very sound. Yet Newton himself poses another explanation for the omission, one that indicates a lasting wound from his earlier bouts with Hooke. When asked later why he omitted calculus from *Principia*, Newton responded that he feared ridicule.

In the meantime, Leibniz had not been idle. By 1674, Leibniz had discovered many results in the field of calculus. Leibniz, like Newton, did not immediately publish his results. His first publication in the field of calculus came during the same year as Newton's letter to Halley, 1684. Unlike Newton, the cause of Leibniz' pause was not apprehension. Leibniz was overcommitted by his many activities. Perhaps the impetus for Leibniz to publish his work was a series of letters between Leibniz and Newton in which each claimed cryptically to be able to solve a set of similar problems and Leibniz wished to claim priority.

Through his published work, Leibniz gained fame in Europe and was considered the sole inventor of calculus for 15 years. Then in 1699, one of Newton's most cherished friends, Nikolas Fatio, accused Leibniz of plagiarism. This marked the beginning of a brutal and pointless brawl that consumed the energies of not only Leibniz and Newton but also others within European scientific circles. Surrounding Newton and Leibniz were men that encouraged their self-righteous instincts. Nationalistic fervor was the source of support for Newton, while a sense of continental superiority encouraged Leibniz' supporters. Both men literally fought on to their grave. Even after the death of Leibniz in 1716, Newton continued to pursue his case against Leibniz for an additional 11 years until his death.

The basis for Newton's accusation of plagiarism rested on Liebniz' 1676 visit to London at the behest of the Royal Academy. The visit did not go too well. In full character Hooke lambasted Leibniz. At any rate, Newton claimed that it was during this visit that Leibniz obtained access to Newton's works. In Newton's mind, Leibniz did nothing more than steal Newton's results, rewrite them in a new notation, and then pass them off as his own. The conduit who passed the results to Leibniz, a man named Collins, was an admirer of Newton who only wished to praise his hero. This is all irrelevant. It is uncertain if the sneak peek contained any material related to calculus, but even if it had, Leibniz had already made his breakthrough in 1675. Another claim of Newton is that in 1678 Newton wrote Leibniz a letter containing applications of calculus. Perhaps so, but the letter merely stated several problems that Newton could solve without explaining methodology and by that time Leibniz could also solve such problems.

Although mostly meaningless, the fight did have one redeeming feature. A famous problem that inspired a whole new discipline in mathematics, the calculus of variations, emerged from the fight. Initially, one of the men who most championed Leibniz' cause was Jacob Bernoulli (1654–1705), a student of Leibniz. Bernoulli was certain that Newton was a complete fraud. He had no reason to believe otherwise, there was no record of Newton's claimed achievements. In a manner similar to the Italian duels, Bernoulli proposed to send Newton a problem that could be solved only by those who truly understood calculus. Newton's certain failure to solve this problem would expose him.

The problem is known as the brachistrone problem. The objective is to find the shape of a curve that allows a body to slide from a higher point to a lower point, not directly beneath the initial point, in the least time. As noted above, this problem would later motivate a branch of mathematics known as the calculus of variations. On continental Europe, apart from Leibniz and Bernoulli, nobody could solve the problem. [Some claim that l'Hopital (1661–1704), an aristocrat who took an interest in mathematics, could also solve the problem. However, as l'Hopital was a benefactor to Bernoulli, independent achievements of l'Hopital have come into question.] It is not known how long it took these men to arrive at a solution. Nevertheless, the fact that they shared solutions indicates that they considered it a significant challenge and one can surmise that each took a considerable time grappling with the problem. The fact that the problem was posed as an open challenge to the European community and nobody else was able to arrive at a solution attests to its difficulty. The problem was posed to Newton, who had been inactive in mathematics for around two decades. According to Newton's niece, the master dispensed with the problem in a single evening between his dinner and bedtime—it was child's play.

Still the fight dragged on. Both Leibniz and Newton, men of great intellect, were stupid enough to attack each other's strengths. Leibniz, with a modest record of achievement in physics, attacked Newton's explanation of planetary motion stating that the concept of gravity was absurd. How could Leibniz have believed he could win an argument in physics with the internationally acknowledged supreme physicist of the times? Newton, whose notation was not as transparent as Leibniz', attacked Leibniz on the grounds that Leibniz' notation deprived its users of geometric insight. In fact, Leibniz' notation was adopted because its simplicity and elegance enable ease of use and allow for keener insight.

The development of calculus through the quest to describe planetary motion is an achievement that cannot be understated. Calculus and the laws of motion initiated modern science, and they are at the root of nearly all other achievements in mathematics and physics in the past 300 years. Certainly, their discovery added to the technical body of knowledge that mankind accumulated. But more importantly, their discovery forever changed the way that the unknown was approached—the unknown would be challenged by the imagination leading to greater discoveries.

The invention of calculus can be viewed in two ways. Chapter 2 stresses the perspective of the invention as an achievement for mankind; it is the culmination of centuries of pursuits that are interwoven with a broader history. The invention of calculus may also be viewed as the achievement of two individuals, both geniuses who applied their talents with a determined spirit. Whichever perspective is taken, a common view is visible. The historical perspective reveals human flaws that resulted in unjust persecutions of men who were deliberating over a subject that ought to have been innocuous. The view of individuals reveals human flaws that cause men of exceptional talent to engage in senseless tomfoolery, just like you and I. The brilliant and the blemish are part of mankind and men.

6.1 STUDIES OF MOTION AND THE FUNDAMENTAL THEOREM OF CALCULUS

As a starting point, this section motivates calculus through an investigation of simple motions. Differential calculus, integral calculus, and the relation between them are introduced from the perspective of the description of motion. Concrete methods for calculating derivatives and integrals follow in later sections.

6.1.1 Constant Velocity and Two Problems of Motion

This section examines the case in which an object moves along a single dimension at constant velocity. A physical setting would be a train moving along a track at constant speed. There are two fundamental problems that we address. The first is, given the object's position, determine its velocity. The second problem inverts the first; given the object's velocity, determine its position. Differential calculus is a generalization of the first problem, while integral calculus is a generalization of the second. The geometry of these problems is emphasized as it is critical to the generalization.

The geometry is set on a Cartesian plane parameterized by t and x or v. The horizontal axis, t, represents time while the vertical axis, x or v, represents the distance that the object has traveled, x, or the velocity of the object, v.

For the first problem, the object's position is described by a function of time:

$$x(t) = at + c \qquad (6.1)$$

A graph of the position is presented in Figure 6.1.

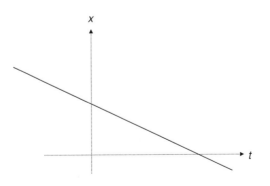

Position, x, at time t

Figure 6.1 Graph of motion with constant speed.

The problem is to find the velocity. The velocity is a ratio of distance to time. With this notion, it is easy to determine the average velocity over a time interval. It is simply the difference in position divided by the difference in time:

$$\text{Average velocity} = \frac{x(t_f) - x(t_i)}{t_f - t_i} \tag{6.2}$$

where t_i and t_f are initial and final times over the time interval of interest. Using the value of $x(t)$ given by equation (6.1), the average velocity is the following:

$$\begin{aligned}
\text{Average velocity} &= \frac{x(t_f) - x(t_i)}{t_f - t_i} \\
&= \frac{(at_f + c) - (at_i - c)}{t_f - t_i} \\
&= \frac{a(t_f - t_i)}{t_f - t_i} \\
&= a
\end{aligned}$$

In this case, the average velocity is independent of the time interval given by t_i and t_f. As such the average velocity is the same as the velocity at all times. Denoting the velocity by $v(t)$, $v(t) = a$. Geometrically, the velocity is the slope of the line given by the graph. Higher velocities correspond with greater slopes, as illustrated in Figure 6.2, where all objects start at the same position. Notice that at any given time the position of the object with higher velocity, which is the same as the slope, is greater than the position of the object with lower velocity.

For the second problem, the object's velocity is given as a function of time and the objective is to determine the position of the object. We start with the simplest case, $v(t) = a$. In this case, the velocity is constant for all times.

One approach toward solving this problem is to appeal to the definition of average velocity and solve for the position. As above, because the velocity is constant,

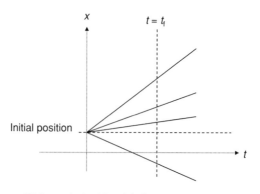

Higher velocity (slope) indicates greater
difference from initial position at a fixed time

Figure 6.2   Position at different speeds.

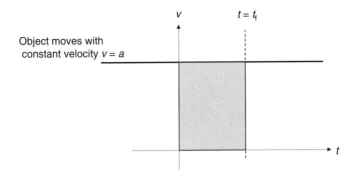

Figure 6.3   Geometric relation between speed and position.

it is the same as the average velocity:

$$\text{Average velocity} = \frac{x(t_f) - x(t_i)}{t_f - t_i} = a$$

$$x(t_f) - x(t_i) = a\,(t_f - t_i)$$

Geometrically, the solution to the second problem is illustrated by Figure 6.3. The velocity is graphed as a horizontal straight line indicating a constant velocity across all time. The difference between the position at times t_i and t_f is the rectangular area between the graph of the velocity and the t axis, Area $= a(t_f - t_i)$.

It is common to express the position at any time in terms of the initial position. Setting t_f to any arbitrary time, $t_f = t$, and t_i to zero yields the following:

$$x(t) - x(0) = at$$
$$x(t) = at + x(0)$$
$$= at + c$$

where c indicates the initial position, $x(0)$.

Note the expressions for velocity and position whether we are first given the position and must solve for the velocity or vice versa.

The geometric interpretations of velocity and displacement, slope and area, are generalized below.

6.1.2   Differential Calculus, Generalizing the First Problem

Next consider a more general motion. While still along one dimension, the position of the object is a general function of time with the property that every point of the function has a unique tangent line. Figure 6.4 illustrates the concept of a unique tangent line; those functions similar to Figure 6.4a are allowable, while the others are not. Note that Figure 6.4b has no tangent line where the function breaks, while the function in Figure 6.4c does not have a unique tangent line at the point where the graph has a corner. The first problem is to determine the velocity of the object at any arbitrary time t.

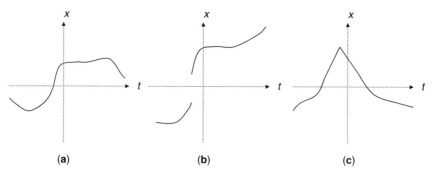

(a)   **(b)**   **(c)**

Figure 6.4   Only part (a) has a tangent line along its entire graph.

The average velocity between times t_i and t_f is given by equation 6.2. The velocity at time t_i, sometimes referred to as the instantaneous velocity, is defined as the result of applying the formula for the average velocity and finding the limiting value as t_f approaches t_i. In symbols, the velocity is defined as follows:

$$v(t_i) = \lim_{t_f \to t_i} \frac{x(t_f) - x(t_i)}{t_f - t_i}$$

The symbol $\lim_{t_f \to t_i}$ is read as the limit as t_f approaches t_i, so the definition of instantaneous velocity is the limiting value of the average velocity as the time interval over which the average is taken shrinks to a point. The question of how the limiting process is evaluated as well as the geometric interpretation of the definition is best illustrated by example. For our first example, suppose $x(t)$ is as follows:

$$x(t) = t^2$$

Let us take $t_i = 1$ and find the corresponding instantaneous velocity. First, construct a table in which t_f assumes values that approach $t_i = 1$. The result is shown in the following table:

t_f	$x(t_f) - x(t_i)$	$t_f - t_i$	$\dfrac{x(t_f) - x(t_i)}{t_f - t_i}$
3	$3^2 - 1 = 8$	2	$\dfrac{8}{2} = 4$
2	$2^2 - 1 = 3$	1	$\dfrac{3}{1} = 3$
$1\dfrac{1}{2}$	$\left(\dfrac{3}{2}\right)^2 - 1 = \dfrac{5}{4}$	$\dfrac{1}{2}$	$\dfrac{5/4}{1/2} = \dfrac{5}{2}$
$1\dfrac{1}{4}$	$\left(\dfrac{5}{4}\right)^2 - 1 = \dfrac{9}{16}$	$\dfrac{1}{4}$	$\dfrac{9/16}{1/4} = \dfrac{9}{4}$
$1\dfrac{1}{8}$	$\left(\dfrac{9}{8}\right)^2 - 1 = \dfrac{17}{64}$	$\dfrac{1}{8}$	$\dfrac{17/64}{1/8} = \dfrac{17}{8}$
$1\dfrac{1}{16}$	$\left(\dfrac{17}{16}\right)^2 - 1 = \dfrac{33}{256}$	$\dfrac{1}{16}$	$\dfrac{33/256}{1/16} = \dfrac{33}{16}$
$1\dfrac{1}{32}$	$\left(\dfrac{33}{32}\right)^2 - 1 = \dfrac{65}{1024}$	$\dfrac{1}{32}$	$\dfrac{65/1024}{1/32} = \dfrac{65}{32}$
$1\dfrac{1}{2^n}$	$\left(\dfrac{1+2^n}{2^n}\right)^2 - 1 = \dfrac{1+2^n}{4^n}$	$\dfrac{1}{2^n}$	$\dfrac{1+2^n/4^n}{1/2^n} = \dfrac{1+2^n}{2^n}$

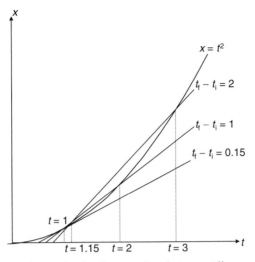

Figure 6.5   As t_f approaches t_i, the chord approaches the tangent line.

Graphically, the limiting operation is viewed in Figure 6.5. As t_f approaches $t_i = 1$, the line between the points $(t_i, x(t_i))$ and $(t_f, x(t_f))$ approaches the tangent line to the curve at t_i. One can surmise, and it is true, that the instantaneous velocity is the slope of the tangent line.

While geometric arguments permit insight, assessing the limiting value requires algebra. It is customary to set $t_f = t_i + \Delta$, where Δ can be either positive or negative. Noting that $t_f - t_i = \Delta$, the instantaneous velocity can be expressed as follows:

$$v(t_i) = \lim_{\Delta \to 0} \frac{x(t_i + \Delta) - x(t_i)}{\Delta}$$

With $t_i = 1$ and $x(t) = t^2$, the instantaneous velocity becomes the following:

$$v(1) = \lim_{\Delta \to 0} \frac{x(1 + \Delta) - x(1)}{\Delta}$$

$$= \lim_{\Delta \to 0} \frac{(1 + 2\Delta + \Delta^2) - (1)}{\Delta}$$

$$= \lim_{\Delta \to 0} \frac{2\Delta + \Delta^2}{\Delta}$$

$$= \lim_{\Delta \to 0} 2 + \Delta$$

$$= 2$$

There is a subtlety in the notation that justifies an explanation. Because of the division by zero, the expression $[(1 + 2\Delta + \Delta^2) - (1)]/\Delta$ does not make sense as a stand-alone expression with $\Delta = 0$. However, it is possible to talk about the limit of the expression as Δ approaches zero. Indeed, as long as Δ is not zero, the algebraic operations that yield the answer $v(1) = 2$ are all allowable and the limit as we proceed toward zero has a well-defined value.

This subtlety expresses itself geometrically. In Figure 6.5, the single point $(t_i, x(t_i)) = (1, 1)$ does not define a unique line; another point, $(t_f, x(t_f))$, is necessary. However, as long as Δ is not equal to zero, two points are available for defining a line whose slope is the average velocity between the points. By allowing t_f to approach t_i (Δ approaches zero) without ever equaling t_i, a sequence of lines that approach the tangent line is established. The sequence of slopes that corresponds with the sequence of lines approaches the instantaneous velocity, which is the same as the slope of the tangent line.

Note that the algebra used to obtain $v(1)$ in the preceding set of equations is not unique to the point $t_i = 1$. The velocity at any general time t can be determined as follows:

$$
\begin{aligned}
v(t) &= \lim_{\Delta \to 0} \frac{x(t + \Delta) - x(t)}{\Delta} \\
&= \lim_{\Delta \to 0} \frac{(t^2 + 2t\Delta + \Delta^2) - (t^2)}{\Delta} \\
&= \lim_{\Delta \to 0} \frac{2t\Delta + \Delta^2}{\Delta} \\
&= \lim_{\Delta \to 0} 2t + \Delta \\
&= 2t
\end{aligned}
$$

As expected, the general expression for the velocity is consistent with the velocity specified at time $t = 1$, $v(1) = 2 \times 1 = 2$.

Leibniz' notation for the instantaneous velocity is the following:

$$
v(t) = \frac{dx}{dt} \quad \text{or} \quad \frac{dx}{dt} = \lim_{\Delta \to 0} \frac{x(t + \Delta) - x(t)}{\Delta} \tag{6.3}
$$

The full reading of dx/dt is the change in x with the change in t; this is a ratio of changes as illustrated in Figure 6.6.

One can also think of d/dt as an operator that when applied to the function x yields a new function. From the above example,

$$
\frac{d}{dt}[x(t)] = \frac{d}{dt}\left(t^2\right) = 2t
$$

That is, applying the operator d/dt to the function $x(t) = t^2$ results in a new function, $dx/dt = 2t$.

Leibniz coined the resulting function the derivative and that is what it been called ever since. Considering $x(t)$ as the position of an object at time t, the derivative of x is the object's velocity.

Remark. While the example of the relation between position and velocity is used to introduce the concept of a derivative, the derivative is much more general. If

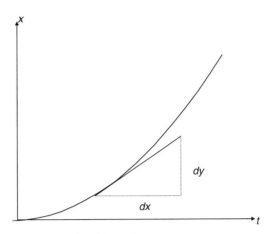

Figure 6.6 The derivative as a ratio of dy to dx.

two quantities y and z are related functionally, $z = f(y)$, then dz/dy represents the ratio of incremental changes in the quantities.

6.1.3 Integral Calculus, Generalizing the Second Problem

We continue to generalize Section 6.1.1. While still constraining motion to one dimension, consider the velocity of the object as a general function $v(t)$. The second problem is to determine the position of the object, $x(t)$, at any arbitrary time t. Section 6.1.1 presents the answer for the case when the velocity is piecewise constant. Calculus proposes an approach for generalizing the piecewise constant solution to more arbitrary functions.

Figure 6.7 illustrates the generalization. Consider the velocity given by the function $v(t)$. We wish to find the distance traveled by the object between times t_i and t_f. Suppose we partition the time interval from t_i to t_f into smaller time segments

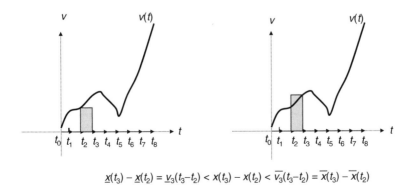

$$\underline{x}(t_3) - \underline{x}(t_2) = \underline{v}_3(t_3 - t_2) < x(t_3) - x(t_2) < \overline{v}_3(t_3 - t_2) = \overline{x}(t_3) - \overline{x}(t_2)$$

Figure 6.7 Underestimate and overestimate between two times.

with end points $t_0, t_1, t_2, \ldots, t_n$ so that $t_i = t_0 < t_1 < t_2 < t_3 < \cdots < t_n = t_f$. Next, let $\underline{v}_j$ be the minimum value of the velocity whenever $t_{j-1} < t < t_j$ and similarly let $\overline{v}_j$ be the maximum value of the velocity whenever $t_{j-1} < t < t_j$. Also, let $\underline{\Delta x}_j$ and $\overline{\Delta x}_j$ denote the differences in position between times t_{j-1} and t_j corresponding to the velocity profiles $\underline{v}_j$ and $\overline{v}_j$ (see Figure 6.7).

By the results of Section 6.1.1, $\underline{\Delta x}_j$ and $\overline{\Delta x}_j$ are calculated as follows:

$$\underline{\Delta x}_j = \underline{v}_j(t_j - t_{j-1}) \qquad \overline{\Delta x}_j = \overline{v}_j(t_j - t_{j-1})$$

Finally, let Δx_j be the difference in position between times t_{j-1} and t_j corresponding to the original velocity profile $v(t)$. Then, because $\underline{v}_j \le v(t) \le \overline{v}_j$ for all t in the interval between t_{j-1} and t_j, the following inequalities hold:

$$\underline{\Delta x}_j \le \Delta x_j \le \overline{\Delta x}_j \tag{6.4}$$

The quantity $x(t_f) - x(t_i)$ is the difference in position from time t_i to time t_f, also referred to as the relative displacement. The relative displacement is equivalent to the sum of all the Δx_j's. Since the inequalities of equation (6.4) hold for each j, the following also holds:

$$\sum_{j=1}^{n} \underline{\Delta x}_j \le \sum_{j=1}^{n} \Delta x_j = x(t_f) - x(t_i) \le \sum_{j=1}^{n} \overline{\Delta x}_j \tag{6.5}$$

As illustrated in Figure 6.8, the left-sided sum is the area between the t axis and the velocity profile given by the $\underline{v}_j$'s, while the right-sided sum is the area between the t axis and the velocity profile given by the $\overline{v}_j$'s. What happens as the partition becomes finer and the size of every cell in the partition approaches zero ($t_j - t_{j-1}$ approaches zero for each j)?

Assuming that the area between the initial velocity curve and the t axis is well defined, as the partition becomes finer, the areas given by $\sum_{j=1}^{n} \underline{\Delta x}_j$ and $\sum_{j=1}^{n} \overline{\Delta x}_j$ approach one another. Then the difference in the initial and final positions is the

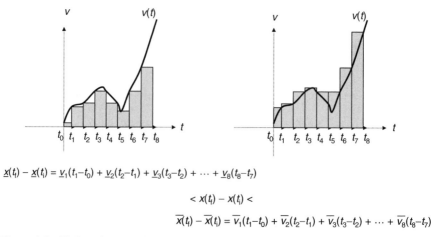

$$\underline{x}(t_f) - \underline{x}(t_i) = \underline{v}_1(t_1-t_0) + \underline{v}_2(t_2-t_1) + \underline{v}_3(t_3-t_2) + \cdots + \underline{v}_8(t_8-t_7)$$

$$< x(t_f) - x(t_i) <$$

$$\overline{x}(t_f) - \overline{x}(t_i) = \overline{v}_1(t_1-t_0) + \overline{v}_2(t_2-t_1) + \overline{v}_3(t_3-t_2) + \cdots + \overline{v}_8(t_8-t_7)$$

Figure 6.8   Underestimate and overestimate of the relative position.

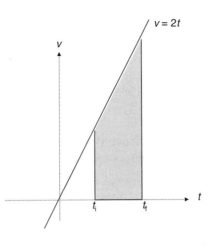

$$x(t_f) - x(t_i) = \text{shaded area under line} = t_f^2 - t_i^2$$

Figure 6.9 Position given by shaded area.

common limiting value, which is the area between the velocity curve and the t axis:

$$x(t_f) - x(t_i) = \lim_{n \to \infty} \sum_{j=1}^{n} \Delta x_j = \lim_{n \to \infty} \sum_{j=1}^{n} \underline{v}_j(t_j - t_{j-1})$$

$$= \lim_{n \to \infty} \sum_{j=1}^{n} \overline{\Delta x}_j = \lim_{n \to \infty} \sum_{j=1}^{n} \overline{v}_j(t_j - t_{j-1}) \qquad (6.6)$$

Note that as the partition becomes finer the number of elements in the partition becomes indefinite. Determining an object's relative displacement, $x(t_f) - x(t_i)$, from the object's velocity profile, $v(t)$, is known as integrating the velocity over the interval from t_i to t_f. We illustrate integration with a simple example.

Consider the case when an object moves with the velocity $v(t) = 2t$ and examine the difference in the object's position between two times t_i and t_f. From the above discussion, the difference in the object's position, $x(t_f) - x(t_i)$, is given by the area under the line, as illustrated in Figure 6.9. The area is the difference in area of the two triangles; the larger triangle has area t_f^2 and the smaller triangle has area t_i^2:

$$x(t_f) - x(t_i) = \tfrac{1}{2}v(t_f)t_f - \tfrac{1}{2}v(t_i)t_i$$

$$= \tfrac{1}{2}2t_f t_f - \tfrac{1}{2}2t_i t_i$$

$$= t_f^2 - t_i^2 \qquad (6.7)$$

Remarks

- The inequalities of equation (6.5) may yield positive as well as negative values for the displacement $x(t_f) - x(t_i)$. This comes about by specifying positive and negative directions of motion. In the case that the displacement is positive, the graph of the velocity tends to be above the time axis. Alternatively, in the case when the displacement is negative, the graph of the velocity tends to fall below the time axis.

- Archimedes' estimation of π squeezes the quantity between upper and lower bounds. In his work *On Floating Bodies* (2002), Archimedes adopts the same approach for estimating the volume of paraboloids and then finds the actual volume by allowing the upper and lower bounds to approach one another.

6.1.4   Relations Between Differentiation and Integration and the Fundamental Theorem of Calculus

The fundamental relationship between velocity and position is that if the position of an object is known over a time span, the velocity can be determined (differentiation); alternatively, if the velocity of the object is known over the time span, the relative displacement can be determined (integration). We explore this in some detail.

Suppose an object is in motion and t_i is set as a fixed time. Let $x(t)$ be the position of the object. Next imagine another coordinate system, $\tilde{x}$, that shifts the original coordinate, $x(t)$, by the object's initial position, $\tilde{x}(t) = x(t) - x(t_i)$, so that $\tilde{x}(t_i) = 0$. Note that in both coordinate systems the velocity is the same. After all, it is the same object with the same motion. Accordingly, the differentiation of the position expressed in either variable results in the same velocity:

$$\frac{d}{dt}[x(t)] = \frac{d}{dt}[\tilde{x}(t)] = v(t) \tag{6.8}$$

Note that $\tilde{x}(t)$ is the relative displacement found by integrating the velocity v through the time interval from t_i to t. Equation (6.8) explicitly shows the relation between differentiation and integration and is the fundamental theorem of calculus.

Let us apply this principle to the examples from Sections 6.1.2 and 6.1.3. In Section 6.1.2, an object's position is given, $x(t) = t^2$. It was found that the velocity is $v(t) = dx/dt = 2t$. In Section 6.1.3, the velocity profile is given, $v(t) = 2t$. As the velocity profiles of these examples are identical, the relative displacement must also be identical. The displacement given by equation (6.7) is indeed consistent with the position, $x(t) = t^2$. [Using the position from $x(t) = t^2$ to $x(t_f) - x(t_i)$ results in equation (6.7).]

Remarks

- Isaac Barrow spent his mathematical career creating a mathematical representation of motion that would relate velocity and position. All that has been presented to this point was known to Isaac Barrow, who built upon the work of Pierre Fermat (1601–1665). Because Newton's legacy overshadows Barrow's, Barrow does not receive his due credit as the discoverer of the fundamental theorem of calculus.

- Newton and Leibniz went well beyond Barrow in many areas. This book presents a glimpse of their contributions, including methods to calculate derivatives and integrals, as well as Newton's application to planetary motion.

- It may be instructive to apply the definition of the derivative to obtain equation (6.8):

$$\frac{d}{dt}[\tilde{x}(t)] = \lim_{\Delta \to 0} \frac{\tilde{x}(t + \Delta) - \tilde{x}(t)}{\Delta}$$

$$= \lim_{\Delta \to 0} \frac{[x(t + \Delta) - x(t_i)] - [x(t) - x(t_i)]}{\Delta}$$

$$= \lim_{\Delta \to 0} \frac{x(t + \Delta) - x(t)}{\Delta}$$

$$= \frac{d}{dt}[x(t)]$$

$$= v(t)$$

6.1.5 Integration, Leibniz' Notation, and the Fundamental Theorem of Calculus

Although there is an equivalence between the area under a curve and the position of an object as demonstrated in Section 6.1.3, Leibniz did not approach calculus from this perspective. For Leibniz, integral calculus was purely the determination of areas under a curve. We follow Leibniz' approach as it is the most natural way to introduce his notation.

Leibniz' starting point was the approximation of the area by sums. In Section 6.1.3, an underestimation and an overestimation are presented, and the estimates converge to the area as the number of partition elements increases. Recall the notation in which the curve is given by $v(t)$:

$$\sum_{j=1}^{n} \underline{\Delta x}_j \le \sum_{j=1}^{n} \Delta x_j = x(t_f) - x(t_i) \le \sum_{j=1}^{n} \overline{\Delta x}_j$$

$$\underline{\Delta x}_j = \underline{v}_j(t_j - t_{j-1}) \qquad \overline{\Delta x}_j = \overline{v}_j(t_j - t_{j-1})$$

$$t_i = t_0 < t_1 < t_2 < t_3 < \cdots < t_n = t_f$$

Note that the middle term, $x(t_f) - x(t_i)$, representing the area under the curve, may be negative (Figure 6.10). Leibniz' interpretation is that when the function $v(t)$ is negative the region between the function and the t axis is assigned the negative value of the corresponding area.

Substituting the values of Δx_j into the sums yields the following:

$$\sum_{j=1}^{n} \underline{\Delta x}_j \le x(t_f) - x(t_i) \le \sum_{j=1}^{n} \overline{\Delta x}_j$$

$$\sum_{j=1}^{n} \underline{v}_j(t_j - t_{j-1}) \le x(t_f) - x(t_i) \le \sum_{j=1}^{n} \overline{v}_j(t_j - t_{j-1}) \qquad (6.9)$$

$$\sum_{j=1}^{n} \underline{v}_j \, \Delta t_j \le x(t_f) - x(t_i) \le \sum_{j=1}^{n} \overline{v}_j \, \Delta t_j$$

where $\Delta t_j = t_j - t_{j-1}$.

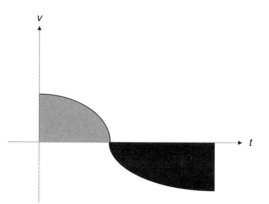

Figure 6.10　Positive (gray) and negative (black) areas of integration.

We consider only curves in which the area is well defined so that the underestimates converge to the overestimates. Accordingly, for ease of exposition, it is only necessary to consider one case. It is possible to choose either underestimates or overestimates, whichever is most convenient. Leibniz then examined the limiting case when the length of all cell intervals, Δt_j, approaches zero. His notation as the limiting case is considered in the following equation:

$$x(t_f) - x(t_i) = \lim_{n \to \infty} \sum_{j=1}^{n} \bar{v}_j \, \Delta t_j = \int_{t_i}^{t_f} v(t) \, dt \qquad (6.10)$$

The final term on the right-hand side is read as "the integral of the function $v(t)$ between t_i and t_f." The symbol $\int$ is known as the integral sign. Leibniz chose a stretched-out S to indicate that this is a summation. The value $\int_{t_i}^{t_f} v(t) \, dt$ is known as the definite integral, indicating that a well-defined area between t_i and t_f is being assessed. Notice the use of "dt" to represent the limiting length of the partitions. This is by design similar to the notation of the derivative, dx/dt, in which "dt" denotes a limiting length of an interval shrinking to zero and dx/dt is the ratio of shrinking intervals. Heuristically, Leibniz considered integration to be the operation of summing up an infinite number of rectangles with height $v(t)$ and base "dt."

Placing this notation into the fundamental theorem of calculus, equation (6.8) results in the following:

$$\frac{d}{dt}[\tilde{x}(t)] = v(t)$$

$$\frac{d}{dt}\left(\int_{t_i}^{t} v(\tau) \, d\tau\right) = v(t) \qquad (6.11)$$

Some attention to the notation is worthwhile. Note that t is considered a variable so that the integral is a function of t and it makes sense to take the derivative with respect to t. Also, note the use of the variable τ following the integral sign. A letter different from t was chosen since t is the uppermost limit of the interval of integration.

Integrals are rarely assessed using the definition equation (6.10). Instead, one assesses integrals using a grab bag of approaches that are sometimes useful and

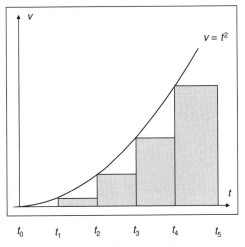

Figure 6.11   Underestimation of the integral by rectangles.

sometimes not (methods are introduced in subsequent sections). Nevertheless, in some instances it is possible to apply the definition, and it is instructive to go through one such example. Consider the case $v(t) = t^2$. Let $t_i = 0$ and t_f be any number greater than zero. Let the interval of integration be partitioned into n cells of equal size; for each cell, $\Delta t_j = t_j - t_{j-1} = t_f/n$. It follows that $t_i = t_0 = 0$, $t_1 = t_f/n$, $t_2 = 2t_f/n$, and in general $t_j = jt_f/n$. Also, for each interval

$$\underline{v}_j = t_{j-1}^2 = \left(\frac{(j-1)t_f}{n}\right)^2 \qquad \overline{v}_j = t_j^2 = \left(\frac{jt_f}{n}\right)^2$$

Placing these values into the inequalities of equation (6.9) results in the following:

$$\sum_{j=1}^{n} \underline{v}_j\, \Delta t_j \le x(t_f) - x(t_i) \le \sum_{j=1}^{n} \overline{v}_j\, \Delta t_j$$

$$\sum_{j=1}^{n} t_{j-1}^2 \frac{t_f}{n} \le x(t_f) - x(t_i) \le \sum_{j=1}^{n} t_j^2 \frac{t_f}{n} \tag{6.12}$$

$$\sum_{j=1}^{n} \left(\frac{(j-1)t_f}{n}\right)^2 \frac{t_f}{n} \le x(t_f) - x(t_i) \le \sum_{j=1}^{n} \left(\frac{jt_f}{n}\right)^2 \frac{t_f}{n}$$

The table below presents values for the left- and right-hand sums for partitions having 1–10 cells. Figure 6.11 corresponds to the table with the number of partitions set at 5 (the first cell has zero area). In both the table and the figure, $t_f = 1$.

Number of cells	5	10	15	20	25	30
Lower sum	0.24	0.285	0.3007	0.3088	0.3136	0.3137
Upper sum	0.44	0.385	0.3674	0.3588	0.3536	0.3502

The table suggests that the left-hand side and the right-hand side converge as the number of cells increases. What happens in the limit as the cell sizes $\Delta t_j = 1/n$

all approach zero (equivalently allow the number of cells, n, to become indefinitely large)? To answer the question, it is necessary to apply some algebra. The induction result of Section 4.5.3 is useful. Recall the result

$$\sum_{j=1}^{n} j^2 = \tfrac{1}{3}n^3 + \tfrac{1}{2}n^2 + \tfrac{1}{6}n$$

It follows that

$$\sum_{j=1}^{n}(j-1)^2 = \sum_{j=0}^{n-1} j^2$$

$$= \tfrac{1}{3}(n-1)^3 + \tfrac{1}{2}(n-1)^2 + \tfrac{1}{6}(n-1)$$

$$= \tfrac{1}{3}n^3 - \tfrac{1}{2}n^2 + \tfrac{1}{6}n$$

Simplifying the left-handed expression of equation (6.12) results in the following:

$$\sum_{j=1}^{n}\left(\frac{(j-1)t_f}{n}\right)^2 \frac{t_f}{n} = \left(\frac{t_f}{n}\right)^3 \sum_{j=1}^{n}(j-1)^2$$

$$= \left(\frac{t_f}{n}\right)^3 \left(\frac{1}{3}n^3 - \frac{1}{2}n^2 + \frac{1}{6}n\right)$$

$$= t_f^3\left(\frac{1}{3} - \frac{1}{2n} + \frac{1}{6n^2}\right)$$

We can now take the limit as n becomes indefinitely large:

$$\lim_{n\to\infty}\sum_{j=1}^{n}\left(\frac{(j-1)t_f}{n}\right)^2 \frac{t_f}{n} = \lim_{n\to\infty} t_f^3\left(\frac{1}{3} + \frac{1}{2n} + \frac{1}{6n^2}\right)$$

$$= \frac{t_f^3}{3}$$

The answer results from insignificance of the terms $1/(2n)$ and $1/(6n^2)$ as n becomes indefinitely large.

A similar calculation on the right-hand side of equation (6.12) results in the same answer.

$$\lim_{n\to\infty}\sum_{j=1}^{n}\left(\frac{jt_f}{n}\right)^2 \frac{t_f}{n} = \lim_{n\to\infty}\left(\frac{t_f}{n}\right)^3 \sum_{j=1}^{n} j^2$$

$$= \lim_{n\to\infty}\left(\frac{t_f}{n}\right)^3 \left(\frac{1}{3}n^3 + \frac{1}{2}n^2 + \frac{1}{6}n\right)$$

$$= \lim_{n\to\infty} t_f^3\left(\frac{1}{3} + \frac{1}{2n} + \frac{1}{6n^2}\right)$$

$$= \frac{t_f^3}{3}$$

Both the left-hand side and the right-hand side of the inequality approach the same value, so the object moves a distance of $\frac{1}{3} t_f^3$ between the times 0 and t_f. Note that when $t_f = 1$ the value of the definite integral is $\frac{1}{3}$. The previous table indicates that the estimates obtained by partitioning indeed approach $\frac{1}{3}$.

Using Leibniz' definition of the integral, we have found

$$\int_0^t \tau^2 \, d\tau = \frac{t^3}{3}$$

But we have found more. The relation between velocity and position as expressed in the fundamental theorem of calculus allows us to also conclude the following:

$$\frac{d}{dt} \left(\int_0^t \tau^2 d\tau \right) = t^2$$

$$\frac{d}{dt} \left(\frac{t^3}{3} \right) = t^2$$

So in addition to the value of the integral, the derivative of the function $\frac{1}{3} t^3$ has been determined. How do we interpret this result in terms of displacement and velocity? If the position of an object is given by the function $x(t) = \frac{1}{3} t^3$, then its velocity is given by $v(t) = t^2$. Conversely, if an object's velocity is given by $v(t) = t^2$ and the position at time $t = 0$ is zero, then the position for any time is given by $x(t) = \frac{1}{3} t^3$.

6.2 MORE MOTION: GOING IN CIRCLES

Copernicus was partial to circles and uniform circular motion. In fact, to Copernicus, one of the more displeasing elements of Ptolemy's universe was the introduction of equants that caused planets to move with non-uniform speed. Copernicus could not eliminate the epicycles because observations of the planets confirmed that they did not move in circular orbits, but he did eliminate equants. Recall that by reason of symmetry circular orbits are compatible with Kepler's view that the sun's force on an object is determined by the object's distance from the sun. By studying circular motion at constant speed, Kepler recognized that the angular momentum of the planet is constant and reasoned that a planet following an elliptic path must respect this property. In this section, we examine circular motion and receive a bonus for the effort. Formulas for the derivative of the sine and cosine functions fall out of the process.

The study of circular motion forces us to generalize the concept of velocity from linear to planar motion. A moving object's position in a plane can be expressed as a vector $\vec{X}(t)$ in which the entries are time-dependent functions, $\vec{X}(t) = (x(t) \ y(t))^{\mathrm{T}}$. (Recall a capital T is used to indicate the transpose of the row into a column vector.) The velocity of the object is defined in a manner similar to that of the velocity

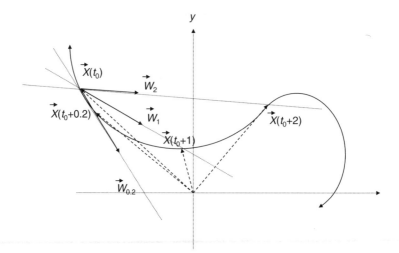

Figure 6.12   Approximating the velocity vector.

in a single dimension:

$$\vec{V}(t) = \lim_{\Delta \to 0} \frac{\vec{X}(t + \Delta) - \vec{X}(t)}{\Delta}$$

$$= \lim_{\Delta \to 0} \left[\begin{pmatrix} x(t + \Delta) \\ y(t + \Delta) \end{pmatrix} - \begin{pmatrix} x(t) \\ y(t) \end{pmatrix} \right] \frac{1}{\Delta}$$

$$= \begin{pmatrix} \dfrac{d}{dt}[x(t)] \\ \dfrac{d}{dt}[y(t)] \end{pmatrix}$$

The equation states that the velocity of a position vector is the derivative of its components. As notation, we use $\vec{V}(t) = (v_x(t)\ v_y(t))^{\mathrm{T}}$; that is,

$$v_x(t) = \frac{d}{dt}[x(t)] \qquad v_y(t) = \frac{d}{dt}[y(t)]$$

Figure 6.12 illustrates the process of determining the velocity vector at a position $\vec{X}(t_0)$. An object's pathway is given along with the position $\vec{X}(t_0)$. A second position, $\vec{X}(t + \Delta)$, is considered. The vector $\vec{W}_\Delta = [\vec{X}(t + \Delta) - \vec{X}(t)]/\Delta$ lies on the line between the two points. The vector $\vec{W}_\Delta$ approaches the velocity vector as Δ approaches zero. Note that, similar to Figure 6.5, the line segments between the points $\vec{X}(t_0)$ and $\vec{X}(t + \Delta)$ approach the tangent line to the curve at the position $\vec{X}(t_0)$. This means that the velocity vector is tangent to the curve. Figure 6.13 illustrates an object's pathway along with the velocity vector at different times along the path. As noted, the velocity vector is always tangent to the curve.

We next determine the position and velocity of an object moving about a circular path centered at the origin at constant speed. The position vector at a point in time is obtained by applying a rotation matrix to the initial position. We assume that the

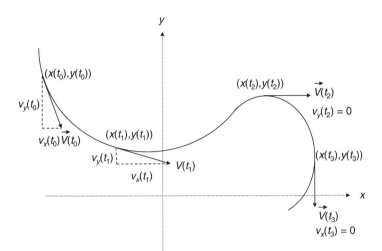

Figure 6.13   Velocity vector always tangent to curve.

initial position is along the x axis:

$$\vec{X}(t) = \begin{pmatrix} \cos[\theta(t)] & -\sin[\theta(t)] \\ \sin[\theta(t)] & \cos[\theta(t)] \end{pmatrix} \begin{pmatrix} r \\ 0 \end{pmatrix}$$

The value r is the radius of the motion. It is necessary to determine $\theta(t)$. Note that, by assumption, the initial angle is zero, $\theta(0) = 0$. As the speed of the object is constant, the rate at which the angle changes, $\frac{d}{dt}\theta(t)$, is also constant, $\frac{d}{dt}\theta(t) = \omega$. It follows from Section 6.1.1, where the relation between constant speed and linear motion is presented, that $\theta(t) = \omega t$ with ω a constant:

$$\vec{X}(t) = \begin{pmatrix} \cos(\omega t) & -\sin(\omega t) \\ \sin(\omega t) & \cos(\omega t) \end{pmatrix} \begin{pmatrix} r \\ 0 \end{pmatrix} = r \begin{pmatrix} \cos(\omega t) \\ \sin(\omega t) \end{pmatrix}$$

The velocity vector is found by differentiating the position vector, but there is an alternative way. The direction of the velocity vector can be determined by noting that the velocity vector is tangent to the circle. Additionally, the length of the velocity vector can be determined by noting that the velocity vector's length is the object's speed. Knowing the direction and the length allows us to specify the velocity vector. The details of the calculation follow.

In Section 4.3.3, the tangent line to a circle is found; the result is that the tangent line is perpendicular to the position vector. As the velocity vector lies along the tangent line, it is also perpendicular to the position vector and points in the direction of the motion. The vector $\vec{W}_0 = (0\ 1)^T$ is perpendicular to the initial position and points in the direction of the velocity. Furthermore, the rotation matrix maintains the angle between two vectors; applying the rotation matrix to the vector $(0\ 1)^T$ results in a vector that is always perpendicular to the position vector $\vec{X}_t$ and always points in the

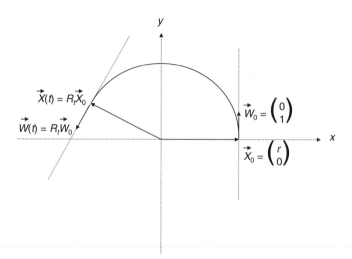

Figure 6.14 Velocity vector for circular motion.

direction of the velocity (Figure 6.14). The vector $W(t) = R_t W_0$ given below points in the direction of the velocity vector:

$$R_t \vec{W}_0 = \begin{pmatrix} \cos(\omega t) & -\sin(\omega t) \\ \sin(\omega t) & \cos(\omega t) \end{pmatrix} \begin{pmatrix} 0 \\ 1 \end{pmatrix}$$

$$= \begin{pmatrix} -\sin(\omega t) \\ \cos(\omega t) \end{pmatrix}$$

$$= \vec{W}(t)$$

The vector $\vec{W}(t)$ has a constant length of 1. Since $\vec{V}(t)$ and $\vec{W}(t)$ point in the same direction, $\vec{V}(t)$ is a multiple of $\vec{W}(t)$, $\vec{V}(t) = s\vec{W}(t)$, for some scalar multiple s. In fact, since the length of $\vec{W}(t)$ is 1, s is the length of $\vec{V}(t)$ and the speed of the object. Also, the distance that the object travels in time t is $r\theta(t) = r\omega t$, giving a constant speed of $s = r\omega$. [Here $r\theta(t)$ is the arc length of the object's path through the angle $\theta(t)$.] The velocity vector can now be specified:

$$\vec{V}(t) = s\vec{W}(t)$$

$$\begin{pmatrix} v_x(t) \\ v_y(t) \end{pmatrix} = s \begin{pmatrix} -\sin(\omega t) \\ \cos(\omega t) \end{pmatrix} = r\omega \begin{pmatrix} -\sin(\omega t) \\ \cos(\omega t) \end{pmatrix}$$

Relating the velocity vector to the derivative of the position yields the following derivatives:

$$\frac{d}{dt}[x(t)] = v_x(t)$$

$$\frac{d}{dt}[r\cos(\omega t)] = -r\omega \sin(\omega t)$$

$$\frac{d}{dt}[y(t)] = v_y(t)$$

$$\frac{d}{dt}[r\sin(\omega t)] = r\omega\cos(\omega t)$$

Setting r and ω to 1 results in the derivatives of the sine and cosine functions:

$$\frac{d}{dt}[\cos(t)] = -\sin(t) \qquad \frac{d}{dt}[\sin(t)] = \cos(t) \qquad (6.13)$$

Remark. The tangency of the velocity vector to the curve can be demonstrated as follows. The slope between two points is given by t and $t + \Delta$:

$$\text{Slope} = \frac{y(t + \Delta) - y(t)}{x(t + \Delta) - x(t)}$$

$$\text{Tangent slope} = \lim_{\Delta \to 0} \frac{y(t + \Delta) - y(t)}{x(t + \Delta) - x(t)}$$

$$= \lim_{\Delta \to 0} \frac{y(t + \Delta) - y(t)\Delta}{x(t + \Delta) - x(t)\Delta}$$

$$= \frac{\frac{d}{dt}[y(t)]}{\frac{d}{dt}[x(t)]}$$

$$= \frac{v_y(t)}{v_x(t)}$$

The above calculation holds provided that $\frac{d}{dt}[x(t)]$ is not zero. In the case that $\frac{d}{dt}[x(t)]$ is zero, the tangent is not defined. This corresponds to a vertical line.

6.3   MORE DIFFERENTIAL CALCULUS

While Barrow discovered the fundamental theorem of calculus and was able to differentiate a few functions, both Newton and Leibniz discovered rules that allow them to differentiate large classes of functions. Using the rules, the problem of differentiation is turned into a rather mechanical process involving algebraic manipulations; to uncover the ellipse, it is necessary to become familiar with the mechanical process. This section first presents the mechanics in a stripped-down, just-the-facts manner. The concept of higher order derivatives is also presented. Finally, there is an excursion that includes a more thorough treatment of the mechanical process of differentiation, the exponential function, and a proof of the identity that Aristarchus uses as presented in Section 6.3.4.

The starting point for the mechanical process is the following table of derivatives for a very short list of functions. In subsequent sections, the table is used to differentiate more complicated functions.

Function	Derivative
$\sin(t)$	$\cos(t)$
$\cos(t)$	$-\sin(t)$
t^n	nt^{n-1}
$c = $ constant	0
e^t	e^t
$\ln(t)$	$1/t$

6.3.1 Differentiation Rules

There are five differentiation rules. These rules are bridges between extremely complicated expressions and the simple functions of the table presented above. The rules
are presented along with examples of their use.

Rule 1. Additive rule

$$\frac{d}{dt}[x(t) + y(t)] = \frac{d}{dt}[x(t)] + \frac{d}{dt}[y(t)]$$

Example: Determine $\frac{d}{dt}[\sin(t) + t^2]$. Using the additive rule and the above table,

$$\frac{d}{dt}[\sin(t) + t^2] = \frac{d}{dt}[\sin(t)] + \frac{d}{dt}(t^2)$$

$$= \cos(t) + 2t$$

Rule 2. Multiplicative constant rule

When the value c is a constant value,

$$\frac{d}{dt}[cx(t)] = c\frac{d}{dt}[x(t)]$$

Example: Determine $\frac{d}{dt}(3t^5)$. *Solution:*

$$\frac{d}{dt}(3t^5) = 3\frac{d}{dt}(t^5)$$

$$= 15t^4$$

Rule 3. Multiplicative rule

$$\frac{d}{dt}[x(t)y(t)] = x(t)\frac{d}{dt}[y(t)] + y(t)\frac{d}{dt}[x(t)]$$

Example: Determine $\frac{d}{dt}[t^3 \cos(t)]$. *Solution:*

$$\frac{d}{dt}[t^3 \cos(t)] = t^3\frac{d}{dt}[\cos(t)] + \cos(t)\frac{d}{dt}(t^3)$$

$$= -t^3 \sin(t) + 3t^2 \cos(t)$$

Rule 4. Quotient rule

$$\frac{d}{dt}\left(\frac{x(t)}{y(t)}\right) = \frac{y(t)\dfrac{d}{dt}[x(t)] - x(t)\dfrac{d}{dt}[y(t)]}{(y(t))^2}$$

Example: Determine

$$\frac{d}{dt}[\tan(t)] = \frac{d}{dt}\left(\frac{\sin(t)}{\cos(t)}\right)$$

Solution:

$$\frac{d}{dt}\left(\frac{\sin(t)}{\cos(t)}\right) = \frac{\cos(t)\dfrac{d}{dt}[\sin(t)] - \sin(t)\dfrac{d}{dt}[\cos(t)]}{\cos^2(t)}$$

$$= \frac{\cos^2(t) + \sin^2(t)}{\cos^2(t)}$$

$$= \frac{1}{\cos^2(t)}$$

$$= \sec^2(t)$$

Rule 5. Chain rule

$$\frac{d}{dt}[x(y(t))] = \frac{d}{dy}[x(y)]\frac{d}{dt}(y(t))$$

Example: Determine $\frac{d}{dt}[\sin(t^3)]$. Set $y = t^3$:

$$\frac{d}{dt}[\sin(t^3)] = \frac{d}{dy}[\sin(y)]\frac{d}{dt}(t^3)$$

$$= -\cos(y)3t^2$$

$$= -3t^2\cos(t^3)$$

6.3.2   Notation and the Derivative at a Specified Point

We have used the notation $\frac{dx}{dt}$ and $\frac{d}{dt}[x(t)]$ to represent the derivative of a function which is another function. The notation $x'(t)$ is also commonly used for the same purpose. The following notation for the evaluation of the derivative at a specified point t_0 is common

$$\frac{dx}{dt}\Big|_{t=t_0} \qquad \frac{d}{dt}(x(t))\,|_{t=t_0} \qquad x'(t_0)$$

Example: Find the derivative of the function $x = \tan(t)$ at the time $t = \pi$.

From the example following rule 4 in Section 6.3.1,

$$\frac{dx}{dt} = \frac{d}{dt}[x(t)] = x'(t) = \sec^2(t)$$

Evaluating the derivative at $t = \pi$ results in the following:

$$\frac{dx}{dt} \big|_{t=\pi} = \frac{d}{dt}(x(t)) \big|_{t=\pi} = x'(\pi) = \sec^2(\pi) = 1$$

6.3.3 Higher Order Differentiation and Examples

The result of differentiating a function is another function. At times it is of interest to differentiate the resulting function, and this is known as taking the second derivative. One can continue to take the third derivative, fourth derivative, and so on, *ad nauseam*. There are many applications where taking a higher order derivative is useful; Newton's laws of motion are based on the second derivative. In this section, notation is introduced through examples.

An additional benefit of the examples is that they illustrate the use of the differentiation rules. As noted above, the rules are bridges toward a simplification that requires only the derivatives of the elementary functions listed in the table at the beginning of the section. The reader is encouraged to identify the rule used as the expression is morphed into a more congenial form.

Example 6.1

$$\frac{d}{dt}(t^3) \qquad \frac{d^2}{dt^2}(t^3) \qquad \frac{d^3}{dt^3}(t^3) \qquad \frac{d^4}{dt^4}(t^3)$$

$$\frac{d}{dt}(t^3) = 3t^2 \quad \frac{d^2}{dt^2}(t^3) = \frac{d}{dt}\left(\frac{d}{dt}(t^3)\right) \quad \frac{d^3}{dt^3}(t^3) = \frac{d}{dt}\left(\frac{d^2}{dt^2}(t^3)\right) \quad \frac{d^4}{dt^4}(t^3) = \frac{d}{dt}\left(\frac{d^3}{dt^3}(t^3)\right)$$

$$\qquad\qquad = \frac{d}{dt}(3t^2) \qquad\qquad = \frac{d}{dt}(6t) \qquad\qquad = \frac{d}{dt}(6)$$

$$\qquad\qquad = 3\frac{d}{dt}(t^2) \qquad\qquad = 6\frac{d}{dt}(t) \qquad\qquad = 0$$

$$\qquad\qquad = 6t \qquad\qquad\qquad = 6$$

Example 6.2

$$\frac{d}{dt}[\sin(5t)] \qquad\qquad\qquad \frac{d^2}{dt^2}[\sin(5t)]$$

$$\frac{d}{dt}[\sin(5t)] = \frac{d}{dy}[\sin(y)]\frac{d}{dt}(5t) \qquad \frac{d^2}{dt^2}[\sin(5t)] = \frac{d}{dt}\left(\frac{d}{dt}[\sin(5t)]\right)$$

$$= \cos(y)5\frac{d}{dt}(t) \qquad\qquad\qquad = \frac{d}{dt}[5\cos(5t)]$$

$$= 5\cos(5t) \qquad\qquad\qquad\qquad = 5\frac{d}{dt}[\cos(5t)]$$

$$\qquad\qquad\qquad\qquad = 5\frac{d}{dy}[\cos(y)]\frac{d}{dt}(5t)$$

$$\qquad\qquad\qquad\qquad = -5\sin(y)5\frac{d}{dt}(t)$$

$$\qquad\qquad\qquad\qquad = -25\sin(5t)\frac{d}{dt}(t)$$

$$\qquad\qquad\qquad\qquad = -25\sin(5t)$$

Example 6.3

(Uses results of Examples 6.1 and 6.2)

$$\frac{d}{dt}\left(t^3 \sin(5t)\right) = t^3 \frac{d}{dt}[\sin(5t)] + \sin(5t)\frac{d}{dt}(t^3)$$
$$= 5t^3 \cos(5t) + 3t^2 \sin(5t)$$

$$\frac{d^2}{dt^2}[t^3 \sin(5t)] = \frac{d}{dt}\left(\frac{d}{dt}[t^3 \sin(5t)]\right) = \frac{d}{dt}\left[5t^3 \cos(5t) + 3t^2 \sin(5t)\right]$$

$$= 5\frac{d}{dt}[t^3 \cos(5t)] + 3\frac{d}{dt}[t^2 \sin(5t)]$$

$$= 5\left(t^3 \frac{d}{dt}[\cos(5t)] + \cos(5t)\frac{d}{dt}(t^3)\right)$$

$$+ 3\left(t^2 \frac{d}{dt}[\sin(5t)] + \sin(5t)\frac{d}{dt}(t^2)\right)$$

$$= -25t^3 \sin(5t) + 15t^2 \cos(5t)$$

$$+ 15t^2 \cos(5t) + 6t \sin(5t)$$

$$= -19t^3 \sin(5t) + 30t^2 \cos(5t)$$

Example 6.4

$$\frac{d}{dt}\left\{\sin[\cos(t^2)]\right\} = \frac{d}{dy}[\sin(y)]\frac{d}{dt}[\cos(t^2)]$$

$$= \cos(y)\frac{d}{dt}[\cos(t^2)]$$

$$= \cos[\cos(t^2)]\frac{d}{dz}[\cos(z)]\frac{d}{dt}(t^2)$$

$$= -\cos[\cos(t^2)]\sin(z)\frac{d}{dt}(t^2)$$

$$= -\cos[\cos(t^2)]\sin(t^2)\frac{d}{dt}(t^2)$$

$$= -2\cos[\cos(t^2)]\sin(t^2)t$$

With a little practice, it is possible to differentiate the above expression quickly to obtain the second derivative. The reader might want to get out paper and pencil and fill in the missing steps to the solution given below:

$$\frac{d^2}{dt^2}\left\{\sin[\cos(t^2)]\right\} = \frac{d}{dt}\left\{-2t\cos[\cos(t^2)]\sin(t^2)\right\}$$
$$= -2\cos[\cos(t^2)]\sin(t^2) - 4t^2 \cos[\cos(t^2)]\cos(t^2)$$
$$- 4t^2 \sin^2(t^2)\sin[\cos(t^2)]$$

6.3.4 Differentiation and the Enquirer

The preceding material in Section 6.3 requests that the reader accept on faith both the initial table of derivatives (except for sine and cosine presented in Section 6.2) and the rules of differentiation. The inquiring mind may wish to understand the basis for the rules of differentiation and derivatives of the elementary functions. This section responds by providing both heuristic and formal arguments.

Additionally, the exponential and logarithmic functions introduced in the table on page 218 are ignored in the subsequent material. While not necessary for pursuing the ellipse, these functions are central to many core concepts that are approached through calculus and are a part of the calculus curriculum that is taught as an international standard. For this reason, it is not possible to write a book that includes calculus without some mention of the exponential and logarithmic functions, so I feel compelled to include the material. This section also includes a demonstration of the identity used by Aristarchus, equation (5.4). The reader may skip this section and still follow the remaining material.

6.3.4.1 Heuristic Understanding of the Addition Rule

$$\frac{d}{dt}[x(t) + y(t)] = \frac{d}{dt}[x(t)] + \frac{d}{dt}[y(t)]$$

A physical example of the additive rule offers some insight. Imagine two trains going down a straight railroad track, A and B, with A in front of B. Also imagine a stationary observer, C, as in Figure 6.15. In this setup, the distance from B to A is given by $x(t)$, the distance from C to B is $y(t)$, and the distance from C to A is $z(t) = x(t) + y(t)$. The additive rule states that given the relative velocity of train A with respect to train B, $\frac{d}{dt}[x(t)]$, and the velocity of train B with respect to the stationary observer C, $\frac{d}{dt}[y(t)]$, the velocity of train A with respect to the stationary observer is the sum of the two given velocities:

$$\frac{d}{dt}[z(t)] = \frac{d}{dt}[x(t) + y(t)] = \frac{d}{dt}[x(t)] + \frac{d}{dt}[y(t)]$$

6.3.4.2 Heuristic Understanding of the Multiplicative Constant Rule

$$\frac{d}{dt}(cx(t)) = c\frac{d}{dt}(x(t))$$

As with the addition rule, intuitive reasoning guides us toward the multiplicative constant rule. Once again using the train analogy, suppose that train B's position from

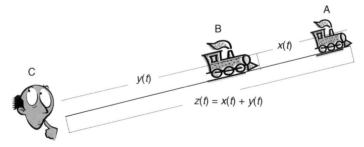

Figure 6.15   Addition rule.

observer C is given by $x(t)$ and train A's position is always c times that of train B. If $y(t)$ is the position of train A as viewed by observer C, $y(t) = cx(t)$. Then train A is moving c times as fast as train B:

$$\frac{d}{dt}[y(t)] = \frac{d}{dt}[cx(t)] = c\frac{d}{dt}[x(t)]$$

6.3.4.3 Heuristic Understanding of the Chain Rule

$$\frac{d}{dt}\{x[y(t)]\} = \frac{d}{dy}[x(y)]\frac{d}{dt}[y(t)]$$

Once again the analysis of a physical problem affords insight. Suppose that an American is standing next to a European and the European continuously records the distance to a moving train in meters. The distance is represented by $y(t)$. Next suppose that the American wishes to have the record of distances in feet, so he has a process that converts meters to feet. The conversion process is represented by $x(y)$, so the record of the distance to the train in feet is given by $x[y(t)]$. To get the speed in feet per second, one finds the speed in meters per second, $\frac{d}{dt}[y(t)]$, and then multiplies it by the ratio indicating the train's change in feet over its change in meters, $\frac{d}{dy}[x(y)]$. Hence,

$$\frac{d}{dt}\{x[y(t)]\} = \frac{d}{dy}[x(y)]\frac{d}{dt}[y(t)]$$

This intuition is more evident using the notation

$$\frac{dx}{dt} = \frac{dx}{dy}\frac{dy}{dt}$$

The expression states that the change in feet per change in time is equal to the change in feet per change in meters multiplied by the change in meters per change in time. The change in meters, dy, cancels on the top and bottom to give the desired answer.

6.3.4.4 Formal Derivation of the Multiplication Rule

$$\frac{d}{dt}[x(t)y(t)] = x(t)\frac{d}{dt}[y(t)] + y(t)\frac{d}{dt}[x(t)]$$

Below, a formal argument of the multiplication rule is presented. A similar argument can be made to formalize the other rules as well. The argument appeals directly to the definition of the derivative, equation (6.3).

Begin by noting the following approximation:

$$x(t + \Delta) \approx x(t) + v(t)\Delta = x(t) + \frac{d}{dt}[x(t)]\Delta$$

Solving for the difference in position $x(t + \Delta) - x(t)$ results in the approximation $x(t + \Delta) - x(t) \approx \frac{d}{dt}[x(t)]\Delta$. The approximation indicates that during a small time increment Δ an object roughly moves through a displacement that is determined solely by its velocity at the beginning of the increment and the length of the increment. This is not an equality because as the object moves from its position at time t to a new position at time $t + \Delta$ the velocity is not necessarily constant but may change. Let us eliminate the approximation using a correction term that we denote by $\xi(t, \Delta)$:

$$x(t + \Delta) = x(t) + \frac{d}{dt}[x(t)]\Delta + \xi(t, \Delta)$$

$$x(t + \Delta) - x(t) - \xi(t, \Delta) = \frac{d}{dt}[x(t)]\Delta$$

Dividing the second equation by Δ and taking the limit as Δ approaches zero results in the following:

$$\lim_{\Delta \to 0} \frac{x(t + \Delta) - x(t) - \xi(t, \Delta)}{\Delta} = \frac{d}{dt}[x(t)]$$

From this one can conclude that $\lim_{\Delta \to 0} \xi(t, \Delta)/\Delta = 0$. This will be useful in the following derivation:

$$\frac{d}{dt}\left[x(t)y(t)\right] = \lim_{\Delta \to 0} \frac{1}{\Delta}\left[x(t + \Delta)y(t + \Delta) - x(t)y(t)\right]$$

$$= \lim_{\Delta \to 0} \frac{1}{\Delta}\left[\left(x(t) + \frac{d}{dt}[x(t)]\Delta + \xi_x(t, \Delta)\right)\right.$$

$$\times \left.\left(y(t) + \frac{d}{dt}[y(t)]\Delta + \xi_y(t, \Delta)\right) - x(t)y(t)\right]$$

$$= \lim_{\Delta \to 0} \frac{1}{\Delta}\left[x(t)y(t) + x(t)\frac{d}{dt}[y(t)]\Delta + x(t)\xi_y(t, \Delta) + y(t)\frac{d}{dt}[x(t)]\Delta\right.$$

$$+ \frac{d}{dt}[y(t)]\frac{d}{dt}[x(t)]\Delta^2 + \frac{d}{dt}[x(t)]\xi_y(t, \Delta) + \xi_x(t, \Delta)y(t)$$

$$+ \left.\xi_x(t, \Delta)\frac{d}{dt}[y(t)]\Delta + \xi_x(t, \Delta)\xi_y(t, \Delta) - x(t)y(t)\right]$$

$$= \lim_{\Delta \to 0} \frac{1}{\Delta}\left[\left(x(t)\frac{d}{dt}[y(t)] + y(t)\frac{d}{dt}[x(t)]\right)\Delta + \frac{d}{dt}[y(t)]\frac{d}{dt}[x(t)]\Delta^2\right.$$

$$+ \xi_x(t, \Delta)\left(y(t) + \frac{d}{dt}[y(t)]\Delta\right) + \xi_y(t, \Delta)\left(x(t) + \frac{d}{dt}[x(t)]\Delta\right)$$

$$+ \left.\xi_x(t, \Delta)\xi_y(t, \Delta)\right]$$

$$= \lim_{\Delta \to 0}\left[\left(x(t)\frac{d}{dt}[y(t)] + y(t)\frac{d}{dt}[x(t)]\right) + \frac{d}{dt}[y(t)]\frac{d}{dt}[x(t)]\Delta\right.$$

$$+ \frac{\xi_x(t, \Delta)\left(y(t) + \frac{d}{dt}[y(t)]\Delta\right)}{\Delta} + \frac{\xi_y(t, \Delta)\left(x(t) + \frac{d}{dt}[x(t)]\Delta\right)}{\Delta}$$

$$+ \left.\frac{\xi_x(t, \Delta)\xi_y(t, \Delta)}{\Delta}\right]$$

$$= \lim_{\Delta \to 0} \left(x(t) \frac{d}{dt}[y(t)] + y(t) \frac{d}{dt}[x(t)] \right) + \lim_{\Delta \to 0} \left(\frac{d}{dt}[y(t)] \frac{d}{dt}[x(t)]\Delta \right)$$

$$+ \lim_{\Delta \to 0} \frac{\xi_x(t, \Delta) \left(y(t) + \frac{d}{dt}[y(t)]\Delta \right)}{\Delta}$$

$$+ \lim_{\Delta \to 0} \frac{\xi_y(t, \Delta) \left(x(t) + \frac{d}{dt}[x(t)]\Delta \right)}{\Delta} + \lim_{\Delta \to 0} \frac{\xi_x(t, \Delta)\xi_y(t, \Delta)}{\Delta}$$

$$= x(t) \frac{d}{dt}[y(t)] + y(t) \frac{d}{dt}[x(t)]$$

The above derivation assumes that both $\frac{d}{dt}[x(t)]$ and $\frac{d}{dt}[y(t)]$ are well defined. The derivation also makes use of an additive property of limits—the limit of sums is the same as the sum of limits—which equates expressions above and to the right of the sixth equality sign. The final step makes use of the fact that a limit of factors is the same as the factor of limits. Within the expression above the final result, the second term disappears since the factor Δ goes to zero. The remaining terms all disappear since the factors $\xi_x(t, \Delta)/\Delta$ and $\xi_y(t, \Delta)/\Delta$ both go to zero. Note that within the limiting expression to the right of the fifth equal sign the terms are lined up as a first-order term in Δ, a second-order term in Δ, and the expressions involving ξ_x or ξ_y. The only term that survives the limiting process is the first-order term in Δ, which is typical for an argument of this type.

6.3.4.5 The Derivative of $x(t) = t^n$ In this section, differentiation of functions of the type $x(t) = t^n$, where n is any real number, is considered. The presentation is in cases, starting with positive integers and proceeding to negative integers and then to rational numbers and finally to any real number.

Case 1. n a positive integer.

The method of induction by the Muslim mathematician Al Karaji is the perfect tool for our purpose. Recall that there are two stages to the method. The first is to demonstrate that the statement is true for some starting value. This stage is already complete; the method is true for the starting value $n = 1$, as demonstrated in Section 6.1. The second stage is to demonstrate that if the identity holds for arbitrary n, then it holds for $n + 1$, showing that under the assumption $\frac{d}{dt}(t^n) = nt^{n-1}$ it is true that $\frac{d}{dt}(t^{n+1}) = (n + 1)t^n$:

$$\frac{d}{dt}(t^{n+1}) = \frac{d}{dt}(tt^n)$$

$$= t\frac{d}{dt}(t^n) + t^n \frac{d}{dt}(t)$$

$$= tnt^{n-1} + t^n$$

$$= nt^n + t^n$$

$$= (n + 1)t^n$$

The second stage is successfully demonstrated and case 1 is complete.

Case 2. $n = -q$ with q a negative integer.

Case 2 proceeds in a manner similar to case 1; the method of induction is used on the integer q. For this case, it is noted that the function t^{-q} is not defined at the point $t = 0$, so the expressions below assume nonzero values for t. There is a little more work to be done; stage 1 is first established with $q = 1$. To accomplish this, the following identity is obtained by long division and placed into the definition of the derivative:

$$\frac{1}{t + \Delta} = \frac{1}{t} - \frac{\Delta}{t^2} + \frac{\Delta^2}{t^2(t + \Delta)}$$

$$\frac{d}{dt}\left(t^{-1}\right) = \lim_{\Delta \to 0} \frac{1}{\Delta}\left(\frac{1}{t + \Delta} - \frac{1}{t}\right)$$

$$= \lim_{\Delta \to 0} \frac{1}{\Delta}\left(\frac{1}{t} - \frac{\Delta}{t^2} + \frac{\Delta^2}{t^2(t + \Delta)} - \frac{1}{t}\right)$$

$$= \lim_{\Delta \to 0} \frac{1}{\Delta}\left(-\frac{\Delta}{t^2} + \frac{\Delta^2}{t^2(t + \Delta)}\right)$$

$$= \lim_{\Delta \to 0}\left(-\frac{1}{t^2} + \frac{\Delta}{t^2(t + \Delta)}\right)$$

$$= \lim_{\Delta \to 0}\left(-\frac{1}{t^2}\right) + \lim_{\Delta \to 0}\frac{\Delta}{t^2(t + \Delta)}$$

$$= -\frac{1}{t^2}$$

$$= -t^{-2}$$

Stage 1 is a success as the result matches the proposed derivative, $\frac{d}{dt}(t^{-q}) = -qt^{-(q+1)}$ when $q = 1$. We proceed to stage 2. Assume that the proposal holds for an arbitrary positive integer value of $q = -n$ and demonstrate that $\frac{d}{dt}(t^{-(q+1)}) = -(q + 1)t^{-(q+2)}$:

$$\frac{d}{dt}(t^{-(q+1)}) = \frac{d}{dt}(t^{-q}t^{-1})$$

$$= t^{-1}\frac{d}{dt}(t^{-q}) + t^{-q}\frac{d}{dt}(t^{-1})$$

$$= -qt^{-1}t^{-(q+1)} - t^{-q}t^{-2}$$

$$= -qt^{-(q+2)} - t^{-(q+2)}$$

$$= -(q + 1)t^{-(q+2)}$$

The second stage is successfully demonstrated and case 2 is complete.

Case 3. $n = 1/q$ with q a positive integer.

Once again the same strategy is taken; apply al-Karaji's method of induction for positive and negative integers. This case is left to the reader.

Case 4. n a rational number.

As n is rational, $n = p/q$ with both p and q being integers. The derivative is found by applying the chain rule.

Let $y(t) = t^p$ and $x(y) = y^{1/q}$. Then $x(y(t)) = (t^p)^{1/q} = t^{p/q}$:

$$\frac{d}{dt}\left(t^{p/q}\right) = \frac{d}{dt}\{x[y(t)]\}$$

$$= \frac{d}{dy}\left[x(y)\right]\frac{d}{dt}\left[y(t)\right]$$

$$= \frac{d}{dy}\left(y^{1/q}\right)\frac{d}{dt}\left(t^p\right)$$

$$= \frac{1}{q}y^{1/q-1}pt^{p-1}$$

$$= \frac{p}{q}t^{p(1/q-1)}t^{p-1}$$

$$= \frac{p}{q}t^{p/q-p+p-1}$$

$$= \frac{p}{q}t^{p/q-1}$$

$$= nt^{n-1}$$

Case 5. n any real number.

For this case, an arbitrary sequence of rational numbers that approaches the real number n is investigated. Since the derivative associated with each rational number follows the general formula, and the rational number approaches n, it can be shown that the derivative approaches the general formula for every real number n.

6.3.4.6 The Quotient Rule

$$\frac{d}{dt}\left(\frac{x(t)}{y(t)}\right) = \frac{y(t)\frac{d}{dt}\left[x(t)\right] - x(t)\frac{d}{dt}\left[y(t)\right]}{\left[y(t)\right]^2}$$

There is some redundancy in the rules of differentiation. The quotient rule follows from the multiplication rule, chain rule, and the derivative of a power. Let $z(y) = y^{-1}$ so that $z(y(t)) = 1/[y(t)]$.

$$\frac{d}{dt}\left(\frac{x(t)}{y(t)}\right) = \frac{d}{dt}(x(t)\{z[y(t)]\})$$

$$= z[y(t)]\frac{d}{dt}[x(t)] + x(t)\frac{d}{dt}\{z[y(t)]\}$$

$$= \frac{\frac{d}{dt}[x(t)]}{y(t)} + x(t)\frac{d}{dy}(z(y))\frac{d}{dt}[y(t)]$$

$$= \frac{\frac{d}{dt}[x(t)]}{y(t)} + x(t)\frac{d}{dy}(y^{-1})\frac{d}{dt}[y(t)]$$

$$= \frac{\frac{d}{dt}[x(t)]}{y(t)} - x(t)(y(t))^{-2}\frac{d}{dt}[y(t)]$$

$$= \frac{\frac{d}{dt}[x(t)]}{y(t)} - \frac{x(t)\frac{d}{dt}(y(t))}{(y(t))^2}$$

$$= \frac{y(t)\frac{d}{dt}[x(t)] - x(t)\frac{d}{dt}[y(t)]}{(y(t))^2}$$

Remark. The quotient rule is not the only redundant rule. The constant rule is also redundant as it is a special case of the multiplication rule.

6.3.4.7 The Exponential Function Let $x(t) = \alpha^t$ for some value of α. To find the derivative, try to use the definition:

$$\frac{d}{dt}(\alpha^t) = \lim_{\Delta \to 0} \frac{\alpha^{t+\Delta} - \alpha^t}{\Delta}$$

$$= \lim_{\Delta \to 0} \alpha^t \frac{\alpha^{\Delta} - 1}{\Delta}$$

$$= \alpha^t \lim_{\Delta \to 0} \frac{\alpha^{\Delta} - 1}{\Delta}$$

$$= \alpha^t \lim_{\Delta \to 0} \frac{\alpha^{\Delta} - \alpha^0}{\Delta}$$

$$= \alpha^t \frac{d}{dt}(\alpha^t)\bigg|_{t=0}$$

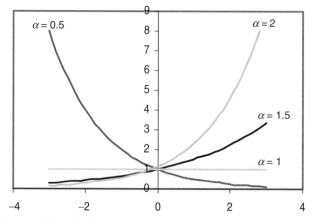

Figure 6.16   Graph of α^t with different α.

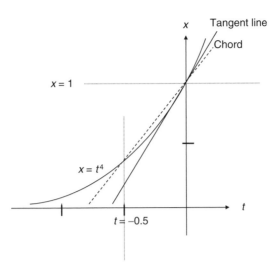

Figure 6.17   Approximation of $d(t^4)/dt$ at $t = 0$.

Is there a value of α so that $\frac{d}{dt}(\alpha^t)\,|_{t=0}$ is 1? If so, then there is a function $x(t) = \alpha^t$ with the property that $\frac{d}{dt}[x(t)] = x(t)$. Figure 6.16 shows the graph of $x(t) = \alpha^t$ for different values of α. Note that, when $\alpha = 1$, x is constant for all t, and $x(t) = 1$, the derivative of $x(t)$ is zero. If α is less than 1, the value $x(t)$ decreases in t, so the derivative of $x(t)$ is negative. Alternatively, as α increases from 1, the derivative of $x(t)$ at $t = 0$ also increases.

Figure 6.17 shows the graph of $x(t) = \alpha^t$ for $\alpha = 4$. The x coordinates at $t = -\frac{1}{2}$ and $t = 0$ are $x(-\frac{1}{2}) = 4^{-1/2} = \frac{1}{2}$ to $x(0) = 4^0 = 1$. The slope of the dashed line between these points is

$$\frac{x(0) - x(-\frac{1}{2})}{0 - (-\frac{1}{2})} = \frac{1 - \frac{1}{2}}{\frac{1}{2}} = 1$$

From the graph one can determine that the slope of the tangent line at $t = 0$ (the solid line) is greater than that of the dashed line. So, for $\alpha = 4$, $\frac{d}{dt}(\alpha^t)\,|_{t=0} > 1$.

When $\alpha = 1$, $\frac{d}{dt}(\alpha^t)\,|_{t=0} = 0$, and when $\alpha = 4$, $\frac{d}{dt}(\alpha^t)\,|_{t=0} > 1$. For some value of α in between 0 and 4, the derivative at 0 should be 1. In fact, such a value exists. Let us denote the value by e. In the next section, we follow Newton's and Leibniz' method for determining e. The method uses the defining property of e; the function $x(t) = e^t$ is its own derivative, $\frac{d}{dt}(e^t) = e^t$.

6.3.4.8 Power Series and the Exponential Function   Power series are polynomials of indefinite order. Both Newton and Leibniz found power series expansions for the exponential function. We retrace their steps.

Assume that e^t can be expanded in a power series:

$$e^t = a_0 + a_1 t + a_2 t^2 + \cdots + a_j t^j + \cdots = \sum_{j=0}^{\infty} a_j t^j$$

Assume next that one can find the derivative of e^t by termwise differentiation of the power series; one applies the sum rule to each term of the series:

$$\frac{d}{dt}(e^t) = \frac{d}{dt}(a_0) + \frac{d}{dt}(a_1 t) + \frac{d}{dt}(a_2 t^2) + \cdots + \frac{d}{dt}(a_j t^j) + \cdots$$

$$= 0 + a_1 \frac{d}{dt}(t) + a_2 \frac{d}{dt}(t^2) + \cdots + a_j \frac{d}{dt}(t^j) + \cdots$$

$$= a_1 + 2a_2 t + \cdots + ja_j t^{j-1} + \cdots$$

$$= \sum_{j=0}^{\infty} (j+1)a_{j+1} t^j$$

By the defining property of e^t, both of these power series must be equal. Two power series are equal only if each term is identical; this means that the coefficients in each power must be the same, $a_j = (j+1)a_{j+1}$, or

$$a_{j+1} = \frac{a_j}{j+1} \tag{6.14}$$

These relations form an infinite set of equations that we next solve.

As a starting point, setting $t = 0$ determines the value for a_0 since all other terms in the power series vanish:

$$e^0 = a_0 = 1$$

It may be helpful to write a few terms and see if a pattern emerges.

$$a_1 = a_0 = 1 \qquad a_2 = \frac{a_1}{2} = \frac{1}{2} \qquad a_3 = \frac{a_2}{3} = \frac{1}{2 \times 3} \qquad a_4 = \frac{a_3}{4} = \frac{1}{2 \times 3 \times 4}$$

The pattern that emerges is $a_j = 1/j!$, where the symbol "!" represents the factorial function over positive integers, $j! = 1 \times 2 \times 3 \times \cdots \times j$. Using induction, one can check that the solution $a_j = 1/j!$ is correct. This solution satisfies equation (6.13) and allows one to match the known quantity $e^0 = a_0 = 1$.

Having found the power series, it is possible to approximate the number e by setting $t = 1$ and summing a reasonable number of terms from the power series. Summing the first six terms gives the following:

$$e \approx 1 + \frac{1}{1!} + \frac{1}{2!} + \frac{1}{3!} + \frac{1}{4!} + \frac{1}{5!} = 1 + 1 + \frac{1}{2} + \frac{1}{6} + \frac{1}{24} + \frac{1}{120} = 2.71666\ldots$$

After 10 terms, the approximation yields $e \approx 2.718282$, which is correct to six digits.

Remarks

- In the seventeenth century, both Newton and Leibniz followed their intuition in deriving the series expansions for the exponential function. The general analysis of power series, which includes a rigorous demonstration of convergence and justification from term-by-term differentiation, was not finalized until two centuries later.

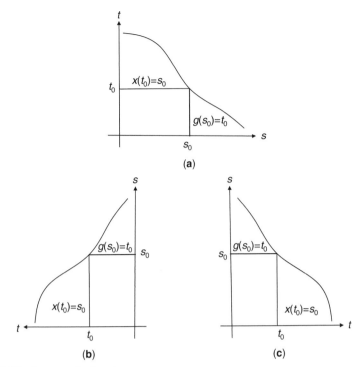

Figure 6.18   Inverse of a function.

- Newton and Leibniz found power series expansions for other functions as well, including the sine and cosine functions.
- The number e is an irrational number.

6.3.4.9 Logarithms

In Section 4.6.4, the inverse of a linear transformation is introduced. As indicated in the remarks following Section 4.6.4, an analogous concept exists for the inverse of a function. Given a function $x(t)$, $g(s)$ is its inverse provided that both the following equalities are satisfied, $g(x(t)) = t$ and $x(g(s)) = s$. Figure 6.18 illustrates the concept of the inverse. In Figure 6.18a, a function $x(t)$ maps the value t_0 to a number s_0 and $g(s)$ maps the number s_0 back to the original value, t_0. The graph can be considered as representing the points $(s, g(s))$ or the points $(t, x(t))$. In the latter case, the independent variable, t, points in the vertical direction, rather than the customary horizontal direction, The reader can graph the function $x(t)$ in its standard manner using two steps. First rotate the axis so that the t axis is horizontal (Figure 6.18b) and then flip the image around the s axis so that the t axis points to the right (Figure 6.18c).

The inverse of the exponential function is known as the logarithm. Figure 6.19a is a graph of the exponential function, while Figure 6.19b is a graph of the logarithmic function. Note that because the exponential of any real number is always positive, the logarithm is defined only over positive real numbers.

The derivative of the logarithm function is obtained by applying the chain rule to the defining property of the logarithm. Start with $\exp[\ln(t)] = t$. Taking the derivative

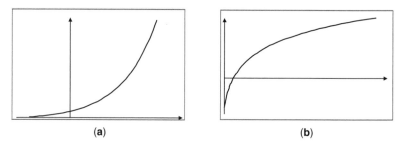

(a)   (b)

Figure 6.19   Exponential and logarithmic functions.

of both sides and simplifying give the result. Let $x(t) = \ln(t)$:

$$\frac{d}{dt}\{\exp[\ln(t)]\} = \frac{d}{dt}(t)$$

$$\frac{d}{dx}[\exp(x)]\frac{d}{dt}[\ln(t)] = 1$$

$$\exp(x)\frac{d}{dt}[\ln(t)] = 1$$

$$\frac{d}{dt}[\ln(t)] = \frac{1}{\exp(x)}$$

$$\frac{d}{dt}[\ln(t)] = \frac{1}{\exp[\ln(t)]}$$

$$\frac{d}{dt}[\ln(t)] = \frac{1}{t}$$

Remark. The method demonstrates a relation between the derivative of a function and the derivative of its inverse:

$$y[x(t)] = t$$

$$\frac{d}{dt}\{y[x(t)]\} = \frac{d}{dt}(t)$$

$$\frac{d}{dx}[y(x)]\frac{d}{dt}[x(t)] = 1$$

$$\frac{d}{dt}[x(t)] = \frac{1}{\frac{d}{dx}[y(x)]}$$

6.3.4.10 A Trigonometric Inequality   As discussed in Section 5.6.2, Aristarchus used the trigonometric inequality $\sin(\alpha)/\alpha > \sin(\beta)/\beta$ (whenever $0 < \alpha < \beta < \pi/2$) to determine bounds on the dimension of the sun. In this section, we demonstrate the inequality. The inequality is identical to the statement that the function $\sin(\theta)/\theta$ is a decreasing function over the interval from 0 to $\pi/2$. Figure 6.20 illustrates the relation

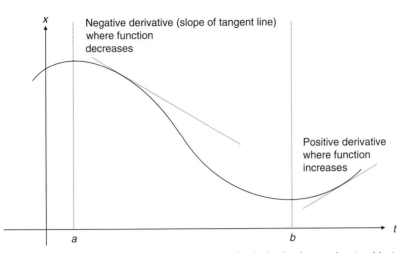

Figure 6.20   A function decreases (increases) where the derivative is negative (positive).

between a decreasing function and the slope of its tangent. The function is decreasing between the values $t = a$ and $t = b$, the same interval where the slope of the tangent is negative. Conversely, the function is increasing wherever $t < a$ and wherever $t > b$. The figure illustrates that it is necessary to demonstrate that $(d/d\theta)[\sin(\theta)/\theta]$ is less than zero in the interval of interest, 0 to $\pi/2$.

$$\frac{d}{d\theta}\left(\frac{\sin(\theta)}{\theta}\right) = \frac{\theta\frac{d}{d\theta}[\sin(\theta)] - \sin(\theta)\frac{d}{d\theta}(\theta)}{\theta^2}$$

$$= \frac{\theta\cos(\theta) - \sin(\theta)}{\theta^2}$$

Within the interval of interest, the derivative $(d/d\theta)[\sin(\theta)/\theta]$ is negative provided that the function $f(\theta) = \theta\cos(\theta) - \sin(\theta)$ is negative. The function $f(\theta)$ is identically zero when θ is zero. If the derivative of the function $f(\theta)$ is negative between 0 and $\pi/2$, then the function $f(\theta)$ decreases from zero and the function $f(\theta)$ is accordingly negative whenever $0 < \theta < \pi/2$ (Figure 6.21):

$$\frac{d}{d\theta}\left[f(\theta)\right] = \frac{d}{d\theta}[\theta\cos(\theta)] - \frac{d}{d\theta}[\sin(\theta)]$$

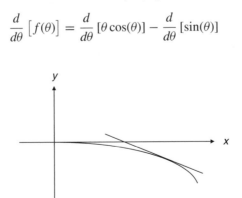

Figure 6.21   A function starting at the origin goes below the x axis with a negative derivative.

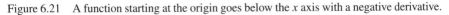

$$= \theta \frac{d}{d\theta}[\cos(\theta)] + \cos(\theta) - \cos(\theta)$$

$$= -\theta \sin(\theta)$$

The derivative of the function $f(\theta)$ is indeed less than zero in the interval of interest, $0 < \theta < \pi/2$. As stated above, $f(\theta)$ decreases from zero and is accordingly negative.

Remark. At the points $t = a$ and $t = b$ in Figure 6.20, the derivative of the function is zero. These points are known as extrema and play a fundamental role in optimization theory.

6.4   MORE INTEGRAL CALCULUS

This is the last of the mathematical sections, and it furnishes the final element of all the mathematics required to reveal the ellipse. The relations between velocity, position, differentiation, and integration are presented in Section 6.1.3. This section follows up with methods of integration as developed by Newton and Leibniz.

6.4.1   The Antiderivative and the Fundamental Theorem of Calculus

In Section 6.3, the fundamental theorem of calculus [equation (6.11)] demonstrates the relationship between integration and differentiation. To determine the definite integral, it is necessary to find a function $x(t)$ whose derivative is equal to the function being integrated, $v(t)$, that is, $\frac{d}{dt}[x(t)] = v(t)$ [see equation (6.8)]. Leibniz' integral sign also indicates this relation:

$$\int v(t)\, dt = x(t) + c \tag{6.15}$$

Notice that the notation differs slightly from that found in Section 6.1.5. Equation (6.10) has an interval associated with the integral sign, $[t_i, t_f]$, whereas equation (6.15) has no associated interval. Additionally, equation (6.15) contains an arbitrary constant, c. The role of the constant is to generalize the result so that it includes all possible functions that satisfy the relation $\frac{d}{dt}[x(t)] = v(t)$; one notes that if $x(t)$ satisfies $\frac{d}{dt}[x(t)] = v(t)$, then so does $x(t) + c$ for any arbitrary constant c. Using the interpretation that x represents position with regard to some arbitrary fixed reference point, v represents velocity, and t represents time, recall that the definite integral is the difference in the relative positions of an object moving with velocity $v(t)$. The choice of c sets the arbitrary point of reference for the position, which can be critical in applications.

The result $x(t) + c$ goes by two names, the antidervative of $v(t)$ and the indefinite integral of $v(t)$. Mathematicians often generate results through intuitive processes; rigorous proofs come later. Leibniz' notation allows one to pursue one's intuition toward correct results, as illustrated in the following equalities:

$$\frac{dx}{dt} = v(t)$$

$$dx = v(t)\,dt$$

$$\int dx = \int v(t)\,dt$$

$$x(t) + c = \int v(t)\,dt$$

So a problem that is originally posed in terms of the derivative is converted to a problem that is posed as an integral.

As noted in Section 6.1.5, the relation between the antiderivative and the definite integral is the following:

$$\int_{t_i}^{t_f} v(t)\,dt = [x(t_f) - c] - [x(t_i) - c] = x(t_f) - x(t_i)$$

The value of c is irrelevant when determining indefinite integrals because relative positions between two points are the same regardless of the reference point against which the two points are measured. Other notation that is commonly used is the following:

$$\int_{t_i}^{t_f} v(t)\,dt = x(t)\,|_{t_i}^{t_f}$$

In this notation, $x(t)\,|_{t_i}^{t_f} = x(t_f) - x(t_i)$.

We close this section with a method for verifying that an integral is properly evaluated. The method is just a restatement of the fundamental theorem, of calculus. If one understands differential calculus, it is easy to check whether or not one has valued an integral correctly, take the derivative of the result, and make sure that the derivative returns the original function being integrated. This fact was made used in Section 6.1.5, where it was shown that $\int_{t_i}^{t_f} 2t\,dt = t^2\,|_{t_i}^{t_f}$.

How can one tell? The derivative of the result t^2 returns the function being integrated, $\frac{d}{dt}(t^2) = 2t$. It is common to refer to the function being integrated as the integrand. Using this parlance, we have verified that the derivative of the integral returns the integrand. This method of verification is used whenever one performs an integral.

6.4.2 Methods of Integration

This section shows some common methods for integrating functions. This section emphasizes commonality between rules of differentiation and methods of integration that surface naturally through the relationship between differentiation and integration. We start by noting that the derivative table at the beginning of Section 6.3 can also be transformed into an integral table by switching columns and applying simple algebra. Verification of the entries may be accomplished by assuring that the derivative of the integral returns the function as discussed in Section 6.4.1.

Function	Integral
$\cos(t)$	$\sin(t)$
$\sin(t)$	$-\cos(t)$
t^n	$\dfrac{t^{n+1}}{n+1}$
e^t	e^t
$1/t$	$\ln(t)$

6.4.2.1 Linearity Rules: Additivity and Multiplication by a Constant   The linearity rules of differentiation, the additivity rule, and the constant-multiplication rule have identical counterparts for integration:

$$\int v(t) + w(t)\, dt = \int v(t)\, dt + \int w(t)\, dt$$

This is noted by taking the derivative of the right-hand side, applying the additivity rule for differentiation, and noting that the result is the integrand on the left-hand side:

$$\frac{d}{dt}\left(\int v(t)\, dt + \int w(t)\, dt\right) = \frac{d}{dt}\left(\int v(t)\, dt\right) + \frac{d}{dt}\left(\int w(t)\, dt\right)$$

$$= v(t) + w(t)$$

$$= \frac{d}{dt}\left(\int v(t) + w(t)\, dt\right)$$

Similarly, for a constant a, $\int av(t)\, dt = a \int v(t)\, dt$.

Example 6.5   POLYNOMIALS

The linearity rules combined with the integral for powers allow for the integration of polynomials:

$$\int 2t^4 + t^3 - 7t^2 - 4t + 3\, dt = \int 2t^4\, dt + \int t^3\, dt - \int 7t^2\, dt - \int 4t\, dt + \int 3\, dt$$

$$= 2\int t^4\, dt + \int t^3\, dt - 7\int t^2\, dt - 4\int t\, dt + 3\int dt$$

$$= \tfrac{2}{5}t^5 + \tfrac{1}{4}t^4 - \tfrac{7}{3}t^3 - 2t^2 + 3t + c$$

In general,

$$\int a_n t^n + a_{n-1}t^{n-1} + a_{n-2}t^{n-2} + \cdots + a_1 t + a_0\, dt$$

$$= \frac{a_n}{n+1}t^{n+1} + \frac{a_{n-1}}{n}t^n + \frac{a_{n-2}}{n-1}t^{n-1} + \cdots + \frac{a_1}{2}t^2 + a_0 t + c$$

6.4.2.2 Chain Rule and the Change of Variables   Let the integrand be the composite function $v(u(t))$ in which $u(t)$ has the corresponding derivative $\frac{du}{dt}$. Then

the following identity holds:

$$\int v(u(t))\,dt = \int v(u)\frac{1}{du/dt}\,du \tag{6.16}$$

To show that the identity holds, the derivative of the right-hand side is taken and shown to be equal to the integrand of the left-hand side. The chain rule must be applied to convert from a derivative with respect to t to one with respect to u:

$$\frac{d}{dt}\left(\int v(u)\frac{1}{du/dt}\,du\right) = \frac{d}{du}\left(\int v(u)\frac{1}{du/dt}\,du\right)\frac{du}{dt}$$

$$= v(u)\frac{1}{du/dt}\frac{du}{dt}$$

$$= v(u(t))$$

For the definite integral, the following holds:

$$\int_{t_i}^{t_f} v(u(t))\,dt = \int_{u(t_i)}^{u(t_f)} v(u)\frac{1}{du/dt}\,du$$

The identity changes the integration from a function over the variable t to a function over the variable u. There are times when it is advantageous to make the change as illustrated by some examples.

Example 6.6

Determine $\int te^{t^2}\,dt$.

Let us change the variable. Let $u(t) = t^2$ and note that $du/dt = 2t$ so that $1/(du/dt) = 1/(2t)$. Applying the change-of-variable formula yields the following result:

$$\int te^{t^2}\,dt = \int te^u\frac{1}{2t}\,du$$

$$= \int \frac{1}{2}e^u\,du$$

$$= \frac{1}{2}e^u + c$$

$$= \frac{1}{2}e^{t^2} + c$$

The result can be checked by verifying that its derivative equals the integrand. Note that when taking the derivative of the result the chain rule must be applied with $u(t) = t^2$:

$$\frac{d}{dt}\left(\frac{1}{2}e^{t^2} + c\right) = \frac{d}{dt}\left(\frac{1}{2}e^{t^2}\right) + \frac{d}{dt}(c)$$

$$= \frac{1}{2}\frac{d}{dt}\left(e^u\right) + 0$$

$$= \frac{1}{2}\frac{d}{du}(e^u)\frac{d}{dt}(u(t))$$

$$= \frac{1}{2}e^u 2t$$

$$= te^{t^2}$$

Example 6.7

Determine $\int_1^2 te^{t^2}\, dt$.

Using the result of the previous example with $u(t) = t^2$, $u(1) = 1$, and $u(2) = 4$ along with the change-of-variable formula for the definite integral yields the following result:

$$\int_1^2 te^{t^2}\, dt = \frac{1}{2}\int_1^4 e^u\, du$$

$$= \frac{1}{2}e^u \;\Big|_1^4$$

$$= \frac{1}{2}(e^4 - e^1)$$

There are several points of interest in this example. First, it is possible to maintain the original range of integration $(1, 2)$ by expressing the result of the indefinite integral in terms of t rather than u:

$$\int_1^2 te^{t^2}\, dt = \tfrac{1}{2}e^{t^2}\;\big|_1^2 = \tfrac{1}{2}\left(e^4 - e^1\right)$$

The example also illustrates the geometry of the variable change. Figure 6.22 presents graphs of the integrands in terms of t and u between their respective ranges of integration. Both $v = t\exp(t^2)$ and $w = \exp(u)$ are plotted on the same set of axes for ease of comparison. The change of variables from t to u stretches out the interval

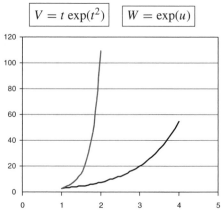

Figure 6.22   Change of variables.

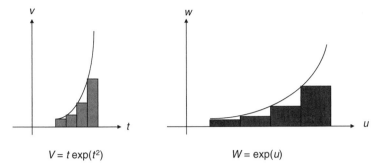

$V = t \exp(t^2)$ $W = \exp(u)$

Figure 6.23   Partitions before and after change of variables.

of integration from $[1, 2]$ to $[1, 4]$. Recall that the integral is the area between the graph and the horizontal axis. If the areas are to be maintained, one must vertically squeeze the graph enough to compensate for the horizontal stretching. As expected, the graph $w = \exp(u)$ is a horizontal expansion and a vertical compression of the graph of $v = t \exp(t^2)$.

Recall from Section 6.1.5 that the integral is defined as a limit of sums of rectangles. Figure 6.23 presents a partition of rectangles for the functions $v = t \exp(t^2)$ and $w = \exp(u)$. Examine a single rectangle of a partition over the function expressed in terms of the t variable and the corresponding rectangle expressed in terms of the u variable, $u_j = u(t_j)$:

$$\text{Area}_{t_j} = v(t_j)(t_{j+1} - t_j) \qquad \text{Area}_{u_j} = w_j(u_{j+1} - u_j)$$

We wish to equate the areas of these rectangles by a judicious choice of w_j:

$$w_j(u_{j+1} - u_j) = v(t_j)(t_{j+1} - t_j)$$

$$w_j = v(t_j)\frac{t_{j+1} - t_j}{u_{j+1} - u_j}$$

The ratio $(t_{j+1} - t_j)/(u_{j+1} - u_j)$ is the squeezing factor that when applied to $v(u_j)$ determines the height of the rectangle, w_j. To find the integral, one examines the limiting case as the bases of each rectangle in the partition approach zero. When the difference between t_j and t_{j+1} is sufficiently small, the ratio $(u_{j+1} - u_j)/(t_{j+1} - t_j)$ is close to the derivative, $(du/dt)\,|_{t_j}$. Equivalently,

$$\frac{t_{j+1} - t_j}{u_{j+1} - u_j} \approx \frac{1}{(du/dt)\,|_{t_j}}$$

The approximation becomes exact in the limit as t_{j+1} approaches t_j:

$$w_j = w(u_j) = \lim_{t_{j+1} \to t_j} v(t_j)\frac{t_{j+1} - t_j}{u_{j+1} - u_j}$$

$$= v(t_j)\frac{1}{(du/dt)\,|_{t_j}}$$

Applying the definition of the integral as the limiting value results in the change-of-variable formula:

$$\int_{t_i}^{t_f} v(t)\,dt = \lim_{n\to\infty} \sum_{j=1}^{n} v(t_j)(t_{j+1} - t_j)$$

$$= \lim_{n\to\infty} \sum_{j=1}^{n} w_j(u_{j+1} - u_j)$$

$$= \lim_{n\to\infty} \sum_{j=1}^{n} v(u_j) \frac{1}{(du/dt)\,|_{t_j}} (u_{j+1} - u_j)$$

$$= \int_{u(t_i)}^{u(t_f)} v(u) \frac{1}{du/dt}\,du$$

$$= \int_{u_i}^{u_f} w(u)\,du$$

where $w(u) = v(u)[1/(du/dt)]$.

Remarks

- Recall that a similar situation arises in Chapter 3 where the surface of a sphere is calculated. A change of variables occurs when a longitudinal arc of the sphere is mapped onto the sphere's central axis running between the poles. In that case, the mapping squeezes the longitudinal arch into a smaller space and a corresponding stretch is necessary. In regions where the map $u(t)$ shrinks the corresponding interval of integration, $1/(du/dt)$ is greater than 1 and there is a corresponding stretch of the integrand. Alternatively, in regions where the map $u(t)$ stretches the corresponding interval of integration, $1/(du/dt)$ is less than 1 and there is a corresponding squeeze of the integrand.
- Leibniz' notation permits additional intuition:

$$v(u(t))\,dt = v(u)\frac{dt}{du}\,du$$

$$v(u(t))\,dt = v(u)\frac{1}{du/dt}\,du$$

$$\int v(u(t))\,dt = \int v(u)\frac{1}{du/dt}\,du$$

- To further cement the correspondence between the chain rule for differentiation and a change of variables for integration, note the following. The chain rule changes differentiation with respect to one variable into differentiation with respect to another by applying a correction factor:

$$\frac{d}{dt}[x(u(t))] = \frac{dx}{du} \times \frac{du}{dt}$$

the correction factor is du/dt. The change of variables formula changes integration over one variable to integration over another by applying a correction factor (stretch or squeeze):

$$\int v(u(t))\, dt = \int v(u)\frac{dt}{du}\, du$$

The correction factor dt/du is the inverse of the correction factor for differentiation, as expected by the inverse relations between differentiation and integration.

6.4.2.3 The Multiplication Rule and Integration by Parts

A very useful method for performing integrals, known as integration by parts, is given by the following formula:

$$\int u(t)\frac{d}{dt}(v(t))\, dt = u(t)v(t) - \int v(t)\frac{d}{dt}(u(t))\, dt$$

where $u(t)$ and $v(t)$ are differentiable functions.

The following example demonstrates the method's use.

Example 6.8

Evaluate $\int_0^{\pi/2} t\cos(t)\, dt$.

Let $\frac{d}{dt}[v(t)] = \cos(t)$ and $u(t) = t$. Then $v(t) = \sin(t)$ and $\frac{d}{dt}[u(t)] = 1$. Applying the integration-by-parts formula gives the following result:

$$\int t\cos(t)\, dt = t\sin(t) - \int \sin(t)\, dt$$

$$= t\sin(t) + \cos(t) + c$$

Evaluating the antiderivative at the end points of the interval of integration establishes the result:

$$\int_0^{\pi/2} t\cos(t)\, dt = [t\sin(t) + \cos(t)]\,|_0^{\pi/2}$$

$$= \left[\tfrac{1}{2}\pi\sin\left(\tfrac{1}{2}\pi\right) + \cos\left(\tfrac{1}{2}\pi\right)\right] - [0\sin(0) + \cos(0)]$$

$$= \tfrac{1}{2}\pi - 1$$

The solution to the indefinite integral is verified by checking that the derivative of the result is equal to the integrand on the left-hand side:

$$\frac{d}{dt}[t\sin(t) + \cos(t)] = \frac{d}{dt}[t\sin(t)] + \frac{d}{dt}\cos(t)$$

$$= \left(t\frac{d}{dt}[\sin(t)] + \sin(t)\frac{d}{dt}(t)\right) - \sin(t)$$

$$= t\cos(t) + \sin(t) - \sin(t)$$

$$= t\cos(t)$$

In the example, the multiplication rule is employed to perform differentiation. It is possible to start with the multiplication rule and arrive at the integration-by-parts formula. Let $u(t)$ and $v(t)$ be differentiable functions:

$$\frac{d}{dt}[u(t)v(t)] = u(t)\frac{d}{dt}[v(t)] + v(t)\frac{d}{dt}[u(t)]$$

$$\int \frac{d}{dt}[u(t)v(t)]\,dt = \int u(t)\frac{d}{dt}[v(t)]\,dt + \int v(t)\frac{d}{dt}[u(t)]\,dt$$

$$u(t)v(t) = \int u(t)\frac{d}{dt}[v(t)]\,dt + \int v(t)\frac{d}{dt}[u(t)]\,dt$$

$$u(t)v(t) - \int v(t)\frac{d}{dt}[u(t)]\,dt = \int u(t)\frac{d}{dt}[v(t)]\,dt$$

6.5   POTPOURRI

This section is the final excursion of the book. The excursion revisits some of the problems that were previously introduced.

6.5.1   Cavalieri's Theorem and the Fundamental Theorem of Calculus

Recall Cavalieri's theorem from Section 3.2.2. The theorem states that two objects with the same cross sections have the same area (see Figure 6.24). One can prove this theorem by using calculus. Referring to Figure 6.24, label the horizontal axis by x. (The axis no longer represents time, so another variable is selected.) Let the lower boundary of object A be given by the function $f_A(x)$ and the upper boundary be given by $g_A(x)$ and the corresponding boundaries for object B be given by $f_B(x)$ and $g_B(x)$.

Using the fact that the cross sections are equal [for all x, $f_A(x) - g_A(x) = f_B(x) - g_B(x)$], equality of the areas of the objects can be demonstrated:

$$\text{Area}_A = \int_{x_i}^{x_f} f_A(x)\,dx - \int_{x_i}^{x_f} g_A(x)\,dx$$

$$= \int_{x_i}^{x_f} \left[f_A(x) - g_A(x)\right]\,dx$$

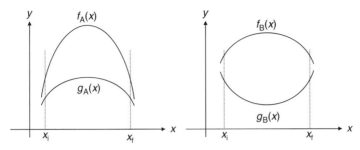

Figure 6.24   Cavalieri's theorem and integration.

$$= \int_{x_i}^{x_f} \left[f_B(x) - g_B(x) \right] \, dx$$

$$= \text{Area}_B$$

A natural question arises for the reverse direction: Can one state anything about the vertical cross sections if the area is known? The answer may be found by reversing the above process. Integration of the vertical cross sections results in the area, and differentiation (the inverse of integration) of the area must result in the vertical cross sections. In the remainder of this section, we take the lower boundary, $g(x)$, as the x axis so that the vertical cross section has height $f(x)$.

Let $A(x)$ be the area of the object between the interval x_i and x, $A(x) = \int_{x_i}^{x} f(\chi) \, d\chi$.

We seek the derivative of $A(x)$. First note the following identity that states that the area associated with the interval initiating at x_i and ending at $x + \Delta$ can be split into two areas, one along the interval from x_i to x and the other along the interval from x to $x + \Delta$ (Figure 6.25):

$$A(x + \Delta) = \int_{x_i}^{x} f(\chi) \, d\chi + \int_{x}^{\Delta} f(\chi) \, d\chi$$
$$= A(x) + \int_{x}^{\Delta} f(\chi) \, d\chi \tag{6.17}$$

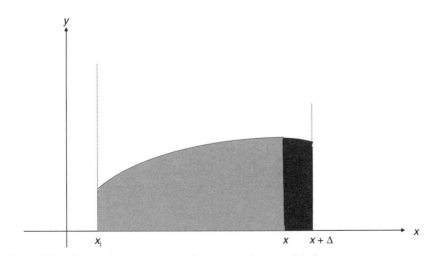

Figure 6.25   Area under the curve equals the sum of gray and dark areas.

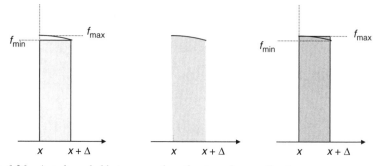

Figure 6.26   Area bounded between underestimate and overestimate.

Let f_{min} be the smallest value of $f(x)$ within the interval from x to $x + \Delta$ and let f_{Max} be the largest value of $f(x)$ in the same interval. Using the inequalities of equation (6.16), upper and lower bounds on the area $\int_x^\Delta f(\chi)\,d\chi$ are given in terms of f_{min} and f_{Max}:

$$f_{min}\Delta \le \int_x^\Delta f(\chi)\,d\chi = A(x + \Delta) - A(x) \le f_{Max}\Delta \qquad (6.18)$$

The inequality states that the area of interest is greater than the rectangle with base delta and height f_{min} but less than the rectangle with base delta and height f_{max} (Figure 6.26).

Assume that $f(x)$ is continuous; it has no breaks so that one can draw its graph without lifting the pen. Then as Δ approaches zero, both f_{min} and f_{Max} approach $f(x)$. Dividing the inequalities of equation (6.17) by Δ and taking the limit result in the fundamental theorem of calculus:

$$f_{min} \le \frac{A(x + \Delta) - A(x)}{\Delta} \le f_{Max}$$

$$\lim_{\Delta \to 0} f_{min} \le \lim_{\Delta \to 0} \frac{A(x + \Delta) - A(x)}{\Delta} \le \lim_{\Delta \to 0} f_{Max}$$

$$f(x) \le \frac{d}{dx}[A(x)] = \frac{d}{dx}\left(\int_{x_i}^x f(\chi)\,d\chi\right) \le f(x)$$

Since the extreme left and extreme right are both the same, the inequalities can be replaced by an equality:

$$\frac{d}{dx}\left(\int_{x_i}^x f(\chi)\,d\chi\right) = f(x)$$

The fundamental theorem of calculus completes the relation between the height of the vertical cross sections and the area. Integration of the vertical cross sections yields the area, and differentiation of the area yields the vertical cross sections.

Remarks

- In the above discussion, corresponding with Cavalieri's theorem only positive values of $f(x)$ have been considered. Integration of a function along an interval with negative values of f corresponds to the negative of the area between the graph of $f(x)$ and the horizontal axis. The fundamental theorem of calculus follows either way with a slight adjustment to the above discussion; both the signs and directions of the inequalities of equation (6.17) are reversed.

- The fundamental theorem of calculus is independent of the initial point x_i that is selected.

6.5.2 Volume of the Sphere and Other Objects with Known Cross-Sectional Areas

In 1615, a wine barrel shortage followed a banner harvest of grapes. The shortage inspired Kepler to determine the volume of a wine barrel. Kepler posed a similar solution to the one that Archimedes presents in his proof for the volume of the sphere. This section presents the sphere and other examples.

Example 6.9   VOLUME OF A SPHERE OF RADIUS R

Consider a sphere depicted on the x, y, and z axes as shown in Figure 6.27a. One way to approximate the volume of the sphere is to partition the x axis into small segments and sum the volumes of discs that are inscribed within the sphere, as depicted in Figure 6.27b.

For a single disc, disc j, the radius of the disk is $r_j = \sqrt{R^2 - x_j^2}$ and the corresponding volume of the disc is $v_j = \pi r_j^2 (x_{j+1} - x_j)$ (Figure 6.28).

The volume of the sphere can be approximated as follows:

$$\text{Volume} \approx \sum_{j=1}^{n} v_j$$

$$= \sum_{j=1}^{n} \pi r_j^2 (x_j - x_{j-1})$$

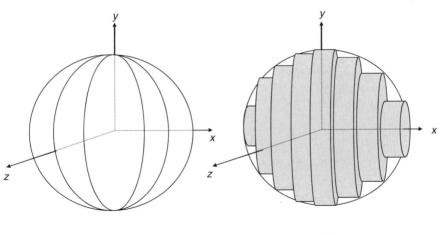

(a)   (b)

Figure 6.27   Approximating sphere with disks.

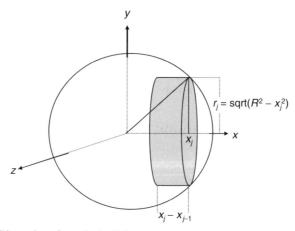

Figure 6.28   Dimensions for a single disk.

$$= \sum_{j=1}^{n} \pi r_j^2 \Delta x_j$$

Taking the limit as the number of partition elements, n, becomes arbitrarily large results in the volume of the sphere. Additionally, taking the limit results in an integral, as in equation (6.10):

$$\text{Volume} = \lim_{n \to \infty} \sum_{j=1}^{n} \pi r_j^2 \Delta x_j$$

$$= \int_{-R}^{R} \pi \, [r(x)]^2 \, dx$$

$$= \pi \int_{-R}^{R} \left(R^2 - x^2 \right) dx$$

$$= \pi \left[R^2 x - \tfrac{1}{3} x^3 \right] \Big|_{-R}^{R}$$

$$= \pi \left[\left(R^3 - \tfrac{1}{3} R^3 \right) - \left(-R^3 - \tfrac{1}{3} - R^3 \right) \right]$$

$$= \frac{4}{3} \pi R^3$$

Example 6.10   WINE BARREL

Position a wine barrel so that its central axis passing through the ends is aligned with the x axis. The cross sections of the barrel are all circles. Assume that the length of the barrel is 1 m and let the radii of the cross-sectional circles be given by $r(x) = \tfrac{1}{3} \left(1 - \tfrac{1}{2} x^2 \right)$ (Figure 6.29).

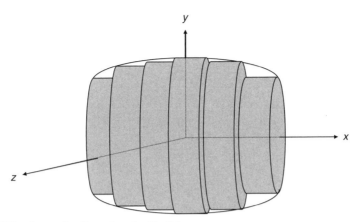

Figure 6.29 Approximating a barrel with disks.

The entire construction that leads to the integral for the volume of the sphere can be replicated for this example:

$$\text{Volume} = \lim_{n \to \infty} \sum_{j=1}^{n} \pi r_j^2 \Delta x_j$$

$$= \int_{-1}^{1} \pi \, [r(x)]^2 \, dx$$

$$= \pi \int_{-1}^{1} \frac{1}{9} \left(1 - x^2 + \frac{x^4}{4} \right) dx$$

$$= \frac{\pi}{9} \left[x - \frac{x^3}{3} + \frac{x^5}{20} \right] \Big|_{-1}^{1}$$

$$= \frac{\pi}{9} \left[\left(1 - \frac{1}{3} + \frac{1}{20} \right) - \left(-1 + \frac{1}{3} - \frac{1}{20} \right) \right]$$

$$= \frac{43\pi}{270}$$

Example 6.11

Volume of an n-dimensional pyramid.

Let an n-dimensional pyramid be given with its base an $(n-1)$-dimensional cube having sides of length x. Additionally, assume that the length of the pyramid is also x. The pyramid is a stack of $(n-1)$-dimensional cubes. If we flip the stack over and lay it upon the y axis, so that it balances on its apex at $y = 0$, the side of the $(n-1)$-dimensional cube at a cross section perpendicular to the y axis is given by the value $s(y) = y$. The volume of the $(n-1)$-dimensional cube is just $\mu(y) = y^{n-1}$. Figure 6.30 illustrates the three-dimensional case $n = 3$ in which $\mu(y)$ represents the cross-sectional area of the squares.

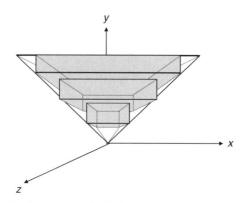

Figure 6.30   Approximating a pyramid with boxes.

The volume of the n-dimensional pyramid is approximated by breaking the pyramid into slices and adding up the volumes of the slices. Let v_j represent the volume of slice j, $v_j = \mu(y_j)(y_j - y_{j-1}) = \mu(y_j)\Delta_{y_j}$:

$$\text{Volume} = \lim_{n \to \infty} \sum_{j=1}^{n} v_j$$

$$= \lim_{n \to \infty} \sum_{j=1}^{n} \mu(y_j)\Delta_{y_j}$$

$$= \int_0^x \mu(y)\,dy$$

$$= \int_0^x y^{n-1}\,dy$$

$$= \left.\frac{y^n}{n}\right|_0^x$$

$$= \frac{x^n}{n}$$

This is the same result as obtained in Section 4.4. Setting $h = x$ in formula (3.23), we note $V_{n-1}(\text{base}) = x^{n-1}$, giving $V_n(\text{pyramid}) = (1/n)x^n$. Knowing that $V_n(\text{pyramid}) = \int_0^x y^{n-1}\,dy$ allows one to conclude $\int y^{n-1}\,dy = y^n/n$, or equivalently $\int y^n\,dy = y^{n+1}/(n+1)$.

CHAPTER 7

EIGHT MINUTES THAT CHANGED HISTORY

Kepler had been working for 5 years trying to decipher Mars' trajectory from Tycho Brahe's observations. He had a correct functional description of the planet's latitude and endeavored to complete the orbital description by determining the longitude. Reconciling the two components of motion required an adjustment that set the pathway off of Tycho's observations by 8 min, $\frac{2}{15}^{\circ}$. Let us put this 8 min in perspective. To launch the shuttle, NASA needs to know the full weight of the payload. Suppose they account for the weight of a crew of five at 900 lb. The crew weighs in on launch date and their combined weight comes to 900 ($^{1}/_{3}$) pounds. Does NASA pull the plug on the mission? Kepler had a fixation with numbers. He once calculated his gestation period within his mother's womb at 224 days 9 h and 53 min. For Kepler, there was only one way to proceed, pull the plug and go back to square one.

One might wonder how the intellectual development of civilization would have proceeded had the earth been an only child with no moon or sister planets. In such a lonely solar system, the geocentric description of the universe would be perfect and a would-be Copernicus may never have questioned it. Fortunately, the sun's sphere of influence contains many spheres. The heliocentric signature is penned by the planets' trajectories as seen by an earthbound observer. The curve of Figure 7.1 indicates the path of the planet Mars as viewed from the earth. A peculiarity that has been at the center of the controversy since observations of the planets have been recorded is retrograde motion of the planets Mars, Jupiter, and Saturn. In Figure 7.1, retrograde motion is seen beneath Aries where Mars initially moves in a dominantly westerly direction, reverses its course, and then resumes its original direction. It is the specifics of Mars' retrograde motion that led Kepler to the ellipse.

The curve of Figure 7.1 gives a series of Mars observations against a background of fixed stars. The time period of observation begins on June 21, 2005 and ends on March 26, 2006, and specific observations are annotated by their dates. An experienced reader may recognize the background constellations against which the observations are made. It is of interest to compare the planet's signature with that of the sun as well as the daily trajectory of a fixed star. The zodiac guides the sun's

The Ellipse: A Historical and Mathematical Journey   by Arthur Mazer
Copyright © 2010 by John Wiley & Sons, Inc.

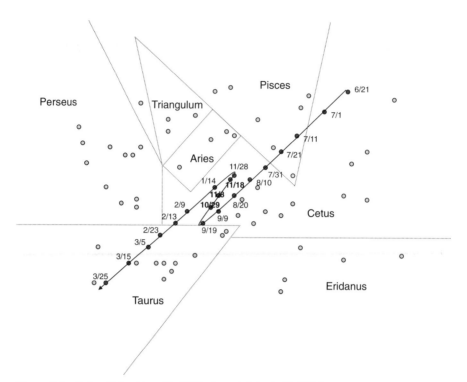

Figure 7.1   Path of Mars June 21, 2005, through March 25, 2006.

movement as it floats easterly through the stars. Figure 7.2 presents the constellation behind the sun throughout the year. The annual trajectory is sinusoidal with its position closest to the North Star in the summer and furthest from the North Star in the winter.

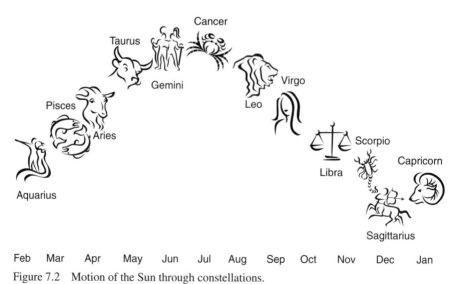

Figure 7.2   Motion of the Sun through constellations.

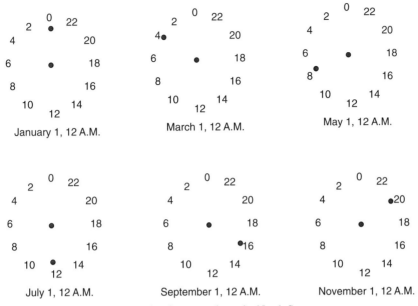

Figure 7.3   Annual movement of a given star about the North Star.

The stars execute a uniform rotation that fixes their position with respect to one another. Figure 7.3 illustrates a star's movement as viewed from the northern hemisphere. The star executes a daily rotation in a counterclockwise direction centered at the North Star. The rotation may be used as a nighttime clock provided that the numbers are ordered in reverse of an actual clock to accommodate the counterclockwise motion. Figure 7.3 superimposes a counterclock on the sky with the North Star at the clock's center and another star that is in the midnight position on New Year's Eve. At 2 A.M. on January 1, the star that at midnight was in the midnight position would be in the 2 o'clock position, and in general, at j o'clock on January 1, the star would be in the j o'clock position. This clock would run fast by about 4 min per day, requiring the clock's reader to adjust the reading. The effect of running fast is illustrated in Figure 7.3. The same star shown at midnight on different dates of the year runs further ahead of the clock as the months pass. In 1 year's time, the star returns to its position directly above the North Star at midnight.

Because the trajectories of the stars and the sun move in a discernible pattern that repeats annually, the geocentric shaman can easily explain these observations using Figure 7.4. The universe is a huge sphere; resting upon its surface are the stars and at its center is the earth. The universal sphere's north–south axis is aligned with that of the earth. Encased within the universal sphere is the sun. The sun rotates about the universal sphere's north–south axis at uniform speed once per day and always rests upon a plane called the ecliptic. The ecliptic forms an angle of 23.4° with the earth's equator and also rotates about the north–south axis. At the winter solstice, on January 21, point a vector from the center of the earth toward the highest latitude upon the equinox and set the sun there. The surface of the universal sphere and all

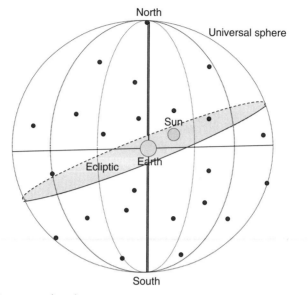

Figure 7.4   The geocentric universe.

the stars affixed to it also rotate about its north–south axis as seen by the star's daily motion. However, the speed of rotation of the universal sphere along with the ecliptic is slightly faster than the daily rotational speed of the sun, $\frac{366}{365}$ times as fast as the sun. Then, the drift of the stars as depicted in Figure 7.3 is visible to an observer because the stars advance slightly each day. Furthermore, if an observation of a star is taken at midnight of some day, a year later the relative position of the stars and sun will realign and the observation will return to its same point. The faster rotation of the universal sphere than the sun also explains the eastern drift of the sun through the zodiac, as depicted in Figure 7.2. Finally, because the ecliptic's daily rotation is slightly faster than the sun's, the sun's latitude performs its sinusoidal oscillation through the stars—also depicted in Figure 7.2.

This is a tidy and convincing portrait of a universe without planets or the moon. The movements of this geocentric model are in nearly perfect accord with observations. There is a slight disagreement between the model and observations. The model predicts equal time frames for the seasons, but in fact in the northern hemisphere the winter is slightly shorter than the summer. This would probably raise some eyebrows, but the model could be salvaged by slightly altering the rate of rotation of the sun. More to the point, the arrangement is simple and pleasing. Apparently many ancient observers found this model particularly pleasing because the orbits of the stars and sun are described by uniform circular motions. The planets upset universal nirvana with their retrograde motion. But the motion cannot be denied so the geocentrist must march on. To maintain tradition and aesthetics, the geocentrist must propose a pathway of circular orbits that explains retrograde motion. A most able advocate is Apollonius, who, as discussed in Chapter 2, proposes epicycles.

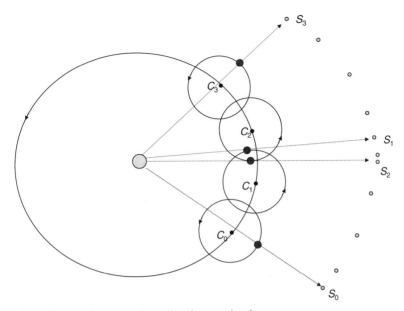

Figure 7.5   Geocentric retrograde motion due to epicycles.

Figure 7.5 illustrates retrograde observations caused by epicyclic motion. The annual motion of the center of the planet's epicycle follows the largest circle. The four smaller circles indicate the position of the epicycle and its center at four different times, C_0 through C_3. The planet's position within its epicycle is also displayed at the same four times. The figure illustrates the associated four observations of the planet from a geocentric earth, and each observation is taken at midnight. The observations indicate the alignment of the stars behind the planet. Typically, the stars that are behind the planet are ordered in a counterclockwise direction associated with the center of the epicycle's counterclockwise motion. However, due to the epicyclic motion, the ordering of the stars behind the planets S_1 and S_2 reverses, this clockwise ordering exhibits retrograde motion.

When Ptolemy attempted to apply Apollonius' epicyclic theory, reality was not accommodating. Rather than reviewing the central precepts, maintaining the earth in the center and explaining any movement in terms of uniform circular motion, Ptolemy made adjustments. Ptolemy added more epicycles so that a planet's motion could be the sum of more than two circular movements. He also shifted the center of the primary circular path of a planet's orbit from the earth to a point close by known as the eccentric and fixed the angular speed of the planet's primary orbit with reference to another point known as the equant. With these final two adjustments, Ptolemy sacrificed the pure notion of the earth being smack-dab in the middle of the universe and uniform circular motions. The sacrifices were accepted because the adjustments were minor, and the earth was still in the middle of the celestial sphere carrying the stars and was only a smidgen off center with respect to the planets. Furthermore, the motions of both the celestial sphere and the sun were still described by uniform rotations, and the adjustments to the planets were slight. While the description was reasonably accurate in terms of matching observations, to many it lost its pre-Ptolmaic beauty and was, as Copernicus put it, not pleasing. The complexity of the system meant that anyone wishing to fully understand its details must put in a significant effort. What is more, no physical explanation of the cause of the motion for any of the heavenly bodies was put forward.

With all its shortcomings, Ptolemy's model was accepted for 1300 years, until the Renaissance, when Copernicus placed his heliocentric model as a competitive alternative to Ptolemy. Copernicus' work receives its well-deserved historical recognition, although the description commonly proffered throughout elementary schools is inaccurate. The common perception is that Copernicus describes circular orbits for the paths of the planets with the sun at its center and that Kepler's ellipses were a minor adjustment. This is not quite right. There is brilliance in Copernicus' work along with some old Ptolemaic baggage. Let us start with the brilliance.

Copernicus' description of the earth's motion matches the historical picture. Copernicus describes the earth as having three motions. The first is uniform circular motion about the sun. The second is the daily rotation about the earth's axis. The third is the most puzzling because it applies to an unusual frame of reference. If a person could stand on the sun while continuously rotating so that he always faces the earth, it would appear to that person that the earth's axis performs a precession. Sometimes the North Pole would be inclined toward that person and sometimes it would be inclined away from that person. Copernicus' third motion is the precession of the earth's axis as seen by the solar inhabitant. From the perspective of someone standing on the North Star always looking down at the earth, there would be no precession of the earth's axis. Unlike the solar cousin, the individual on the North Star would not have to continuously rotate to keep the earth in sight and the earth's axis would be fixed—no motion. Since Copernicus describes his first motion from the perspective of the individual on the North Star, which is a reasonable proposition, it is a bit perplexing that he even introduces a third motion. Perhaps Copernicus wishes to state this motion so that he can emphasize the relation between the orientation of the earth's axis and the seasons. This Copernicus does with great skill.

By replacing geocentrism with heliocentrism, Copernicus is able to correct a serious flaw in the Ptolemaic universe. A natural question that arises in the Ptolemaic system is the ordering of each heavenly body by their distance from the earth. The Ptolemaic answer is Moon, Mercury, Venus, Sun, Mars, Jupiter, Saturn, and finally the stars. The answer is wrong because the question makes no sense. Since the earth and the planets are in motion, there is no fixed ordering of the planets and the sun in terms of their distance from the earth. Copernicus correctly poses the question of ordering in terms of distance from the sun. He then gets the right answer for the six planets known at the time, Mercury, Venus, Earth, Mars, Jupiter, and Saturn. Not only does Copernicus correctly infer the solar ordering, but he also used a similar three-bodied trigonometric relations between the earth, the sun, and a planet that Aristarchus applied between the sun, earth, and moon to estimate each planet's distance to the sun relative to the earth's distance from the sun. It must be noted that Copernicus correctly viewed the earth as a planet, while others would have maintained that there are only five planets. As for the moon, Copernicus correctly describes it as a satellite of the earth.

Copernicus describes the orbits of the planets as nearly circular. A natural question is the time period of revolution for each planet, each planet's year. Once again Copernicus addresses this issue using trigonometry and presents a method for calculating each planet's year. He gives some crude estimates from data that he had assembled. The method would later be used by Kepler with better data. One more success of Copernicus is that he deduces the presence of gravity on all heavenly bodies, just as it is evident on the earth, and correctly concludes that it is due to gravity that the heavenly bodies are round. Copernicus' concept of gravity is primitive in that he considers it a surface effect unable to act across space between heavenly bodies.

These accomplishments are considerable, but to convince a skeptical audience, Copernicus would have to pass the same test as Ptolemy. The predictions of his model would have to match astronomical observations. At first glance, Copernicus' arrangement offers some promise. One can replicate retrograde motion in a helio-centric model with the planets executing circular obits. Figure 7.6 illustrates how this arises. The earth moves about its orbit quicker than a planet at greater distance from the sun. The stars behind the planet vary providing the signature seen in Figure 7.1. (Compare this illustration with the geocentric illustration, Figure 7.5.) While promising, the circles do not yield signatures that match observations. Copernicus was stuck in the same position as Ptolemy. Not knowing what else to do, Copernicus applied the same epicyclic medicine. Copernicus had an issue with equants, finding them most displeasing, That left Copernicus with epicycles that he applied liberally. In fact, Copernicus' description has more epicycles than Ptolemy's and, like Ptolemy, he had no explanation for their cause.

One has to wonder about the data that both Copernicus and Ptolemy worked with. The observations were taken over a period of history that dates back to Ptolemy. The precise dates of these measurements can be questioned and the accuracy of any measurement must also be called into question. Referring to inaccuracies, in today's parlance, one would describe the data set as noisy. One could argue that it was a

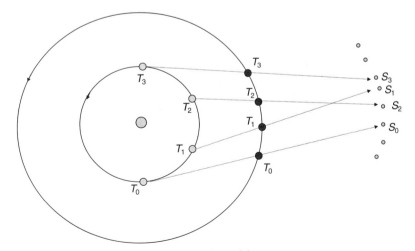

Figure 7.6   Retrograde motion in a heliocentric model.

rather meaningless exercise to try to match any model to a noisy data set. Or, on the contrary, one could confer success upon an incorrect model by stating that the observations, not the model, are in error. There was a remarkably clean and very populated data set back in Copernicus' time. This was the data set that had been continuously and meticulously gathered over many centuries by the Chinese using techniques that proffered very accurate measurements. But, needless to say, Europeans did not have access to the Chinese data. For Europeans to make further headway on the arrangement of the solar system, they would need their own clean data set. Is it a freak happenstance that Tycho Brahe, Europe's preeminent astronomical observer, dwelled on earth at just the moment that Europeans had the wherewithal to put a clean data set to use, or was it the aftermath of the Renaissance that allowed a man with Tycho's aptitude and determination to pursue his interest?

Half of Tycho's nose was lopped off by the sword of a classman with whom Tycho had brawled. Tycho later used an artificial replacement made of gold and silver. Reflecting this episode, Tycho's life is both bizarre and gilded. Tycho was born into one of the most connected families of the Danish Kingdom; the family had several members serving the inner court of the monarchy. At the age of two, Tycho was kidnapped by his childless uncle, Joergen Brahe. An arrangement was struck whereby Tycho was coparented by his biological parents and his adopted parents, sharing the households of both. Despite all efforts to guide Tycho down the path of a political career in the service of the king, as all four of his brothers had, Tycho stubbornly pursued his interest in astronomy. Unlike his brothers who were fully parented by their biological parents, perhaps Tycho felt the influence of his father's sister-in-law, his comother, who had an intellectual disposition. Or perhaps the strange parental arrangement left Tycho a bit different from the others. Or perhaps Tycho was just different. Whatever the case, he was a maverick who could afford to pursue his dreams. With the assistance of connections made through his family as well as resources inherited from his father's death, Tycho was able to construct several instruments for taking astronomical observations. Using these observations, Tycho at age 27 published a book, *De Nova Stella* (1573) that gained him notoriety.

The sovereign of Denmark, King Frederick II, was lucky to be alive as Tycho passed into adulthood. In 1565, when Tycho was 21 years old, King Frederick had fallen off a bridge into icy waters whereby a loyal member of the king's inner court sacrificed himself by diving into the waters and rescuing the king, while the loyal servant subsequently contracted pneumonia and died. The savior was Tycho's own uncle-cofather Joergen and the Brahe family received its share of gratitude. King Frederick seemed to have taken a personal liking to Tycho, despite or perhaps because of Tycho's maverick ways. When Tycho refused to accept the gift of four castles and fiefdoms because of the political duties associated with their operation, the King offered to build Tycho an astronomical observatory on the island of Hven so that Tycho could fully engage himself as an astronomer. Tycho accepted the offer. With his bride, a commoner whose social status was several strata beneath Tycho's, along with a retinue of servants, Tycho moved to Hven. The King was forthcoming in providing sources of income from a fiefdom. Following the path of Copernicus, the King granted Tycho the position of Church Canon from which Tycho gleaned additional revenues.

At Hven, over a 21-year period, Tycho assembled his treasured data set. Despite being publicly financed, the data in no measure belonged to the public. There were observations that Tycho did publish. But Tycho safeguarded as personal property the observations of the planets and their signatures that encoded the correct structure of the universe. Even on the island of Hven, Tycho was the only individual with access to all observations. There were three competing theories on the universe's structure, geocentric, heliocentric, and Tycho's own halfway system in which the planets are satellites of the sun, while the sun revolves about the earth. While Tycho was the most able observer in all of Europe, he and the individuals in his retinue did not have the theoretical skills to tease the data into revealing the universe's correct structure. This matter was made worse by Tycho's prejudice that the data should only be used to demonstrate the correctness of Tycho's own Ptolemaic–Copernican compromise.

There are many avenues that history could have followed. It is not hard to envision a scenario in which Tycho Brahe's observations languish in obscurity. The course of history flirted with this outcome as Tycho was entrenched on Hven. Something would need to perturb Tycho's isolation so that his data could fall into the hands of someone who could actually make use of it. In 1588, the death of King Frederick and ascension of his son, Christian IV, instigated a change. Relations between Tycho and the young king deteriorated to the point that King Christian dispossessed Tycho of Hven. Tycho responded by seeking a better offer elsewhere. It took some time to reach a breaking point and find another suitor, but by 1599, Tycho arranged a deal with Emperor Rudolf of the Habsburg Empire. He would be the imperial mathematician and would be given a residence that was larger than his castle at Hven. The move to the outskirts of Vienna was not an easy task. It was not until 1600 that Tycho resided in his new location.

Tycho was aware of his limitations as a theorist. He desired to use his catalog of observations to demonstrate his halfway system. To accomplish this, Tycho would need to bring a theorist of superior intellect into his inner circle. As Tycho accidentally stumbled into the neighborhood of the most brilliant mathematician in Europe, Johannes Kepler was reluctantly on his doorstep in desperate circumstances. While Tycho was recruited into the Habsburg Empire with extensive offerings, Johannes Kepler was no more than a refugee. Archduke Ferdinard, emperor Rudolf's cousin proclaimed and enforced a decree evicting Protestants from well-delineated Catholic areas in which Kepler happened to reside and work. Remaining faithful to his Lutheran beliefs, Kepler, the man who would later be excommunicated from the Lutheran Church, refused to convert to Catholicism, despite real earthly advantages that a conversion would confer upon Kepler and his family. As Kepler drew the line and stood on the other side, he and his family were evicted. Kepler's connections at Tubingen refused to offer assistance in obtaining a well-deserved university position either in Tubingen or elsewhere. He was homeless, unemployed, and rejected by the Lutheran community that he had remained faithful to. Apart from his unfortunate circumstances, there was another cause for Kepler to seek out Tycho. Kepler continued his research into God's order and passionately pursued theories relating musical harmonies to the structure of the universe. He was certain that Tycho's data would enable him to substantiate his views.

Kepler initially stayed as Tycho's guest. The circumstances of both men brought out the worst in each. Outfitting his new observatory to the standards of Hven was an overwhelming task that frustrated Tycho. Cost overruns, unfulfilled promises of support from the emperor, difficulties with the transport of equipment, and a homesick staff that took flight all conspired against Tycho. Regarding Kepler, Tycho was fearful that Kepler was a closet Copernican, not fully on board with the Tychonic compromise. Tycho was right because Kepler was not exactly in the closet. On his side, Kepler was fearful that his circumstances were being taken advantage of and he was not being accorded the status he deserved. Inadequate morsels were tossed his way in the form of material and intellectual substance. On the material side, like a student he was given room and board. But he was without salary, as a man of his accomplishment deserved, and he needed a salary to support his family. On the intellectual side, Tycho's prized observations were withheld. An emotionally charged Kepler brought the situation to a head, resulting in a public shouting match between the nobleman and the commoner that was witnessed by Tycho's own staff and so it also seemed ending the possibility that the empiricist and the theorist could together resolve the structure of the universe.

Humans are not governed by a simple set of physical laws. Drop an object from your hand a thousand times and surely enough of it falls toward the ground each time. Drop it one more time and surely enough of it falls again. Witness a commoner public shout at nobility as though the commoner has higher authority a thousand times and each time the commoner will discover very quickly that he is mistaken. Witness it one more time and then wonder just what happened as the two men come to agreement. It took some intervention by a mutual friend, but after a period of time, Tycho offered Kepler the terms of employment that Kepler was seeking and Kepler accepted. They might have had problems in the personal arena but wished to cohabitate within the scientific arena. Luckily, their relationship was not based on politics.

Recruiting Kepler was the last gift to science that Tycho would make. Full reconstruction of a functioning observatory at the outskirts of Prague never occurred. In fact, Tycho found himself in a position that he would have recoiled from in his youth. Emperor Rudolf called upon Tycho as his personal astrologer and, finding Tycho's counsel useful, forced Tycho to take up residence away from his observatory near the imperial palace in Prague. (That Tycho, a protestant, was able to adroitly maneuver between political columns of Catholics attests to his ability in this role.) It is in this residence that Tycho died in a bizarre set of circumstances that are legendary.

As legend goes, Tycho died of urethral infection instigated by his insistence on upholding etiquette and remaining at his seat at a banquet when in truth he really had to go bad. This description does not fully satisfy the modern forensic pathologist and there has been an ongoing investigation that began in 1991. Based upon chemical analysis of beard samples, the conclusion of forensic experts is that Tycho died of mercury poisoning. There are two camps as to how Tycho ingested the mercury. One camp proposes that Tycho prepared his own medications that included mercury and unwittingly poisoned himself. Another camp states that he was murdered, and this assertion has precipitated a whodunit hunt. Suspects include envious advisers to Rudolf and a distant cousin of Tycho's who was sent as an assassin by King Christopher of Denmark.

Included in the list of suspects, but one that nobody takes seriously, is Kepler. The only supporting evidence is that Kepler benefited from the death, which cannot be argued. In the aftermath of Tycho's death, Tegnagel, Tycho's ambitious employee turned son-in-law, did his best to preserve the family inheritance. This was an enormous task given the holdings of Tycho. Kepler correctly assumed that it would be a while before anyone would notice a small portion of missing observations. During a visit to Tycho's estate, Kepler took his opportunity and absconded with the Mars observations. The newly named imperial mathematician, Johannes Kepler, pored over these observations for 2 years before Tegnagel noticed they were missing. That it took so long to notice the missing observations testifies to the dysfunctional state of the observatory from the time of Tycho's death. Indeed from then on Tycho's instruments gathered cobwebs.

For 5 years Kepler worked to establish his first theory of orbital shape. Upon finding an 8-min discrepancy with Tycho's data, Kepler delightfully torched 5 years of work and began anew. It was not Kepler's first nor last demonstration of having the Scarecrow's capacity to pick himself up, put himself together, and start all over again. This time, he did so with the Scarecrow's most cheery disposition. Kepler describes his decision in the following words:

> *Since the divine benevolence has vouchsafed us Tycho Brahe, a most diligent observer, from whose observations the eight minute error in this ptolemaic computation is shown, it is fitting that we with thankful mind both acknowledge and honor this benefit from God. For it is in this that we shall find at length the true form of the celestial motions ...*

Just as Tycho's 8 min caught up with Kepler, so did Tegnagel. Upon discovering that Kepler was in possession of the Mars observations, Tegnagel demanded that the data be returned. Kepler forfeited the data, interrupting his research, and then turned to another seemingly unrelated pursuit, optics. As noted in Chapter 4, it is through Kepler's work in optics that he becomes thoroughly familiar with the conic sections, in particular the ellipse. While depriving Kepler of the data, Tegnagel inadvertently steered Kepler in the right direction. But Kepler would have nothing to do with it. Concerning the possibility of elliptical orbits, in a letter to a friend, Kepler writes that the problem would have been solved by Archimedes or Apollonius if the true path were an ellipse. No little anguish would result from this assumption as Kepler convinced himself out of the correct answer before ever giving it a try.

During the 2-year interlude in which Kepler wrote his work *The Optical Part of Astronomy*, Tycho's invaluable Mars observations found company with Tycho's invaluable observations of other planets and stars, all filed and forgotten. Tycho repeatedly said on his deathbed, "Let me not seem to have lived in vain." While Tegnagel may have been acting on behalf of the legal heirs to Tycho's property, Kepler was the heir to Tycho's legacy. After Tycho's departure, Kepler had the will and the imagination to fuel Tycho's fire; Tegnagel did not and never even tried. Through negotiations between Kepler and Tegnagel, Kepler regained possession of the Mars data and was ready for round two.

Having convinced himself that the path was oval, but not an ellipse, Kepler proceeded to consider alternatives. Kepler (1993) pursued several alternatives that occupied him for perhaps 2 years and also occupy 10 chapters of his 70-chapter book *New Astronomy*. To give an abridged version, all failed. What is striking is that Kepler actually uses an ellipse to establish estimations for his oval shapes. But he is only willing to use the ellipse as a device to assist with calculations, not yet recognizing that the contrivance is the actual prize.

Once more, Kepler cheerfully played the scarecrow, accepting the failure in fine humor; it is at this point that Kepler compares himself to a dog through the quote that is given in Chapter 2. But then what they say in the proverb, "A hasty dog bears blind pups", happened to me. Kepler's further elaboration of his efforts indicates a sense of frustration that accompanied his good humor:

> *While I am thus celebrating a triumph over the motions of Mars, and fetter him in the prison of tables and the leg-irons of eccentric equations, considering him utterly defeated, it is announced in various places that the victory is futile, and war is breaking out again with full force. For while the enemy was in the house as a captive, and hence lightly esteemed, he burst all the chains of the equations and broke out of the prison of the tables. That is, no method administered geometrically under the direction of the opinion of ch. 45 was able to emulate in numerical accuracy the vicarious hypothesis of chapter 16 (which has true equations derived from false causes)... I send new reinforcements of physical reasoning in a hurry to the scattered troops and old stragglers, and informed with all diligence, stick to the trail without delay in the direction whither the captive has fled. (New Astronomy (1993), p. 508)*

Kepler considers himself at war, but throughout the battles, he never waivers from his original purpose. Philosophizing about the errors evident in all his attempts, Kepler writes the following:

> *But, my good man, if I were concerned with results, I could have avoided all this work, being content with the vicarious hypothesis. Be it known therefore, that these errors are going to be our path to the truth. (New Astronomy (1993), p. 494)*

At long last, Kepler comes to the realization that the path is an ellipse. Once the ellipse has been accepted, the specifics can be determined and it can be put to the test and proved with brevity. These tasks do not take considerable time and only require ten chapters of *New Astronomy* (recall there are 70 chapters in all). The accomplishment is a testimony to both Kepler and Tycho Brahe. Though his Tychonic compromise proved to be incorrect, Kepler's work memorializes Tycho. The faith that Kepler placed in Tycho's observations is particularly noteworthy. Despite underlying tensions, there was considerable mutual respect. As for Kepler, this work along with his work in optics is his best. While Kepler's previous work demonstrates mathematical acumen, it is not based on science but has an element of mysticism to it. Kepler is at his best when he confronts the empirical data of Tycho in an honest manner. His faith in Tycho not only is a testament to Tycho but also allows Kepler to mature as a scientist. Along with Kepler's growth, science takes a leap forward.

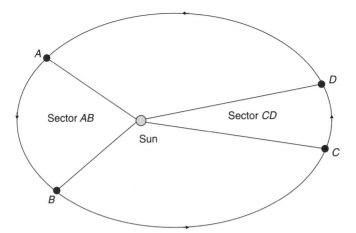

Figure 7.7   Kepler's law relating sector areas to path time.

Kepler's work culminates in two laws of planetary motion; he would add a third at a later time. It is worth reiterating the original two laws because of their relation to one another. The first law states that the ellipse is the shape of a planet's trajectory. The second law concerns the area of pie-shaped sectors, as illustrated in Figure 7.7. Each sector has three corner points, two are the planet's position at two different times and the final corner point is the sun. One boundary of the sector is the pathway of the orbit between the planet's corner points, while the other two boundaries are straight lines between the planet's corner positions and the sun. The second law states that the area of the sector is proportional to the amount of time required for the planet to move from the initial corner point to the latter corner point. In the illustration, the planet moves faster when it is closer to the sun so the time required to move from point A to B is the same as that required to move from point C to D. According to Kepler's law, the areas of the two corresponding sectors are equal. Kepler proposed his second law perhaps 3 years before he had settled on the ellipse as the planet's path.

The next scientific work of historical significance is Galileo's (2001) *Dialogue*. The attention afforded the trial of Galileo is disproportionate to that afforded Galileo's scientific works to an extent that Galileo's true role in science is often overlooked. Galileo's interest was in the study of motion. His insight guided him toward conclusions that formed the basis of Newton's laws, but his descriptions were qualitative. Clearly, Galileo understood inertia and had a sense of the relation between force and motion. With his understanding he had a good intuition of an object's motion when subject to a force, but without mathematical relations he was unable to perform mathematical analysis. Galileo used his intuition to guide his experiments in motion, and through data collected from the experiments, Galileo was able to describe an object's trajectory.

Galileo was not an astronomer and had no particular training in astronomy. Unlike Kepler, it is unclear that he ever read Copernicus' (1995) *On Revolutions* and most probable that he never read Kepler's (1993) *New Astronomy*. Prompted by his success at constructing telescopes, Galileo became interested in astronomy when he was nearing 50. Galileo writes his argument in favor of Copernican astronomy, *Dialogues*, not from the perspective of an astronomer, but from the perspective of his understanding of motion. True to form, while Galileo is not able to quantify his statements regarding motion, he nevertheless brings insight demonstrating exceptional instinct.

What conclusion can be drawn from the fact that one of Galileo's greatest follies also demonstrates his brilliant insight? Galileo's argument that tides are caused by the effects that the earth's revolution about the sun has upon the seas that daily rotate about the north–south axis of the earth cost him dearly. He was so enamored by his argument that he would stop nothing short of using it to convert the pope to geocentrism. In Chapter 2, the consequences of this decision that laid the groundwork for Galileo's later persecution are described. But the venture was not only a personal folly but also a scientific one.

Let us examine a bit more closely Galileo's proposition. Consider the dynamics of two particles, one that is pushed toward and one that is pushed away from the center of rotation of a rotating object. The perspective from which to view the particles is that of an observer standing at the center of the object while rotating with the object. The particle pushed toward the center appears to drift ahead of the revolving object, while the particle pushed away from the center appears to drift behind the revolving object. The cause of the drift is the conservation of angular momentum. The angular velocity of the particle increases as the particle's distance toward the center decreases, while the particle's angular velocity decreases as the particle's distance from the center increases. Galileo attributes the tides to the tendency of the oceans to drift as they recede from or approach the sun in accordance with the earth's rotation.

There is an element that Galileo overlooks. The tendency to drift must be compared with the rotation of the earth and the two must not be compatible. Furthermore, it must be shown that if the two are incompatible, the overall effect on the motion is enough to cause the tides. Newton's laws of motion along with his calculus offer a means to perform this analysis but had not sufficiently matured to tackle such a problem until over a century after Newton. Galileo proposed his hypothesis before Newton's birth; the tools were not available. Galileo was in a similar situation long before his interests turned to astronomy. While studying the motion of falling bodies, without Newton's tools Galileo was unable to perform a mathematical analysis of a body's trajectory, so he turned to experiment to acquire empirical data. Similarly, without mathematical constructs empirical data are necessary to support Galileo's tidal hypotheses. Nature's own empirical data display a huge embarrassing hole. If the cause were as Galileo claims, there would be a rhythmic daily pattern to the tides that just does not exist. The data reject the theory, but Galileo could not. He did not have Kepler's disposition to accept nature's verdict, so he never bothered to ask for it. Perhaps it was this disposition that caused him to recklessly engage the pope in an issue that could only spell trouble.

So what makes the folly a success? Examine Kepler's second law and its consequences. The speed of a planet moving toward its center of revolution, the sun, increases while the speed of a planet moving away from its center of revolution decreases. Both men have found the essence of conservation of angular momentum. Kepler finds this through intuition and verifies it using Tycho's data. He is also able to furnish a quantitative description. Galileo arrives at his conclusion through a combination of experience and pure intuition. He does not verify it with any data. While he is unable to set forth any quantitative law, he is on to something quite remarkable. What Galileo intuits would later be called the Coriolis force, named after the mathematician Gaspard-Gustav Coriolis (1792–1843), who quantified Galileo's intuition using Newton's laws of motion. While Galileo was wrong concerning the tides, the Coriolis force due to the earth's rotation (not the revolution about the sun) does impact ocean currents and weather systems.

When Newton was born, not only mathematics but also physics was on the brink of discovery. Copernicus had introduced the seeds of the concept of gravity and correctly used it to explain the spherical shape of stars, planets, the sun, and the moon. Kepler had gone further, correctly asserting that a body's gravity reaches across space and furthermore attributing planetary motion to a force emanating from the sun. Galileo had qualitatively described inertia and the effect of forces on objects. Newton inherited this sketch and was able to complete it into a vivid portrait. His portrait in which his invention of calculus delineates the laws of motion is one of the greatest intellectual feats in history and arguably has had a broader influence on mankind's development than any accomplishment of any other man.

The accomplishment does not come without its blemish. The description of Robert Hooke as nasty and cantankerous is apt. Newton harbored a hatred for Hooke that displays an equally nasty nature. While Hooke was an object of Newton's ill will from the time of Newton's publication in optics, Leibniz would be a later target. But while the battle with Leibniz only consumed Newton's energy with no productive result, Newton's hatred for Hooke was motivational. Exactly how Halley inspired Newton to abandon alchemy and return to physics is unknown. But it is possible that during Halley's 1684 visit to Newton, Halley, knowing Newton's hatred of Hooke, hinted that Hooke was on the verge of demonstrating the elliptical orbit of planets and this hint was enough to prompt Newton to complete his work and claim priority.

Newton would have taken a threat from Hooke seriously. Several years earlier, during a brief interlude from alchemy, Newton and Hooke corresponded. Newton had sent Hooke a confidential letter concerning the kinematics of an imaginary object under the influence of earth's gravity that could penetrate the earth's surface and move freely within the earth's interior. The letter contained an error and, as expected from a man of Hooke's disposition, Hooke used the error to publicly humiliate Newton. Confidentiality be damned, Hooke displayed the letter at a meeting of the Royal Society, pointed out the error, and proceeded to give his own analysis of the kinematics. Hooke's response contains his correct view that the force on an object under the influence of gravity falls off inversely to the square of the distance to the object. Along with the bouts over optics, this painful episode resided in Newton's psyche and a skillful Halley could well have used it.

The episode also may have been more than a motivator. Hooke later claimed priority over the use of the inverse square law and prevailed upon Halley to request that Newton recognize Hooke's contribution in Newton's (1995) upcoming publication, *Principia*. Newton responded that Hooke's claim was nonsense and threatened to abandon the project. Hooke was dispensable; Newton was not, and so the issue never resurfaced. Nevertheless, there is merit to Hooke's claim. Had Newton formulated the inverse square law during his astonishing research days only to have forgotten it, in which case Hooke merely prodded Newton's memory, or did Hooke have a legitimate claim to priority over the discovery? It is a secret that Newton took to his grave. Either way Newton must be smirking.

Eighteen hundred years prior to Newton's *Principia*, Euclid (2002) opens his seminal work, *The Elements*, in the following manner.

Definition 7.1   *A line is that which has no part.*

Twenty-two additional definitions follow and then Euclid stepwise moves to axioms and theorems. His work resonates with Newton as Newton in a similar "let's get down to business" tone begins *Principia* in the following manner:

Definition 7.2   *The quantity of matter is the measure of the same, arising from its density and bulk conjunctly.*

Eight definitions follow and then Newton stepwise moves to axioms (the three laws of motion) and theorems. It is Euclid's well-deserved legacy that Newton chose Euclid's work as a template. The success of both *The Elements* and *Principia* is such that Euclid's work is a template for nearly all modern mathematics and theoretical physics. There are no confessions, references to dogs, battles with an enemy, requests for sympathy, or invocations of an almighty entity; it is all humorless and impersonal and it is perfect for its purpose. Kepler was the last of his kind—perhaps one of a kind.

The previous chapters present all the mathematics needed to demonstrate elliptic planetary orbits. In this chapter, the mathematics is joined with Newton's laws of motion to unveil Kepler's ellipse. The starting point is a presentation of Newton's laws. From there we investigate Galileo's parabola, which affords two benefits. First the simpler problem of Galileo's parabola facilitates an understanding of the laws of motion as well as the mechanism used to determine an object's trajectory. Second, the implementation of Newton's program, which successfully models Galileo's experiments, furnishes confirmation of the program's correctness. Once that stage has been reached, we apply Newton's program to planetary motion and find the ellipse. Our presentation is a bit more ambitious than Newton's original presentation in *Principia*. Since Newton refrains from the use of calculus, his goal is limited to demonstrating that the ellipse satisfies the laws of motion. The ellipse is a starting point and it is shown to be correct. In our approach, we do not begin with the final shape of the trajectory but use calculus to generate the ellipse.

With this chapter's introduction, the book's narrative ends. How does one conclude a messy story that includes Greek tragedies, backstabbings, nose loppings, crisped corpses, sacked cities, plagues, fragile egos, papal bastards, enigmatic geniuses, loose cannons, quirky noblemen, idealistic paupers, rock bed faith, excommunication, stubborn conviction, unrequited dedication, and blissful discovery—a story that as a fictional work would be criticized as devoid of reality? I have decided to end it by recalling my favorite part. There is lots to choose from, but for me the best is when after 2 years of struggling with the Mars data Kepler discovers the 8-min error and shortly afterward Tegnagel demands that the data be returned to him.

The back-to-back setbacks do not stop Kepler, and he returns to his work in optics. Through this work, Kepler becomes thoroughly familiar with the ellipse. In the end, Kepler's recognition of a data point within the Mars data that matches his previous work in optics sparks his realization that God chose the ellipse as the path of the planets. Kepler's back-to-back setbacks become his blessing.

7.1 NEWTON'S LAWS OF MOTION

Newton unveils his theory of motion in *Principia*. Reflecting the practice of scholarly works of the times, the original publication is in Latin. The influence of Euclid's *Elements* is visible in Newton's *Principia* as Newton applies the axiomatic process to physics. Newton's axioms are his laws of motion; he reveals three laws that are central to the rest of his analysis. This section presents the laws. It is tempting to present them through a direct translation and be done with it. Unfortunately, this leaves a false impression; the laws must be interpreted within the context of other material. Accordingly, in addition to providing a translation of the laws, an interpretation is presented.

1. Every body perseveres in its state of rest, or of uniform motion in a right (straight) line, unless it is compelled to change that state by forces impressed thereupon.

2. The alteration of motion is ever proportional to the motive force impressed and is made in the direction of that straight line in which that force is impressed.

3. To every action there is always opposed an equal reaction, or the mutual actions of two bodies upon each other are always equal and directed to contrary parts.

Suppose that one were to present these axioms to Euclid. What might his reaction be? Most certainly, Euclid would consider this attempt at an axiomatic process to be amateurish. As the foundations of the deductive process, axioms are independent statements. There is a serious flaw when one axiom can be deduced from the remaining axioms. At first glance, this flaw appears in Newton's laws; the first axiom appears as a consequence of the second. Indeed, one can prove the first from the second as follows. From the second axiom, if a body experiences no impressed force, it remains at rest or its motion is unaltered; this is a rewording of the first axiom.

To be sure, Newton was no amateur. His first axiom is far from a blunder; it addresses relativity of laws of motion. The concept of relativity is something that Galileo dwelled upon at length. Galileo's concern was that motion appears different to different observers. Furthermore, there is not a preferred observer. Imagine two objects in space moving past one another at uniform velocity. Two observers, one on each object, each concludes that the object upon which he resides is at rest while the other object is moving past him. A third observer not on either object may perceive that both objects are in motion. Any analysis of motion must be applicable to different observers.

Like Galileo, Newton was fully cognizant of the concept of relativity, and in developing his mathematical analysis, Newton wished to take the concept into account. Newton did not philosophize over the issue as Galileo did; rather he quickly addresses the issue through his first law, which frames the conditions under which his general analysis is applicable. Prior to applying his method, Newton demands that the observer obey his first law. If an observer views an object altering its course without any force impressed upon it, then the observer violates the first law and cannot use Newton's system. Observers obeying Newton's first law are said to be observers in an inertial frame of reference. Inertial frames of reference are at rest or move with constant velocity.

Once the setting for the analysis is established, Newton can present a description of the relation between force and motion. This is the second law; the alteration of an object's motion as observed by an observer in an inertial frame is proportional to the force impressed upon the object. It is the translation of this law into a mathematical expression and the analysis of the resulting expression using the methods of calculus that distinguish Newton from his predecessors. Whereas his predecessors could merely philosophize about motion, Newton could calculate. He could determine an object's future trajectory knowing only its current state.

In the privacy of his own company, Newton employed his notation of calculus to express the second law of motion mathematically, but not in *Principia*. Indeed, the first publication of the second law of motion in terms of calculus is given by Johann Lambert after Newton's death. The law relates the alteration of the motion to force. The law speaks not of motion, which Newton quantifies as the momentum, *mass × velocity*, but of an alteration of motion. The corresponding mathematical expression is not the momentum but the derivative of the momentum. The law speaks of proportionality to the impressed force. Putting these ideas together results in the following equations:

$$\frac{d}{dt}(m\vec{v}) = \vec{F}$$

$$m\frac{d}{dt}(\vec{v}) = \vec{F}$$

$$m\frac{d^2}{dt^2}(\vec{x}) = \vec{F}$$

In this set of equations, m is the mass of the moving object, $\vec{v}$ is the velocity with respect to an observer in an inertial frame, $\vec{x}$ is the position with respect to an observer in an inertial frame, $\vec{F}$ is the force impressed upon the object, and t is the time. The quantity

$$\frac{d}{dt}(\vec{v}) = \frac{d^2}{dt^2}(\vec{x})$$

is known as the acceleration. Notice the relation between acceleration, mass, and force. A low-mass object accelerates at a quicker pace than a high-mass object when subject to the same force. Also, notice that the position, velocity, and force are all vectors in three-dimensional Euclidean space.

Newton considered the mass of an object to be fixed. If not, the equations become

$$m\frac{d}{dt}(\vec{v}) + \vec{v}\frac{d}{dt}m = \vec{F}$$

It was not until Einstein challenged the notion that mass remains constant that the second term was considered. For motions that are perceived in everyday experiences, Newton's view remains valid and Newton's analysis is at the foundation of all but highly specialized investigations. For the remainder of this chapter, the mass of all objects is considered constant.

The analysis of planetary motion looks at a two-body system, the sun and a planet. With the third law, our program, seeking Kepler's ellipse with the sun fixed at a focus of the ellipse, appears to be in jeopardy. After all, the third law states that if the sun pulls on the planet, the planet pulls on the sun with an equal force. And if the planet pulls on the sun, how in an inertial reference frame can the sun remain at rest at the focus of the ellipse? Indeed, the sun cannot remain at rest in an inertial reference frame, and it appears that Newton's program is bound for failure by his own laws.

In Chapter 2, there was a similar sense of despair concerning the broader goal of quantifying anything. Recall that the cause of concern is the discovery of the sparseness of numbers that we can actually represent—rational numbers. The question that was asked is that if most physical measurements such as position, speed, and force are irrational values, does not the fact that our measurements are limited to rational values doom any analysis? We saw that the rational numbers, while sparse, were sufficient because rational numbers could be used to come within any required error tolerance.

It is the same concept of error tolerance that allows us to apply Newton's analysis to planetary motion. While the sun may not be exactly fixed with respect to an inertial frame, its movement is so slight that fixing it introduces an error that is inconsequential. The ellipse of Kepler is not the precise orbit, but it is oh so close, just as a rational number can be made oh so close to an irrational number.

That the sun's movement is nearly fixed is a consequence of Newton's second law,

$$\frac{d}{dt}(\vec{v}) = \frac{1}{m}\vec{F}$$

The mass of the sun is orders of magnitude larger than that of the planet; $1/m$ is so huge for the sun that the acceleration is inconsequential. Accordingly, our analysis considers the sun's position as fixed and analyzes the motion of the planet.

Remarks

- The more general problem known as the two-body problem allows the ratio of the masses of the bodies to be any value. For the two-body problem, the inertial frame of interest is usually taken to be the center of mass of the bodies and each body is in motion with respect to the center of mass. Newton's analysis applies, although the mathematical solutions become somewhat complicated. Kepler's ellipse is a limiting case when the ratio of the masses of the larger object to the smaller object becomes indefinite.

- The third law is equivalent to the statement that the motion of the center of mass of a system of objects obeys the first law; if the system is not subject to external forces, the motion of its center of mass is unchanged. As an illustration of the concept, consider a two-object system, A and B. The center of mass of two objects, A and B, is given by the following equation:

$$m_A \vec{x}_A + m_B \vec{x}_B$$

where m_A is the mass of object A and $\vec{x}_A$ is its position, and similarly for object B. Suppose that an imaginary observer who does not interfere with the motion of either object makes his observations of the motions of the bodies with respect to a fixed center of mass that we may arbitrarily set at the origin:

$$m_A \vec{x}_A + m_B \vec{x}_B = 0$$

Taking the derivative of the center of mass, which is by assumption fixed, results in the following:

$$\frac{d}{dt}(m_A \vec{x}_A + m_B \vec{x}_B) = 0$$

$$m_A \frac{d}{dt}(\vec{x}_A) + m_B \frac{d}{dt}(\vec{x}_B) = 0$$

Taking the second derivative produces the following result:

$$m_A \frac{d^2}{dt^2}\vec{x}_A + m_B \frac{d^2}{dt^2}\vec{x}_B = 0$$

$$\vec{F}_A + \vec{F}_B = 0$$

$$\vec{F}_A = -\vec{F}_B$$

where $\vec{F}_A$ and $\vec{F}_B$ represent the forces on bodies A and B, respectively. As there are no external forces, these are the forces that the bodies impress upon each other. These forces are of equal magnitude but in opposite direction just as the third law demands. By performing the steps in reverse, one demonstrates that the third law implies that the first law applies to the system's center of mass.

7.2 GALILEAN CHECKPOINT

Galileo studied the motion of projectiles near the earth's surface and found that the pathway of a tossed object is a parabola. Galileo's explanation of the parabolic shape was a precursor to Newton's laws of motion. Galileo reasoned that the projectile is attracted toward the earth's surface in a vertical direction, and hence its motion in the vertical direction is influenced by the attraction. However, as there is no force influencing its motion in the horizontal direction, Galileo reasoned that the projectile's

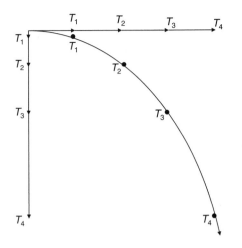

Vertical, horizontal, and composite positions at times T_1, T_2, T_3, and T_4

Figure 7.8   Motion of thrown object.

horizontal motion proceeds in an unaltered state. Figure 7.8 shows the separate horizontal and vertical components of an object thrown horizontally off of an elevated platform. The separate motions are then coupled with a parabola as the resulting shape.

There are several admirable features of Galileo's analysis. He is able to separate motion into two components, one that is influenced by a force and one that is not. This allows each component to be assessed independently, simplifying the problem. Also, by correctly describing the horizontal motion, Galileo arrived at Newton's first law before Newton. While Galileo presents a qualitative description, he is unable to propose quantitative laws. Galileo required experimental observations to determine the precise nature of the vertical component of motion. The experiments are described in Chapter 2. These data were collected and analyzed just as Kepler had analyzed Tycho's data. Below we apply Newton's program to determine the trajectory.

Let a stationary observer sitting comfortably on the ground record the position of a tossed object. Align the axes of the inertial frame such that the y axis points in the vertical direction and the x axis points in the direction of the horizontal motion. Newton's second law of motion gives the following equations:

$$\frac{d}{dt}(v_x) = 0 \qquad \frac{d}{dt}(v_y) = \frac{1}{m}F_y$$

where

- v_x is the horizontal velocity
- v_y is the vertical velocity
- F_y is the vertical force of gravity
- m is the mass of the object

As the tossed object remains near the surface of the earth, the vertical force F_y is considered to be a constant. Anyone lifting two weights of different mass can attest to the fact that the forces of both weights are proportional to their mass. Accordingly, we set $F_y = -mg$, where g is the constant of proportionality. The constant value g has been measured and depends upon the units of measurement. In metric units, the value of g is approximately 9.8 m/s^2. The negative sign indicates that the accelerating force F_y acts downward in the negative y direction. Using $F_y = -mg$ admits the following equations for the motion of the object:

$$\frac{d}{dt}(v_x) = 0 \qquad \frac{d}{dt}(v_y) = -g \tag{7.1}$$

The position of the tossed object can be determined by integrating the equations of motion. Furthermore, consistent with the independence between the horizontal and vertical components of motion, the equations may be solved independently. We begin with the horizontal motion.

Because only a constant has zero derivative, $v_x(t)$ must be a constant value denoted by v_{x0}, $v_x(t) = v_{x0}$. Note that $v(t) = \frac{d}{dt}x$ furnishes the following result:

$$\frac{d}{dt}(x) = v_{x0}$$

$$x(t) = \int v_{x0}\, dt \tag{7.2}$$

$$= v_{x0}t + x_0$$

In the above equation, x_0 is both the initial horizontal position of the object and the constant of integration.

Next we consider the motion in the vertical direction:

$$\frac{d}{dt}(v_y) = -g$$

$$v_y(t) = -\int g\, dt \tag{7.3}$$

$$= -gt + v_{y0} = \frac{d}{dt}(y)$$

Once again the constant of integration v_{y0} has a physical interpretation. It is the object's initial vertical velocity. The velocity is written as a derivative and the resulting equation is integrated to obtained the vertical position:

$$\frac{d}{dt}(y) = -gt + v_{y0}$$

$$y(t) = \int -gt + v_{y0}\, dt \tag{7.4}$$

$$= -\tfrac{1}{2}gt^2 + v_{y0}t + y_0$$

Another constant of integration, y_0, results. This is the initial height of the object. We can conclude that the initial velocity (v_{x0}, v_{y0}) and the initial position

(x_0, y_0) determine the position $(x(t), y(t))$ at any time and this position is given by equations (7.3) and (7.4).

One way to demonstrate the parabolic trajectory is to solve for the time variable t in equation (7.3) and substitute the result into equation (7.4):

$$x = v_{x0}t + x_0 \qquad t = \frac{x - x_0}{v_{x0}}$$

$$y = -\frac{1}{2}gt^2 + v_{y0}t + y_0 \qquad\qquad (7.5)$$

$$= -\frac{1}{2}g\left(\frac{x - x_0}{v_{x0}}\right)^2 + v_{y0}\left(\frac{x - x_0}{v_{x0}}\right) + y_0$$

It is noted that y is quadratic in x, so the shape of the trajectory is a parabola.

Example 7.1   SHOT PUT

A standard shot put has a mass of 7.26 kg (weighs 16 lb). Suppose that the height of an athlete's hand as he releases the shot put is 2 m. In addition, the athlete releases the shot put at an angle of $45°$ with a speed of 30 m/s. How far does the shot put travel before striking the ground?

The initial conditions are stated in the problem:

$$(x_0, y_0) = (0, 2) \qquad (v_{x0}, v_{y0}) = (30\cos 45°, 30\sin 45°) \approx (14.14, 14.14)$$

The terminal condition, when the shot put strikes the ground, is $y = 0$. Placing the initial conditions and terminal condition into the quadratic expression of equation (7.5) results in the following equation for the terminal horizontal distance:

$$y = -\frac{1}{2}g\left(\frac{x - x_0}{v_{x0}}\right)^2 + v_{y0}\left(\frac{x - x_0}{v_{x0}}\right) + y_0$$

$$0 = -\frac{9.8}{225}x^2 + x + 2$$

$$\approx 0.044x^2 + x + 2$$

Solving for x using the quadratic equation yields the result $x \approx [-1 - \sqrt{1 - 8(0.044)}]/[-2(0.044)] \approx 20.51$ m.

Note that the mass of the shot put does not directly enter into the solution, but it affects the initial velocity that the shot putter is capable of at release time. As of July 2009, the world record is 23.12 m.

Remark. While we have been casual concerning the units associated with the terms, they are important. A very useful quality check on any equation is to assure that the units on each side of the equation match.

7.3   CONSTANTS OF MOTION AND ENERGY

A constant of motion is a function of an object's position, velocity, or both that does not change even as the object moves. For example, in the tossed object of Section 7.2, the term $v_x(t)$ is a constant of the motion; the equations (7.1) inform us that $v_x(t)$ is a constant of motion because

$$\frac{d}{dt}(v_x(t)) = 0$$

The constant of motion $v_x(t)$ allows for separation between horizontal and vertical components of motion and results in a simple solution to the horizontal component. In general, constants of motion, also known as conserved quantities, simplify solutions. A universal constant of motion for all mechanical systems obeying Newton's laws of motion is the energy. This section develops the energy for the tossed-object problem and shows that indeed it is a constant of the motion. The method is then generalized to determine the energy for a general system.

7.3.1   Energy of a Tossed Object

First consider the simple case of the tossed object discussed in the previous section, an object that vertically drops. The equations of motion for such a case are the following:

$$\frac{d}{dt}(y) = v_y \qquad \frac{d}{dt}(v_y) = -g$$

Let us look for a constant of the motion of the following form:

$$E_m(y, v_y) = K_m(v_y) + P_m(y)$$

where $K_m(v_y)$ and $P_m(y)$ are functions of their associated variable.

Note that because each of the variables is a function of time, $y = y(t)$, and $v_y = v_y(t)$, both $K_m(v_y)$ and $P_m(y)$ are also functions of time. Using the sum rule and the chain rule for differentiation and substituting for the values of $\frac{d}{dt}(v_y)$ and $\frac{d}{dt}(y)$ when appropriate, it is possible to determine the time derivative of $E_m(y, v_y)$:

$$\frac{d}{dt}\left[E_m(y(t), v_y(t))\right] = \frac{d}{dt}\left[K_m(v_y(t))\right] + \frac{d}{dt}\left[P_m(y(t))\right]$$

$$= \frac{d}{dv_y}\left[K_m(v_y)\right]\frac{d}{dt}(v_y) + \frac{d}{dy}\left[P_m(y)\right]\frac{d}{dt}(y)$$

$$= -g\frac{d}{dv_y}\left[K_m(v_y)\right] + v_y\frac{d}{dy}\left[P_m(y)\right] \qquad (7.6)$$

To ensure that E_m is a constant of the motion, set its time derivative to zero:

$$\frac{d}{dt}\left[E_m(y(t), v_y(t))\right] = -g\frac{d}{dv_y}\left[K_m(v_y)\right] + v_y\frac{d}{dy}\left[P_m(y)\right] = 0$$

$$\qquad\qquad (7.7)$$

$$g\frac{d}{dv_y}\left[K_m(v_y)\right] = v_y\frac{d}{dy}\left[P_m(y)\right]$$

The left-hand side of equation (7.7) is a product of a constant and a function in v_y so the right-hand side must also be of the same form. Equating the function in v_y yields the following equation:

$$\frac{d}{dv_y}\left[K_m(v_y)\right] = v_y$$

Integrating furnishes a solution for K_m:

$$K_m(v_y) = \int v_y \, dv_y$$

$$= \frac{1}{2}v_y^2$$

The remaining right-hand factor of equation (7.7) must be equated with the remaining factor on the left-hand side:

$$\frac{d}{dy}[P_m(y)] = g$$

Integrating with respect to the variable y establishes a solution of P:

$$P_m(y) = \int g \, dy$$

$$= gy$$

Having found K_m and P_m, the constant of motion E_m is available:

$$E_m(y, v_y) = K_m(v_y) + P_m(y) = \frac{1}{2}v_y^2 + gy$$

Multiplying the constant of motion E_m by the mass furnishes the energy of the system:

$$E = \frac{1}{2}mv_y^2 + mgy$$

The energy is split into two components, one that depends only upon velocity v_y and one that depends only upon position y. The term depending upon velocity, $\frac{1}{2}mv_y^2$, is known as the kinetic energy while the term depending only upon position, mgy, is known as the potential energy.

Remarks

- The conserved quantity emerges regardless of the constants of integration that are chosen. It most common to set the constants of integration to zero.
- It is possible to verify that the energy is indeed a constant of the motion by taking the time derivative and ensuring that it is equal to zero. It is necessary to use the chain rule in the exact same manner as in equations.

7.3.2 Energy of a System Moving in a Single Dimension

The approach that results in the energy for the dropped object is applicable to any of Newton's equations of motion in a single dimension provided that the force is a function of position only. Below, the approach is repeated for a general force $F(y)$. The equations of motion are as follows:

$$\frac{d}{dt}(y) = v_y \qquad \frac{d}{dt}(v_y) = \frac{F(y)}{m}$$

A constant of motion $E_m = E_m(y, v_y)$ is determined and its derivative set to zero:

$$E_m(y(t), v_y(t)) = K(v_y(t)) + P_m(y(t))$$

$$\frac{d}{dt}\left[E_m(y(t), v_y(t))\right] = \frac{d}{dt}\left[K_m(v_y(t))\right] + \frac{d}{dt}\left[P_m(y(t))\right]$$

$$= \frac{d}{dv_y}\left[K_m(v_y)\right]\frac{d}{dt}(v_y) + \frac{d}{dy}\left[P_m(y)\right]\frac{d}{dt}(y)$$

$$= \frac{d}{dv_y}\left[K_m(v_y)\right]\frac{F(y)}{m} + v_y\frac{d}{dy}\left[P_m(y)\right]$$

$$= 0$$

From the above relations, the following equality holds:

$$-\frac{d}{dv_y}\left(K_m(v_y)\right)\frac{F(y)}{m} = v_y\frac{d}{dy}\left(P_m(y)\right)$$

Each side of the equality is a multiple of two factors. The first factor is a function of v_y and the second factor is a function of y. Equating the factors and solving through integration result in solutions for K_m and P_m:

$$\frac{d}{dv_y}\left[K_m(v_y)\right] = v_y$$

$$K_m = \int v_y \, dv_y$$

$$= \frac{1}{2}v_y^2$$

$$\frac{d}{dy}\left[P_m(y)\right] = -\frac{F(y)}{m}$$

$$P_m = -\frac{1}{m}\int F(y)\, dy$$

Placing the expressions for K_m and P_m into the expression for E_m results in the following expression for E_m:

$$E_m = K_m + P_m$$

$$= \frac{v_y^2}{2} - \frac{1}{m}\int F(y)\, dy$$

Multiplying through by m furnishes the energy with the first term the kinetic energy and the second term the potential energy:

$$E = \frac{mv_y^2}{2} - \int F(y)\, dy$$

Remarks

- Direct verification that the energy is a constant of motion is performed by ensuring that the time derivative is zero. To take the time derivative, the chain rule is necessary.

- In the above calculations, the energy has been derived from the force. In applications, it is often the case that the energy is known and the force is derived from the energy. From the expression of the energy, it is seen that the force is the negative of the derivative of the potential energy,

$$F = -\frac{d}{dy}(\textit{potential energy})$$

Knowing the force, the equations of motion can be determined, so the equations of motion are available from the energy. For example, the potential energy for the tossed object in one dimension is Potential energy $= mgy$. From this we find the force,

$$F = -\frac{d}{dy}(mgy) = -mg$$

- The process above is generalizable to motions in three dimensions. The generalization requires further definitions that would take us along more excursions. The final target, Kepler's ellipse, is within reach; we forgo the excursion and head toward the ellipse. There may be an objection to this approach. After all, planetary orbits are not along a single direction; a planet moves in an ellipse that lies in a plane. If energy is a critical requirement for solving the problem of planetary motion, a generalization to at least two dimensions should be considered. While the objection has merit, nature is most accommodating. Notice that the tossed-object problem of Galileo can be split into separate and independent components, motion along the horizontal direction and motion along the vertical direction. Each component can be solved separately, so the solution to the motion in two dimensions yields two separate one-dimensional problems. In the next section, it is shown that a similar separation occurs for planetary motion.

7.4 KEPLER AND NEWTON: ARISTARCHUS REDEEMED

7.4.1 Polar Coordinates

Kepler states that the force between the sun and the planet is directed along the line between their centers. Fixing the inertial frame of an observer so that the origin is at the center of the sun, the law states that the force acting on the planet is always in the

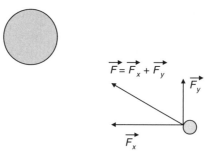

Figure 7.9   Sun pulls planet directly toward it.

radial direction (Figure 7.9). The force is a vector with two components, F_x aligned with the x axis and F_y aligned with the y axis. Using the relations between polar and Euclidean variables, $r\cos(\theta) = x$, $r\sin(\theta) = y$, and $r = \sqrt{x^2 + y^2}$, the components of force are $F_x = \cos(\theta)F(r)$ and $F_y = \sin(\theta)F(r)$, where $F(r)$ is the force expressed in terms of a point's distance from the sun.

Newton's equations of motion may be expressed as follows:

$$\frac{d}{dt}(x) = v_x$$

$$\frac{d}{dt}(y) = v_y \tag{7.8}$$

$$\frac{d}{dt}(v_x) = \frac{1}{m}F_x = \frac{1}{m}\cos(\theta)F(r)$$

$$\frac{d}{dt}(v_y) = \frac{1}{m}F_y = \frac{1}{m}\sin(\theta)F(r)$$

The result is difficult to deal with as there is no obvious way to integrate the derivatives so that the underlying coordinates are available. As the force is directed only in the radial component, it is natural to describe the system entirely in polar coordinates. This is the immediate task.

7.4.1.1 The Position Variables and Velocity   In Cartesian coordinates, the relation between position variables (x, y) and velocity variables (v_x, v_y) is straightforward; the velocity is the derivative of the position. There is an intuitive approach toward determining the relation between the polar position variables (r, θ) and the velocity expressed in polar variables. The intuitive approach is given below. With a trust-but-verify attitude toward intuition, a more detailed demonstration is also given.

We begin with some notation. Let $\vec{v}_{xy}$ denote the velocity vector expressed in standard Cartesian coordinates and $\vec{v}_{\text{polar}}$ denote the same velocity vector expressed in polar coordinates. The components expressed in polar coordinates are $\vec{v}_{\text{polar}} = (v_r, v_\theta)^{\text{T}}$.

Let us begin by assuming that an object's motion is in a radial line emanating from the origin. Then, just as with Cartesian coordinates, the derivative of the radial position coordinate r is the velocity in the radial direction v_r; $\frac{d}{dt}(r) = v_r$. Next assume that the object's motion is circular about the origin with constant speed, v_θ. Let $\Delta\theta$ be the angle change in time Δt. The distance traveled in time Δt is given by $r\,\Delta\theta$. To determine the speed, divide the distance traveled by the time:

$$\frac{r\,\Delta\theta}{\Delta t} = v_\theta$$

Taking the limit as Δt approaches zero establishes the relation between the time derivative of the angle and the angular component of the velocity:

$$\lim_{\Delta t \to 0} \frac{r\,\Delta\theta}{\Delta t} = r\frac{d}{dt}(\theta) = v_\theta$$

Sometimes we divide by r to obtain

$$\frac{d}{dt}(\theta) = \frac{1}{r}v_\theta$$

The relations between derivatives of the position variables and the velocity have been found for two special cases, radial and circular motion:

$$\frac{d}{dt}(r) = v_r \qquad \frac{d}{dt}(\theta) = \frac{1}{r}v_\theta$$

Consider any arbitrary motion as a composite of these cases and the relations hold, so says our intuition. We next follow a more detailed path.

Recall from Section 5.8 that the components of $\vec{v}_{\text{polar}}$ are obtained by applying a rotation matrix $R(-\theta)$ as follows:

$$\begin{pmatrix} v_r \\ v_\theta \end{pmatrix} = R(-\theta)\begin{pmatrix} v_x \\ v_y \end{pmatrix}$$

$$R(-\theta) = \begin{pmatrix} \cos(-\theta) & -\sin(-\theta) \\ \sin(-\theta) & \cos(-\theta) \end{pmatrix} = \begin{pmatrix} \cos(\theta) & \sin(\theta) \\ -\sin(\theta) & \cos(\theta) \end{pmatrix}$$

Also recall that the inverse of the matrix, $R(-\theta)$, is the matrix $R(\theta)$. This allows one to obtain the Cartesian representation $\vec{v}_{xy}$ if the polar representation is known:

$$\begin{pmatrix} v_x \\ v_y \end{pmatrix} = R(\theta)\begin{pmatrix} v_r \\ v_\theta \end{pmatrix}$$

$$R(\theta) = \begin{pmatrix} \cos(\theta) & -\sin(\theta) \\ \sin(\theta) & \cos(\theta) \end{pmatrix}$$

To determine the relation between polar coordinates and the velocity, begin by expressing the Cartesian position $(x, y)^{\text{T}}$ in polar coordinates:

$$\begin{pmatrix} x(t) \\ y(t) \end{pmatrix} = \begin{pmatrix} r(t)\cos[\theta(t)] \\ r(t)\sin[\theta(t)] \end{pmatrix}$$

Next take the time derivative of each side. The right-hand side requires the use of the multiplication rule and the chain rule:

$$\begin{pmatrix} x(t) \\ y(t) \end{pmatrix} = \begin{pmatrix} r(t)\cos[\theta(t)] \\ r(t)\sin[\theta(t)] \end{pmatrix}$$

$$\begin{pmatrix} \frac{d}{dt}[x(t)] \\ \frac{d}{dt}[y(t)] \end{pmatrix} = \begin{pmatrix} \frac{d}{dt}\{r(t)\cos[\theta(t)]\} \\ \frac{d}{dt}\{r(t)\sin[\theta(t)]\} \end{pmatrix}$$

$$\begin{pmatrix} v_x(t) \\ v_y(t) \end{pmatrix} = \begin{pmatrix} \cos[\theta(t)]\frac{d}{dt}[r(t)] + r(t)\frac{d}{dt}\{\cos[\theta(t)]\} \\ \sin[\theta(t)]\frac{d}{dt}[r(t)] + r(t)\frac{d}{dt}\{\sin[\theta(t)]\} \end{pmatrix}$$

$$= \begin{pmatrix} \cos[\theta(t)]\frac{d}{dt}[r(t)] + r(t)\frac{d}{d\theta}[\cos(\theta)]\frac{d}{dt}[\theta(t)] \\ \sin[\theta(t)]\frac{d}{dt}[r(t)] + r(t)\frac{d}{dt}\{\sin[\theta(t)]\} \end{pmatrix}$$

$$= \begin{pmatrix} \cos[\theta(t)]\frac{d}{dt}[r(t)] + r(t)\frac{d}{d\theta}[\cos(\theta)]\frac{d}{dt}[\theta(t)] \\ \sin[\theta(t)]\frac{d}{dt}[r(t)] + r(t)\frac{d}{d\theta}[\sin(\theta)]\frac{d}{dt}[\theta(t)] \end{pmatrix}$$

$$= \begin{pmatrix} \cos[\theta(t)]\frac{d}{dt}[r(t)] - r(t)\sin(\theta)\frac{d}{dt}[\theta(t)] \\ \sin[\theta(t)]\frac{d}{dt}[r(t)] + r(t)\cos(\theta)\frac{d}{dt}[\theta(t)] \end{pmatrix}$$

$$= R(\theta)\begin{pmatrix} \frac{d}{dr}[r(t)] \\ r(t)\frac{d}{dt}[\theta(t)] \end{pmatrix} \tag{7.9}$$

Expressing the relationship of equation (7.9) purely in terms of polar coordinates and simplifying result in the required relationship:

$$R(\theta)\begin{pmatrix} \frac{d}{dt}[r(t)] \\ r(t)\frac{d}{dt}[\theta(t)] \end{pmatrix} = \begin{pmatrix} v_x(t) \\ v_y(t) \end{pmatrix}$$

$$\begin{pmatrix} \frac{d}{dt}[r(t)] \\ r(t)\frac{d}{dt}[\theta(t)] \end{pmatrix} = R(-\theta)\begin{pmatrix} v_x \\ v_y \end{pmatrix}$$

$$\begin{pmatrix} \frac{d}{dt}[r(t)] \\ r(t)\frac{d}{dt}[\theta(t)] \end{pmatrix} = \begin{pmatrix} v_r(t) \\ v_\theta(t) \end{pmatrix}$$

Equivalently,

$$\frac{d}{dt}(r) = v_r \qquad \frac{d}{dt}(\theta) = \frac{v_\theta}{r} \tag{7.10}$$

7.4.1.2 The Velocity Variables and Acceleration

To complete the polar description of Newton's equations, it is necessary to express the acceleration in polar coordinates. Begin by taking the derivative of $\vec{v}(t)_{\text{polar}} = R(-\theta(t))\vec{v}(t)_{xy}$. Below, we apply the product rule, which works for matrix functions just as it does for standard functions:

$$\frac{d}{dt}\left[\vec{v}(t)_{\text{polar}}\right] = \frac{d}{dt}\left[R(-\theta(t))\vec{v}(t)_{xy}\right]$$

$$= \left(\frac{d}{dt}\left[R(-\theta(t))\right]\right)\vec{v}(t)_{xy} + R(-\theta(t))\frac{d}{dt}\left[\vec{v}(t)_{xy}\right] \qquad (7.11)$$

The derivative of the matrix R is obtained by taking the derivative of each component:

$$\frac{d}{dt}\left[R(-\theta(t))\right] = \begin{pmatrix} \frac{d}{dt}\{\cos[\theta(t)]\} & \frac{d}{dt}\{\sin[\theta(t)]\} \\ -\frac{d}{dt}\{\sin[\theta(t)]\} & \frac{d}{dt}\{\cos[\theta(t)]\} \end{pmatrix}$$

Use of the chain rule results in the following expression for the derivative of the matrix:

$$\frac{d}{dt}\left[R(-\theta(t))\right] = \begin{pmatrix} \frac{d}{d\theta}[\cos(\theta)]\frac{d}{dt}(\theta) & \frac{d}{d\theta}[\sin(\theta)]\frac{d}{dt}(\theta) \\ -\frac{d}{d\theta}[\sin(\theta)]\frac{d}{dt}(\theta) & \frac{d}{d\theta}[\cos(\theta)]\frac{d}{dt}(\theta) \end{pmatrix}$$

$$= \begin{pmatrix} -\sin(\theta)\frac{d}{dt}(\theta) & \cos(\theta)\frac{d}{dt}(\theta) \\ -\cos(\theta)\frac{d}{dt}(\theta) & -\sin(\theta)\frac{d}{dt}(\theta) \end{pmatrix}$$

$$= \frac{d}{dt}(\theta)\begin{pmatrix} -\sin(\theta) & \cos(\theta) \\ -\cos(\theta) & -\sin(\theta) \end{pmatrix}$$

Substituting the values of $\frac{d}{dt}v_x$, $\frac{d}{dt}v_y$ from Newton's equations of motion, equation (7.8), and the expression for $\frac{d}{dt}[R(-\theta(t))]$ into the expression for $\frac{d}{dt}[\vec{v}(t)_{\text{polar}}]$, equation (7.11), results in the following equation for $\frac{d}{dt}[\vec{v}(t)_{\text{polar}}]$:

$$\frac{d}{dt}\left[\vec{v}(t)_{\text{polar}}\right] = \frac{d}{dt}\left[R(-\theta(t))\right]v(t)_{xy} + R(-\theta(t))\frac{d}{dt}\left[v(t)_{xy}\right]$$

$$\begin{pmatrix} \frac{d}{dt}(v_r) \\ \frac{d}{dt}(v_\theta) \end{pmatrix} = \frac{d}{dt}(\theta)\begin{pmatrix} -\sin(\theta) & \cos(\theta) \\ -\cos(\theta) & -\sin(\theta) \end{pmatrix}\begin{pmatrix} v_x \\ v_y \end{pmatrix}$$

$$+ \begin{pmatrix} \cos(\theta) & \sin(\theta) \\ -\sin(\theta) & \cos(\theta) \end{pmatrix}\begin{pmatrix} \frac{d}{dt}(v_x) \\ \frac{d}{dt}(v_y) \end{pmatrix}$$

$$= \frac{d}{dt}(\theta)\begin{pmatrix} -\sin(\theta) & \cos(\theta) \\ -\cos(\theta) & -\sin(\theta) \end{pmatrix}\begin{pmatrix} v_x \\ v_y \end{pmatrix}$$

$$+ \begin{pmatrix} \cos(\theta) & \sin(\theta) \\ -\sin(\theta) & \cos(\theta) \end{pmatrix}\begin{pmatrix} \frac{1}{m}\cos(\theta)F(r) \\ \frac{1}{m}\sin(\theta)F(r) \end{pmatrix} \qquad (7.12)$$

The vector $\vec{v}_{xy}$ is next expressed in polar coordinates:

$$\begin{pmatrix} v_x \\ v_y \end{pmatrix} = R(\theta) \begin{pmatrix} v_r \\ v_\theta \end{pmatrix}$$

$$= \begin{pmatrix} \cos(\theta) & -\sin(\theta) \\ \sin(\theta) & \cos(\theta) \end{pmatrix} \begin{pmatrix} v_r \\ v_\theta \end{pmatrix}$$

$$= \begin{pmatrix} v_r \cos(\theta) + v_\theta \sin(\theta) \\ -v_r \sin(\theta) + v_\theta \cos(\theta) \end{pmatrix}$$

Placing the expression for $\vec{v}_{xy}$ into equation (7.12) and simplifying establish a polar expression for the acceleration:

$$\begin{pmatrix} \dfrac{d}{dt}(v_r) \\ \dfrac{d}{dt}(v_\theta) \end{pmatrix} = \dfrac{d}{dt}(\theta) \begin{pmatrix} -\sin(\theta) & \cos(\theta) \\ -\cos(\theta) & -\sin(\theta) \end{pmatrix} \begin{pmatrix} v_x \\ v_y \end{pmatrix}$$

$$+ \begin{pmatrix} \cos(\theta) & \sin(\theta) \\ -\sin(\theta) & \cos(\theta) \end{pmatrix} \begin{pmatrix} \dfrac{1}{m}\cos(\theta)F(r) \\ \dfrac{1}{m}\sin(\theta)F(r) \end{pmatrix}$$

$$= \dfrac{d}{dt}(\theta) \begin{pmatrix} -\sin(\theta) & \cos(\theta) \\ -\cos(\theta) & -\sin(\theta) \end{pmatrix} \begin{pmatrix} v_r \cos(\theta) + v_\theta \sin(\theta) \\ -v_r \sin(\theta) + v_\theta \cos(\theta) \end{pmatrix}$$

$$+ \begin{pmatrix} \cos(\theta) & \sin(\theta) \\ -\sin(\theta) & \cos(\theta) \end{pmatrix} \begin{pmatrix} \dfrac{1}{m}\cos(\theta)F(r) \\ \dfrac{1}{m}\sin(\theta)F(r) \end{pmatrix}$$

$$= \dfrac{d}{dt}(\theta) \begin{pmatrix} v_\theta \\ -v_r \end{pmatrix} + \begin{pmatrix} \dfrac{F(r)}{m} \\ 0 \end{pmatrix}$$

$$\begin{pmatrix} \dfrac{d}{dt}(v_r) \\ \dfrac{d}{dt}(v_\theta) \end{pmatrix} = \dfrac{d}{dt}(\theta) \begin{pmatrix} v_\theta \\ -v_r \end{pmatrix} + \begin{pmatrix} \dfrac{F(r)}{m} \\ 0 \end{pmatrix} \tag{7.13}$$

Using the expression for angular velocity, equation (7.10) yields the two acceleration equations:

$$\frac{d}{dt}(v_r) = \frac{v_\theta^2}{r} + \frac{F(r)}{m} \qquad \frac{d}{dt}(v_\theta) = \frac{-v_r v_\theta}{r} \tag{7.14}$$

Equations (7.10) and (7.14) are the equations of motion in polar coordinates.

Remarks

- To convince oneself that the product rule for differentiation applies to equation (7.11), expand the terms and compare the results.
- A more direct path from equation (7.12) to equation (7.13) is available for those who are familiar with matrix multiplication:

$$\begin{pmatrix} \dfrac{d}{dt}(v_r) \\ \dfrac{d}{dt}(v_\theta) \end{pmatrix} = \dfrac{d}{dt}(\theta) \begin{pmatrix} -\sin(\theta) & \cos(\theta) \\ -\cos(\theta) & -\sin(\theta) \end{pmatrix} \begin{pmatrix} v_x \\ v_y \end{pmatrix}$$

$$+ \begin{pmatrix} \cos(\theta) & \sin(\theta) \\ -\sin(\theta) & \cos(\theta) \end{pmatrix} \begin{pmatrix} \dfrac{1}{m}\cos(\theta)F(r) \\ \dfrac{1}{m}\sin(\theta)F(r) \end{pmatrix}$$

$$= \dfrac{d}{dt}(\theta) \begin{pmatrix} -\sin(\theta) & \cos(\theta) \\ -\cos(\theta) & -\sin(\theta) \end{pmatrix} \begin{pmatrix} \cos(\theta) & -\sin(\theta) \\ \sin(\theta) & \cos(\theta) \end{pmatrix} \begin{pmatrix} v_r \\ v_\theta \end{pmatrix}$$

$$+ \begin{pmatrix} \cos(\theta) & \sin(\theta) \\ -\sin(\theta) & \cos(\theta) \end{pmatrix} \begin{pmatrix} \dfrac{1}{m}\cos(\theta)F(r) \\ \dfrac{1}{m}\sin(\theta)F(r) \end{pmatrix}$$

$$= \dfrac{d}{dt}(\theta) \begin{pmatrix} 0 & 1 \\ -1 & 0 \end{pmatrix} \begin{pmatrix} v_r \\ v_\theta \end{pmatrix} + \begin{pmatrix} \dfrac{F(r)}{m} \\ 0 \end{pmatrix}$$

$$= \dfrac{d}{dt}(\theta) \begin{pmatrix} v_\theta \\ -v_r \end{pmatrix} + \begin{pmatrix} \dfrac{F(r)}{m} \\ 0 \end{pmatrix}$$

7.4.2 Angular Momentum

Integration of equations (7.10) and (7.14) so that the location variables are determined explicitly still appears difficult. A simplification occurs using another constant of the motion, the angular momentum. Angular momentum is the quantity mrv_θ. This section shows that in a Newtonian system angular momentum is conserved by demonstrating that the time derivative of the angular momentum is zero. Afterward, the equivalence with Kepler's second law is shown.

7.4.2.1 *Conservation of Angular Momentum* We begin by taking the time derivative of the quantity rv_θ:

$$\frac{d}{dt}(rv_\theta) = r\frac{d}{dt}(v_\theta) + v_\theta\frac{d}{dt}(r)$$

Next substitute for the values of $\frac{d}{dt}(r)$ and $\frac{d}{dt}(v_\theta)$ from Newton's equations in polar coordinates and simplify to demonstrate that rv_θ is a conserved quantity:

$$\frac{d}{dt}(rv_\theta) = r\frac{d}{dt}(v_\theta) + v_\theta\frac{d}{dt}(r)$$

$$= -r\frac{v_r v_\theta}{r} + v_\theta v_r \qquad \text{equation (7.10)}$$

$$= 0$$

Conservation of angular momentum allows one to set the quantity rv_θ to a constant value and then express the angular component of velocity, v_θ, in terms of the variable r:

$$rv_\theta = k$$
$$v_\theta = \frac{k}{r} \qquad (7.15)$$

This relation is critical for revealing the ellipse.

Notation. From this point onward, justification for a single equality in a sequence of equalities is at times given on the right of the equality of interest. For example, in the equalities preceding equation (7.15), equation (7.10) is used to obtain the second equality.

Remarks

- A point of interest is that up to now the force function $F(r)$ has not been specified. The results then apply to any system for which the force function is constant along all fixed radial values. In the case of planetary motion, let the planet be at a certain point with a fixed radial distance from the sun and measure the force. Now move the object to any other point having the same distance from the sun. If one were to measure the force on the planet, it would be the same at both points. Any force satisfying this property defines a Newtonian system in which angular momentum is conserved.

- The case in which there is no force at all satisfies the property identified in the previous remark. Fixing any arbitrary point as the origin, an object at points that are equidistant from the origin experiences the same force along all those points, zero. This means that angular momentum is conserved. A point moving in such a force field moves in a straight line with constant velocity. One can demonstrate using Cavalieri's theorem that Kepler's second law holds for such motion.

7.4.2.2 Kepler's Law and Angular Momentum

We close this section by demonstrating the equivalence between Kepler's second law and conservation of angular momentum. This is the final excursion of the book, and it is completely un-necessary for uncovering the ellipse. However, I made so much hullabaloo about this point throughout the book that I feel obligated to formally demonstrate it.

Kepler's second law states that the area swept out by the line between the sun and the planet between times t_0 and t is proportional to the length of the time interval (Figure 7.7). Kepler's law may be stated as an equation:

$$A(t) = \frac{1}{2}k(t - t_0) \qquad (7.16)$$

where

- $A(t)$ is the area of interest
- k is a constant value
- and t_0 is the initial time, which may be set to zero

Note that differentiation of equation (7.16) results in

$$\frac{d}{dt}[A(t)] = \frac{k}{2}$$

which is another way to express Kepler's law.

To establish equivalence between Kepler's law and conservation of momentum, it is necessary to show that each statement may be derived from the other. First, it is shown that conservation of angular momentum follows from Kepler's second law. Toward this end, we demonstrate that assuming

$$\frac{d}{dt}[A(t)] = \frac{k}{2}$$

leads one to the conclusion that the constant quantity mk is the angular momentum, or equivalently $k = rv_\theta$. Afterward, we demonstrate the converse statement, assuming the angular momentum is constant, $rv_\theta = k$, it follows that

$$\frac{d}{dt}[A(t)] = \frac{k}{2}$$

For the first part, assume

$$\frac{d}{dt}[A(t)] = \frac{k}{2}$$

and consider Figure 7.10. For ease of exposition, t_0 has been set to zero and the planet's initial position is along the x axis. The symbols $A(t)$ correspond to the area of the sector within the ellipse enclosed by r_0 and r_t, while $A(t + \Delta)$ corresponds to the area of the sector within the ellipse enclosed by r_t and $r_{t+\Delta}$; $\underline{A}(t + \Delta) - A(t)$ corresponds to the area of the triangular region that is colored in turquoise, and

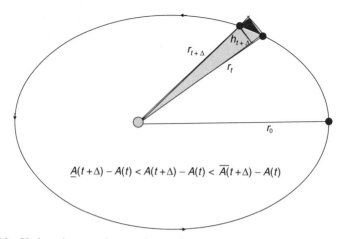

$$\underline{A}(t+\Delta) - A(t) < A(t+\Delta) - A(t) < \overline{A}(t+\Delta) - A(t)$$

Figure 7.10 Underestimate and overestimate of sector between t and $t + \Delta$.

$\overline{A}(t + \Delta) - A(t)$ corresponds to the area of the circular sector with radius enclosed by r_t and $r_{t+\Delta}$ with radius r_t. The following inequalities result from these definitions for the corresponding areas, $\overline{A}(t)$, resulting in the following equality:

$$\underline{A}(t + \Delta) - A(t) = \tfrac{1}{2}r_t h(t + \Delta) \leq A(t + \Delta) - A(t) \leq \tfrac{1}{2}r_t^2 \Delta_\theta(t)$$
$$= \overline{A}(t + \Delta) - A(t)$$

where $\Delta_\theta(t)$ is the angle between r_t and $r_{t+\Delta}$. Note that $\Delta_\theta(t) = \theta(t + \Delta) - \theta(t)$.

The inequalities are valid for all times in an interval between t and $t + \Delta$ for a small increment Δ. As such, across the inequalities, it is possible to divide through by Δ and take the limit to determine a derivative:

$$\frac{\underline{A}(t + \Delta) - A(t)}{\Delta} \leq \frac{A(t + \Delta) - A(t)}{\Delta} \leq \frac{\overline{A}(t + \Delta) - A(t)}{\Delta}$$

$$\lim_{\Delta \to 0} \frac{\underline{A}(t + \Delta) - A(t)}{\Delta} \leq \lim_{\Delta \to 0} \frac{A(t + \Delta) - A(t)}{\Delta}$$

$$\leq \lim_{\Delta \to 0} \frac{\overline{A}(t + \Delta) - A(t)}{\Delta}$$

$$\lim_{\Delta \to 0} \frac{A(t + \Delta) - A(t)}{\Delta} \leq \frac{d}{dt}[A(t)] = \frac{k}{2}$$

$$\leq \lim_{\Delta \to 0} \frac{\overline{A}(t + \Delta) - A(t)}{\Delta} \qquad \text{assumption } \frac{d}{dt}[A(t)] = \frac{k}{2}$$

$$(7.17)$$

Next, determine the left-hand expression of equation (7.17):

$$\lim_{\Delta \to 0} \frac{A(t + \Delta) - A(t)}{\Delta}$$

$$= \lim_{\Delta \to 0} \frac{r_t h(t + \Delta)}{2\Delta}$$

$$= \lim_{\Delta \to 0} \frac{r_t^2 \sin[\theta(t + \Delta) - \theta(t)]}{2\Delta} \qquad h = \sin[\Delta_\theta(t)]$$

$$= \lim_{\Delta \to 0} \left[\frac{r_t^2 \sin[\theta(t + \Delta)] \cos[\theta(t)]}{2\Delta} \right.$$

$$\left. - \frac{r_t^2 \sin[\theta(t)] \cos[\theta(t + \Delta)]}{2\Delta} \right] \qquad \text{trigonometric identities}$$

$$= \frac{r_t^2}{2} \left[\cos[\theta(t)] \lim_{\Delta \to 0} \frac{\sin[\theta(t + \Delta)]}{\Delta} \right.$$

$$\left. - \sin[\theta(t)] \lim_{\Delta \to 0} \frac{\cos[\theta(t + \Delta)]}{\Delta} \right] \qquad \text{simplification of limits}$$

$$= \frac{r_t^2}{2} \left[\cos[\theta(t)] \frac{d}{dt} \{\sin[\theta(t)]\} \right.$$

$$\left. - \sin[\theta(t)] \frac{d}{dt} \{\cos[\theta(t)]\} \right] \qquad \text{definition of derivative}$$

$$= \frac{r_t^2}{2} \left[\cos[\theta(t)] \frac{d}{d\theta} [\sin(\theta)] \right.$$

$$\left. - \sin[\theta(t)] \frac{d}{d\theta} [\cos(\theta)] \right] \frac{d}{dt}(\theta) \qquad \text{chain rule}$$

$$= \frac{r_t^2}{2} \left[\cos^2[\theta(t)] + \sin^2[\theta(t)] \right] \frac{d}{dt}(\theta) \qquad \text{derivative of trignometric functions}$$

$$= \frac{r_t^2}{2} \frac{d}{dt}(\theta) \qquad \text{Pythagorean identity}$$

$$= \frac{r_t}{2} v_\theta(t) \qquad \text{equation (7.10)}$$

Finally, evaluate the right-hand expression of expression (7.17):

$$\lim_{\Delta \to 0} \frac{\overline{A}(t + \Delta) - A(t)}{\Delta} = \lim_{\Delta \to 0} \frac{r_t^2 [\theta(t + \Delta) - \theta(t)]}{2\Delta}$$

$$= \frac{r_t^2}{2} \lim_{\Delta \to 0} \frac{\theta(t + \Delta) - \theta(t)}{\Delta} \qquad \text{simplification of limits}$$

$$= \frac{r_t^2}{2} \frac{d}{dt}(\theta) \qquad \text{definition of derivative}$$

$$= \frac{r_t}{2} v_\theta(t) \qquad \text{equation (7.10)}$$

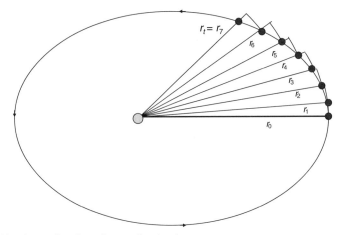

Figure 7.11 Approximation of sector by circular sectors.

Replacing the left-hand and right-hand expressions into the inequality establishes the following inequality:

$$\tfrac{1}{2}rv_\theta \leq \tfrac{1}{2}k \leq \tfrac{1}{2}rv_\theta$$

Noting that the left-hand and right-hand expressions are equal furnishes the desired result, $rv_\theta = k$.

With the above result, we have accomplished the first step, demonstrating that conservation of momentum follows from Kepler's law. We next move to the final step; the assumption of conservation of angular momentum results in Kepler's law.

Consult Figure 7.11. As in Figure 7.10, $A(t)$ is the area of the region within the ellipse enclosed by r_0 and r_t. Approximating the area by a sum of sector areas results in the following:

$$A(t) \approx \sum_{j=0}^{n} A_j$$

$$= \sum_{j=1}^{n} \frac{r_j^2}{2} \Delta_\theta$$

In the above expression, the radius of sector j is r_j and each sector has the same angle Δ_θ. The actual area, $A(t)$, is determined by taking the limit as the sector angle approaches zero. Simplifying the expression yields the result:

$$A(t) = \lim_{\Delta_\theta \to 0} \sum_{j=1}^{n} \frac{r_j^2}{2} \Delta_\theta$$

$$= \int_{0}^{\theta_f} \frac{[r(\theta)]^2}{2} d\theta \qquad \text{definition of integral}$$

$$= \frac{1}{2} \int_{t_0}^{t} [r(\tau)]^2 \frac{d\theta}{d\tau} d\tau \quad \text{change of variables, angle to time}$$

$$= \frac{1}{2} \int_{t_0}^{t} r(\tau) v_\theta(\tau) \, d\tau \quad \text{equation (7.10)}$$

$$= \frac{1}{2} \int_{t_0}^{t} k \, d\tau \quad \text{conservation of angular momentum}$$

$$= \frac{k}{2}(t - t_0)$$

The final equality is Kepler's law, which does indeed follow from the assumption of conservation of angular momentum.

7.4.3 The Ellipse

In this section, we reveal the ellipse. Our path toward the ellipse follows six steps that are first outlined. Afterward, the steps are executed. The objective is to find the shape of the trajectory by determining a relation between the radial and angular components of the trajectory.

Step 1. Find an equation for $\frac{d}{dr}(\theta)$.
Using conservation of momentum, it is possible to determine an equation for $\frac{d}{dr}(\theta)$. The result is an equation in both r and v_r.

Step 2. Separation of radial component of motion.
The equations governing motion along the radial direction are written only in terms of the radial variables r and v_r.

Step 3. Find the associated energy and use it to establish a relation between r and v_r.
Another constant of motion, the energy, is determined for the equations governing radial motion. This generates a relation between r and v_r.

Step 4. Simplify the result of step 1 using the relationship between r and v_r from step 3.
Substituting for v_r the result of step 1 establishes an equation of the form $\frac{d}{dr}(\theta) = h(r)$, where $h(r)$ is a function of r only.

Step 5. Determine θ by integration.
Integrate the result of step 4 to determine a relation between θ and r.

Step 6. Transform the relation of step 5 to reveal the ellipse.
The map is marked. Let us begin.

Step 1. Find an equation for $\frac{d}{dr}(\theta)$.
Using conservation of momentum, it is possible to determine an equation for $\frac{d}{dr}(\theta)$. The result is an equation in both r and v_r:

$$mrv_\theta = k \quad \text{equation (7.15)}$$

$$v_\theta = \frac{k}{mr}$$

$$r\frac{d}{dt}[\theta(r)] = \frac{k}{mr} \quad \text{equation (7.10)}$$

$$\frac{d}{dt}[\theta(r)] = \frac{k}{mr^2}$$

$$\frac{d}{dr}(\theta)\frac{d}{dt}(r) = \frac{k}{mr^2} \qquad \text{chain rule}$$

$$\frac{d}{dr}(\theta)v_r = \frac{k}{mr^2} \qquad \text{equation (7.10)}$$

$$\frac{d}{dr}(\theta) = \frac{k}{mv_rr^2}$$

Step 2. Separation of radial component of motion.

The equations governing motion along the radial direction are written only in terms of the radial variables r and v_r:

$$\frac{d}{dt}(v_r) = \frac{v_\theta^2}{r} + \frac{F(r)}{m} \qquad \text{equation (7.14)}$$

$$= \frac{k^2}{m^2r^3} + \frac{F(r)}{m} \qquad \text{equation (7.15)}, v_\theta = \frac{k}{mr}$$

Step 3. Find the associated energy and use it to establish a relation between r and v_r. Another constant of motion, the energy, is determined for the equations governing radial motion. This establishes a relation between r and v_r. The system, is a Newtonian system:

$$\frac{d}{dt}(r) = v_r$$

$$\frac{d}{dt}(v_r) = \frac{k^2}{m^2r^3} + \frac{F(r)}{m} \qquad \text{result of step 2}$$

Using the procedure described in Section 7.3, the energy for this system can be determined:

$$E = m(K_m(v_r) + P_m(r))$$

$$\frac{d}{dt}(E) = m\left(\frac{d}{dt}[K_m(v_r)] + \frac{d}{dt}[P_m(r)]\right) \qquad \text{product rule}$$

$$= m\left(\frac{d}{dv_r}(K_m)\frac{d}{dt}(v_r) + \frac{d}{dr}(P_m)\frac{d}{dt}(r)\right) \qquad \text{chain rule}$$

$$= m\left(\frac{d}{dv_r}(K_m)\left(\frac{k^2}{m^2r^3} + \frac{F(r)}{m}\right) + \frac{d}{dr}(P_m)v_r\right) \qquad \text{system equations}$$

$$= 0 \qquad \text{conservation of energy}$$

The following relations for K and P assure that the energy is fixed ($dE/dt = 0$):

$$\frac{d}{dv_r}(K_m) = v_r \qquad \frac{d}{dr}(P_m) = -\left(\frac{k^2}{m^2r^3} + \frac{F(r)}{m}\right)$$

Integrating both equations furnishes the kinetic and potential components:

$$K_m = \int v_r \, dv_r \qquad P_m = -\int \left(\frac{k^2}{m^2 r^3} + \frac{F(r)}{m} \right) dr$$

$$= \tfrac{1}{2} v_r^2$$

At this point, we introduce the force $F(r)$. According to the inverse square law, the sun pulls the planet with a force that is proportional to the inverse square of the distance between the two bodies, $F(r) = -\gamma/r^2$. Newton reasoned that the constant of proportionality is the product of the mass of the planet, m, and that of the sun, M, along with another constant known as the universal constant of gravity, g; $\gamma = mMg$. Placing this force into the integral for P_m and performing the integration result in an expression for the potential component (the constant of integration has been set to zero):

$$P_m = \int \left(-\frac{k^2}{m^2 r^3} + \frac{Mg}{r^2} \right) dr = \frac{1}{2} \left(\frac{k}{mr} \right)^2 - \frac{Mg}{r}$$

With the kinetic and potential components, the energy is determined:

$$E = \frac{m v_r^2}{2} + \frac{k^2}{2mr^2} - \frac{mMg}{r}$$

Since the energy is a constant value, v_r can be solved in terms of r:

$$v_r = \sqrt{\frac{2E}{m} + \frac{2Mg}{r} - \left(\frac{k}{mr} \right)^2}$$

Step 4. Simplify the result of step 1 using the relationship between r and v_r from step 3.

Substituting the result of step 3 into the result of step 1 establishes an equation of the form $\frac{d}{dr}(\theta) = h(r)$, where $h(r)$ is a function of r only:

$$\frac{d}{dr}(\theta) = \frac{k}{m v_r r^2} \qquad\qquad \text{step 1}$$

$$= \frac{k}{mr^2 \sqrt{\dfrac{2E}{m} + \dfrac{2Mg}{r} - \left(\dfrac{k}{mr} \right)^2}} \qquad\qquad \text{step 3}$$

Step 5. Determine θ by integration.

Integrate the result of step 4 to determine a relation between θ and r:

$$\frac{d}{dr}(\theta) = \frac{k}{mr^2 \sqrt{\dfrac{2E}{m} + \dfrac{2Mg}{r} - \left(\dfrac{k}{mr} \right)^2}} \qquad\qquad \text{step 4}$$

$$\theta = \int \frac{k}{mr^2 \sqrt{\dfrac{2E}{m} + \dfrac{2Mg}{r} - \left(\dfrac{k}{mr} \right)^2}} \, dr$$

$$= \int \frac{1}{r^2 \sqrt{\left(\frac{m}{k}\right)^2 \left(\frac{2E}{m} + \frac{2Mg}{r} - \left(\frac{k}{mr}\right)^2\right)}} dr$$

$$= \int \frac{1}{r^2 \sqrt{\frac{2Em}{k^2} + \frac{2Mg}{r} \left(\frac{m}{k}\right)^2 - \frac{1}{r^2}}} dr$$

Change the variables to simplify the integrand. Let $s = 1/r$, then $ds = -\frac{1}{r^2} dr$:

$$\theta = \int \frac{1}{r^2 \sqrt{\frac{2Em}{k^2} + \frac{2Mg}{r} \left(\frac{m}{k}\right)^2 - \frac{1}{r^2}}} dr$$

$$= - \int \frac{1}{\sqrt{\frac{2Em}{k^2} + 2Mg \left(\frac{m}{k}\right)^2 s - s^2}} ds$$

Complete the square of the radical:

$$\frac{2Em}{k^2} + 2Mg \left(\frac{m}{k}\right)^2 s - s^2 = \left(\frac{Mgm^2}{k^2}\right)^2 + \frac{2Em}{k^2} - \left(s - \frac{Mgm^2}{k^2}\right)^2$$

$$= \left[\left(\frac{Mgm^2}{k^2}\right)^2 + \frac{2Em}{k^2}\right]\left[1 - \frac{\left(s - Mgm^2/k^2\right)^2}{\left(Mgm^2/k^2\right)^2 + 2Em/k^2}\right]$$

Substitute the completed square into the integral and simplify:

$$\theta = - \int \frac{1}{\sqrt{2Em/k^2 + 2Mg\,(m/k)^2\,s - s^2}} ds$$

$$= -\frac{1}{\sqrt{2}} \int \frac{1}{\sqrt{\left[\left(\frac{Mgm^2}{k^2}\right)^2 + \frac{2Em}{k^2}\right]\left[1 - \frac{\left(s - Mgm^2/k^2\right)^2}{\left(Mgm^2/k^2\right)^2 + 2Em/k^2}\right]}} ds$$

$$= -\frac{1}{\sqrt{\left(\frac{Mgm^2}{k^2}\right) + \frac{2Em}{k^2}}} \int \frac{1}{\sqrt{1 - \frac{\left(s - Mgm^2/k^2\right)^2}{\left(Mgm^2/k^2\right)^2 + 2Em/k^2}}} ds$$

(7.18)

Once again change the variables to simplify the integrand:

$$\cos(u) = \frac{s - Mgm^2/k^2}{-\sqrt{\left(Mgm^2/k^2\right)^2 + 2Em/k^2}}$$

$$-\sin(u)\,du = \frac{ds}{-\sqrt{\left(Mgm^2/k^2\right)^2 + 2Em/k^2}}$$

$$du = \frac{1}{\sqrt{\left(Mgm^2/k^2\right)^2 + 2Em/k^2}}\,\frac{ds}{\sin(u)}$$

$$= \frac{1}{\sqrt{\left(Mgm^2/k^2\right)^2 + 2Em/k^2}}\,\frac{ds}{\sqrt{1 - \cos^2(u)}} \qquad \text{Pythagorean identity}$$

$$= \frac{1}{\sqrt{\left(\dfrac{Mgm^2}{k^2}\right)^2 + \dfrac{2Em}{k^2}}}\,\frac{ds}{\sqrt{1 - \dfrac{\left(s - Mgm^2/k^2\right)^2}{\left(Mgm^2/k^2\right)^2 + 2Em/k^2}}}$$

$$\int du = \frac{1}{\sqrt{\left(\dfrac{Mgm^2}{k^2}\right)^2 + \dfrac{2Em}{k^2}}}\int \frac{ds}{\sqrt{1 - \dfrac{\left(s - Mgm^2/2k^2\right)^2}{\left(Mgm^2/k^2\right)^2 + 2Em/k^2}}}$$

$$= -\theta \qquad\qquad\qquad \text{equation (7.18)}$$

$$u = -\theta$$

The constant of integration has been set to zero. The constant determines the orientation of the axes with respect to the solution, so it is not relevant to the shape of the trajectory.

Tracing back through the variable changes allows one to find the relation between r and θ:

$$-\theta = u$$

$$\cos(-\theta) = \cos(u)$$

$$\cos(\theta) = \cos(u)$$

$$\cos(\theta) = \frac{s - Mgm^2/k^2}{-\sqrt{\left(Mgm^2/k^2\right)^2 + 2Em/k^2}}$$

$$-\sqrt{\left(\frac{Mgm^2}{k^2}\right)^2 + \frac{2Em}{k^2}}\,\cos(\theta) = s - \frac{Mgm^2}{k^2}$$

$$-\sqrt{\left(\frac{Mgm^2}{k^2}\right)^2 + \frac{2Em}{k^2}}\,\cos(\theta) + \frac{Mgm^2}{k^2} = s$$

$$-\sqrt{\left(\frac{Mgm^2}{k^2}\right)^2 + \frac{2Em}{k^2}}\,\cos(\theta) + \frac{Mgm^2}{k^2} = \frac{1}{r}$$

$$\left(-\sqrt{\left(\frac{Mgm^2}{k^2}\right)^2 + \frac{2Em}{k^2}}\,\cos(\theta) + \frac{Mgm^2}{k^2}\right) r = 1$$

Step 6. Transform the relation of step 5 and reveal the ellipse.

The ellipse is coded into the last equation of step 5 in polar coordinates. In this step, we display the ellipse in Cartesian coordinates. To clarify the process, the following notation is used:

$$\alpha = \frac{Mgm}{2E} \qquad -\beta^2 = \frac{k^2}{2Em} \tag{7.19}$$

Placing the notation in the result of step 5, the ellipse is uncovered through an algebraic process that is presented in an example of Section 5.8. In that section, the polar coordinates are determined from the Cartesian coordinates. Here we work in reverse:

$$\left(-\sqrt{\left(\frac{Mgm^2}{k^2}\right)^2 + \frac{2Em}{k^2}\cos(\theta) + \frac{Mgm^2}{k^2}}\right)r = 1$$

$$\left(-\sqrt{\left(\frac{\alpha}{\beta^2}\right)^2 - \frac{1}{\beta^2}\cos(\theta) - \frac{\alpha}{\beta^2}}\right)r = 1$$

$$-\frac{\alpha}{\beta^2}r = 1 + \sqrt{\left(\frac{\alpha}{\beta^2}\right)^2 - \frac{1}{\beta^2}}r\cos(\theta)$$

$$\left(\frac{\alpha}{\beta^2}\right)^2 r^2 = \left(1 + \sqrt{\left(\frac{\alpha}{\beta^2}\right)^2 - \frac{1}{\beta^2}}r\cos(\theta)\right)^2$$

$$\left(\frac{\alpha}{\beta^2}\right)^2 (x^2 + y^2) = 1 + 2\sqrt{\left(\frac{\alpha}{\beta^2}\right)^2 - \frac{1}{\beta^2}}x + \left(\left(\frac{\alpha}{\beta^2}\right)^2 - \frac{1}{\beta^2}\right)x^2$$

$$\frac{1}{\beta^2}x^2 - 2\sqrt{\left(\frac{\alpha}{\beta^2}\right)^2 - \frac{1}{\beta^2}}x + \left(\frac{\alpha}{\beta^2}\right)^2 y^2 = 1$$

$$\frac{1}{\beta^2}\left(x^2 - 2\beta^2\sqrt{\left(\frac{\alpha}{\beta^2}\right)^2 - \frac{1}{\beta^2}}x\right) + \left(\frac{\alpha}{\beta^2}\right)^2 y^2 = 1$$

$$\left(x^2 - 2\sqrt{\alpha^2 - \beta^2}x\right) + \frac{\alpha^2}{\beta^2}y^2 = \beta^2$$

$$\left(x - \sqrt{\alpha^2 - \beta^2}\right)^2 - (\alpha^2 - \beta^2) + \frac{\alpha^2}{\beta^2}y^2 = \beta^2$$

$$\left(x - \sqrt{\alpha^2 - \beta^2}\right)^2 + \frac{\alpha^2}{\beta^2}y^2 = \alpha^2$$

$$\frac{\left(x - \sqrt{\alpha^2 - \beta^2}\right)^2}{\alpha^2} + \frac{y^2}{\beta^2} = 1$$

At long last, the ellipse is revealed. But wait, Apollonius scores! Greek mathematicians set a precedent in their willingness to investigate topics with no apparent application. This precedent finds remarkable payouts again and again. Appolonius demonstrates no application for the conic sections that he revered, but the ellipse weaves itself into planetary motion. Not only the ellipse—take a good look at the final result. From equation (7.19), when $E < 0$ ($\beta^2 > 0$), the solution is the ellipse with major axis α, minor axis β, centered at the point $(x, y) = (\sqrt{\alpha^2 - \beta^2}, 0)$, with the left-hand focal point, the sun, at the origin (see Figure 7.7). But when $E > 0$ ($\beta^2 < 0$), the solution is the hyperbola. As we further pan the solution, another nugget falls out. When $E = 0$, we have the following:

$$\left(-\sqrt{\left(\frac{Mgm^2}{k^2}\right)^2 + \frac{2Em}{k^2}\cos(\theta) + \frac{Mgm^2}{k^2}}\right) r = 1$$

$$\frac{Mgm^2}{k^2} r[1 - \cos(\theta)] = 1$$

$$\frac{Mgm^2}{k^2} r = \frac{Mgm^2}{k^2} r \cos(\theta) + 1$$

$$\left(\frac{Mgm^2}{k^2} r\right)^2 = \left(\frac{Mgm^2}{k^2} r \cos(\theta) + 1\right)^2$$

$$\left(\frac{Mgm^2}{k^2}\right)^2 \left(x^2 + y^2\right) = \left(\frac{Mgm^2}{k^2}\right)^2 x^2 + 2\frac{Mgm^2}{k^2} x + 1$$

$$\left(\frac{Mgm^2}{k^2}\right)^2 y^2 = 2\frac{Mgm^2}{k^2} x + 1$$

$$\frac{Mgm^2}{2k^2} y^2 = x + \frac{k^2}{2Mgm^2}$$

$$x = \frac{Mgm^2}{2k^2} y^2 - \frac{k^2}{2Mgm^2}$$

The equation describes a parabola opening up on the x axis. All of Appolonius' conic sections are solutions to Newton's equations.

Since the energy controls the solution, a look at the energy is in order:

$$E = \frac{mv_r^2}{2} + \frac{k^2}{2mr^2} - \frac{mMg}{r}$$

$$= \frac{mv_r^2}{2} + \frac{mr^2v_\theta^2}{2} - \frac{mMg}{r} \qquad \text{equation (7.15)}$$

The first two terms are the total kinetic energy, while the final term is the gravitational potential, which increases from negative infinity to zero as the planet's distance from the sun goes from zero to infinity. When the gravitational potential dominates, the orbit is the ellipse. Alternatively, when the kinetic energy dominates, the planet drifts off along a hyperbolic trajectory. Finally, when the kinetic energy and the gravitational potential are in balance, the planet's trajectory is a parabola.

EPILOGUE

Behind my home in Arcadia is a pathway leading up to Mount Wilson. It is quite a strenuous hike with a 4500-ft elevation gain over 7 miles. The pathway is very diverse; it begins along a barren stretch, passes on through a forest, and then there are a series of switchbacks that take you up the final assault. Once atop Mt. Wilson, there is a broad vista. The highest peak of the San Gabriel, Mt. Baldy, is visible to the east and on a clear day one can see the San Bernadino Mountains beyond Mt. Baldy. To the west, the Pacific Coast Range is visible along with the ocean. Indeed, in every direction there is a natural landmark that piques at one's curiosity and is very alluring. It's a full day's hike, morning to evening, to climb up and down the Mt. Wilson trail. The distant landmarks are noted for another day.

There are many directions that we could explore with our understanding of calculus. Optimization theory is nearly within our grasp and we could follow the brachistrone problem into the calculus of variations. With a bit more effort, we could look into differential equations, partial differential equations, and more general dynamic systems. Alternatively, we could follow the route of physicists and review a broader set of mechanical systems, optics, and electrodynamics. And more in the distance lies relativity and quantum mechanics. But this book ends at the ellipse, leaving other challenges for another day.

Would it not be nice if all those who contributed toward this achievement could gather together at the top of Mount Wilson, survey the trail that brought us to the peak, and have a chat. Since I have entered into a fantasy world, I will make it the best of all fantasies and assume that the contributors do not display the more flawed aspects of their dispositions. Aristarchus would possibly take a few jovial shots at his fellow Greeks for their ridicule while he was right all along. And Archimedes might retort that if Aristarchus had left the specifications of the planetary orbit to him just as Kepler had left the specifications to Newton, then he, Archimedes, could have proved its elliptic path. Apollonius would be able to brag that he knew there was something fundamental to the conic sections and history proved him right. Ptolemy might be embarrassed but he could point out that his system very accurately describes the position of the planets and has endured for 1300 years. Copernicus might thank Kepler for having faith in his work and salvaging the truth. Kepler and Galileo, though different in character, might share a special bond for having lived during troubled times and having felt the consequences of the troubled times in a very personal way. Newton would have to recognize Leibniz' contribution to calculus as Leibniz' notation is the

The Ellipse: A Historical and Mathematical Journey   by Arthur Mazer
Copyright © 2010 by John Wiley & Sons, Inc.

current standard. And Leibniz would have to concede Newton's historical role in both mathematics and physics.

While not physically present, these men are with us through the legacy that they have left. They may have had different personalities with some flaws, but all of them were truth seekers who laid out their work for us to review with full honesty and openness. We have all been beneficiaries of their achievements. Unfortunately, it is not only the truth seekers who accompany us, but also among us are the ignorant, such as the soldier who speared Archimedes; the intolerant, such as the men who burned the library at Alexandria; the sadistic, such as the men of the Inquisition who tortured and executed Bruno; the hoodwinkers, such as the men who presented Aristotle and their own interpretation of the bible as the final truth about everything; the bigots, such as the men who cast out and persecuted minorities; and the self-aggrandizers, such as those who claimed to be the sole interpreters of God's will. As we set our own journeys, we are free to choose who within our legacy will be our guides. Choose wisely.

BIBLIOGRAPHY

Al-Khwarizmi. *Algebra*. Translated by Frank Rosen. Publisher unknown. London, 1831.

Apollonius. *Conics*. Translated by William Donahue. Santa Fe, 1997.

Appelbaum, Wilbur. *Encyclopedia of the Scientific Revolution: From Copernicus to Newton*. Garland, London, 2008.

Archimedes. *The Works of Archimedes*. Translated by Thomas Heath. Dover, New York, 2002.

Arnold, Vladimir. *Geometrical Methods in the Theory of Ordinary Differential Equations*. Springer, New York, 1983.

Bardi, Jason. *The Calculus Wars*. Thunder's Mouth Press, New York, 2006.

Beckmann, Petr. *A History of Pi*. St. Martin's Press, New York, 1976.

Boorstin, Daniel. *The Discoverers*. Vintage Books, New York, 1985.

Bottazini, Umberto and Van Egmond, Warren. *The Higher Calculus: A History of Real and Complex Analysis from Euler to Weierstrass*. Springer-Verlag, New York, 1986.

Boyer, Carl. *A History of Mathematics*. Revised by Uta Merzbach. Wiley, New York, 1991.

Brecht, Bertolt. *Life of Galileo*, Penguin, New York, 2008.

Burman, Edward. *The Inquisition: The Hammer of Heresy*. Dorset, London, 1984.

Cardano, Girolamo. *Ars Magna*. Translated by Richard Witmer. Dover, New York, 2007.

Connor, James. *Kepler's Witch*. HarperCollins, New York, 2005.

Copernicus, Nicolaus. *On the Revolutions of Heavenly Spheres*. Translated by Charles Wallis. Prometheus Books, New York, 1995.

Crowe, Michael. *Theories of the World from Antiquity to the Copernican Revolution*. Dover, New York, 1990.

Cullen, Christopher. *Astronomy and Mathematics in Ancient China*. Cambridge University Press, Cambridge, 2008.

Descartes, Rene. *The Geometry of Rene Descartes*. Dover, New York, 1954.

Estep, William. *Renaissance and Reformation*. Grand Rapids, 1986.

Euclid. *The Elements*. Translated by Thomas Heath. Green Lion Press, Santa Fe, 2002.

Ferguson, Kitty. *Tycho and Kepler*. Walker and Company, New York, 2002.

Feynman, Richard, Leighton, Robert, and Sands, Matthew. *The Feynman Lectures on Physics*. Addison Wesley Longman, New York, 1970.

Galileo, Galilei. *Dialogue Concerning the Two Chief World Systems*. Translated by Stillman Drake. University of California Press, New York, 2001.

The Ellipse: A Historical and Mathematical Journey   by Arthur Mazer
Copyright © 2010 by John Wiley & Sons, Inc.

Gingerich, Owen. *The Book Nobody Read: Chasing the Revolutions of Nicolaus Copernicus.* Walker and Company, New York, 2004.

Gleick, James. *Isaac Newton.* Pantheon, New York, 2003.

Gullberg, Jan. *Mathematics: From the Birth of Numbers.* W.W. Norton and Company, New York, 1997.

Hawking, Stephen editor. *On the Shoulders of Giants.* Running, Philadelphia, 2002.

Heath, Thomas. *A History of Greek Mathematics, Volumes I and II.* Dover, New York, 1921.

———. *The Copernicus of Antiquity, MacMillan,* New York, 1920

Hofstadter, Douglas. Godel, Escher, Bach: An Eternal Golden Braid. Basic Books, New York, 1979.

Kepler, Johanes. *New Astronomy.* Translated by William Donahue. Cambridge University Press, Cambridge, 1993.

———. *The Optical Part of Astronomy.* Translated by William Donahue. Green Lion Press, Santa Fe, 2000.

———. *Epitome of Copernican Astronomy and Harmonies of the World.* Translated by Charles Wallis. Prometheus, New York, 1995.

Kuhn, Thomas. *The Copernican Revolution.* Shambhala, Boston, 1991.

Koestler, Arthur. *The Sleepwalkers: A History of Man's Changing Vision of the Universe.* Penguin, New York, 1990.

Laubenbacher, Reinhard and Pangelley, David. *Mathematical Expeditions: Chronicles by the Explorers.* Springer, New York, 2000.

Leibniz, Gottfried. *The Early Mathematical Manuscripts of Leibniz.* Translated by J. M. Child. Cosimo Classics, New York, 2008.

Litvinoff, Barnet. *1492.* Little, Brown Book Group, New York, 1991 .

Marsden, Jerrold and Ratiu, Tudor. *Introduction to Mechanics and Symmetry.* Springer, New York, 2002.

Martin, Thomas. *Ancient Greece.* Yale University Press, New Haven, 1996.

Newton, Isaac. *Principia.* Translated by Andrew Motte. Prometheus Books, New York, 1995.

O'Connel, Marvin. *The Counter Reformation, 1550-1610.* Harper and Row, New York, 1974.

Penrose, Roger. *The Road to Reality: A Complete Guide to the Laws of the Universe.* Knopf, New York, 2005.

Ptolemy. *Almagest.* Translated by G. J. Toomer. Princeton University Press, Princeton, 1998.

Saliba, George. *Islamic Science and the Making of the European Renaissance.* M.I.T. Press, Boston, 2007.

Shuckburgh, Evelyn. *A History of Rome to the Battle of Actium.* MacMillan and Company, New York, 1894.

Sobel, Dava. *Galileo's Daughter.* Walker and Company, New York, 1999.

Spivak, Michael. *Calculus.* Cambridge University Press, Cambridge, 2006.

Van Helden, Albert. *Measuring the Universe: Cosmic Dimensions from Aristarchus to Halley.* University of Chicago Press, Chicago, 1986.

Voelkel, James. *The Composition of Kepler's Astronomia Nova.* Princeton University Press, Princeton, 2001.

INDEX

The Ellipse: A Historical and Mathematical Journey by Arthur Mazer
Copyright © 2010 by John Wiley & Sons, Inc.